5 95

Technical choice
innovation and economic
growth

Technical choice innovation and economic growth

Essays on American and British experience in the nineteenth century

PAUL A. DAVID
Professor of Economics, Stanford University

CAMBRIDGE UNIVERSITY PRESS

Published by the Syndics of the Cambridge University Press
Bentley House, 200 Euston Road, London NW1 2DB
American Branch: 32 East 57th Street, New York, N.Y.10022

Library of Congress Catalogue Card Number: 74–76583

Hardback ISBN 0 521 20518 2
Paperback ISBN 0 521 09875 0

First published 1975

Printed by R. R. Donnelley & Sons Company, U.S.A.

To JANET

Contents

Acknowledgements

The writing of this book was inspired by Alexander Gerschenkron, in more than one sense. His wit and intellectual energy as a teacher generated the heady excitement that drew me first, as an undergraduate, to study economic history. His vast erudition set an unattainable standard for youthful scholarship, but the challenge nonetheless served to keep me laboring in this vineyard until the encouraging appearance of some fruits of my toil. And, many years later, it was he who first put in my mind the thought of gathering this harvest.

To Henry Rosovsky I remain grateful for wise counsel and concrete suggestions about the organization of the volume, as well as a gift of extra secretarial assistance which he generously arranged for me to have during my 1972–3 sojourn as a visiting member of the Economics Department at Harvard University. In writing the Introduction I have been fortunate to receive much predictably good advice from Moses Abramovitz, as well as from Stanley L. Engerman and Simon Kuznets. Alas, not all of it have I been able to absorb.

On the occasions when the individual essays were first being prepared to appear in print, I undertook the customary and pleasant chore of trying to express my gratitude for the intellectual and financial help I had obtained in the course of each piece of research. Rather than gathering and rearranging those tokens of my indebtedness here, I have allowed them to stand in their original places within Chapters 2–6. For permitting me to make use of previously published papers, which appear here in substantially unaltered form – new material having been interpolated, and significant editorial corrections having been made in only a few indicated passages – I am obliged to the *Economic History Review*, *Explorations in Economic History*, the *Journal of Economic History*, and also to John Wiley and Sons, Inc. of New York, and Methuen & Co., Ltd. of London.

The opening essay of the collection, however, appears in print here for the first time. Hence this would seem the appropriate place for me to record that the ideas it draws together derive from many sources and have percolated through my mind for an unusually long time. To be precise, some of those notions have been with me since 1962, when economic historians' interests in Anglo-American technological divergences were suddenly raised from a quiet simmer to a furious boil by the publication of H. J. Habakkuk's now celebrated book on the subject. My debts to earlier and subsequently published works on this and closely related topics are amply enumerated in the footnote citations within Chapter 1 itself. Yet, it remains quite impossible for me at this point to identify and label the origins of the many insights I gleaned over the years in recurring discussions of the subject with friends, students and colleagues.

I can recollect, however, having benefited greatly from opportunities to try out some of my present views in a still formative stage during more recent conversations with Moses Abramovitz, Richard A. Easterlin, James N. Rosse, S. Berrick Saul, Warren C. Sanderson, and particularly, in a continuing and for me most fruitful dialogue, with Peter Temin. I am grateful for their company, their patience and their intelligence. Stephen J. DeCanio, Stanley L. Engerman, Mordecai Kurz, Nathan Rosenberg and Paul A. Samuelson, each have put me deeply in their debt by their willingness to read and comment extensively on the essay in a preliminary form. I believe the present version to have been improved significantly in consequence of their penetrating criticisms and constructive suggestions, but the responsibility for such defects of analysis or exposition as may remain surely rests with me.

During the writing of the first essay, and in the preparation of the volume for publication, I have received financial support from the National Science Foundation under NSF Grant GS–36837. David Galenson was a great aid to me in the editing and proof-reading of the whole manuscript, and in the compilation of the bibliographical references. Norma Block loyally rendered secretarial help far above and beyond the normal call of duty.

Looking back through all these lines of acknowledgement, I am only too aware of the paradox that nowhere in them does Janet David's name appear. Yet, it is most doubtful that without her I would have found it possible to bring any of these studies to completion, or have regained the fortitude to commence the next. I remain touched and deeply grateful to recall how unwillingly and with what communicated sense of personal loss, on each occasion she relinquished her husband to the obsessions of this work, and ultimately reclaimed him from it. The dedication lovingly returns a book taken from her.

1974 P.A.D.

Introduction: Technology, history and growth

The essays brought together here are concerned with technological innovations of the nineteenth century. They consider the comparative utility of alternative *concepts* of technical progress as guides for historical research, and the *preconceptions* harbored by economic historians under the recent influence of H. J. Habakkuk's thesis regarding the differentiating characteristics of new methods and prevailing practices of production in America and Britain during that century. They present evidence concerning the exploitation of 'learning opportunities' arising in actual production experience, and the consequent endogenous *generation* of productivity-raising designs for new plant and equipment or modifications of existing industrial operations. They deal with the determinants of the timing, speed and ultimate extent of the *diffusion* of process innovations – a necessary step in studying the latters' impact, inasmuch as new methods often found no immediate commercial applications, and quite typically failed to win instantaneous universal adoption even when introduced with commercial success. They address the problem of discerning the more extended economic *ramifications* of discrete 'inventions,' and of evaluating the full contribution made to the growth of an economy by the addition of a new technique to the previous array of production methods.

My arrangement of the chapters in four main parts – Concepts and preconceptions, Generation, Diffusion, and Ramifications – is meant to suggest the particular aspect of the broader subject with which each is most concerned. But some overlapping of these areas of focus within individual essays is inevitable, particularly in a collection formed as this is from studies that were independently conceived and executed.

A common emphasis upon micro-economic phenomena pervades the historical inquiries upon which the present group of essays report. Elsewhere I have tackled problems of estimating and explaining past trends of productivity considered at very high levels of aggregation, and there the discussion ran in terms of the experience of the 'agricultural sector,' the 'private domestic economy,' and equivalent abstractions. In these studies, by contrast, the effort to understand the role of technological choices and changes in the modern economic development of the United States and Great Britain is pursued within the restricted context of the experiences of particular branches of industry, or firms, and even of individual manufacturing establishments.

This unifying microcosmic focus of the whole volume reflects a theoretical orientation developed in the first, and most recent of the essays. It is argued that many questions about the forces determining the pace and the direction of factor-saving 'biases' in the generation and diffusion of technological change can be examined more usefully by conceiving of innovation in terms of (the original development or modification of) existing specific production processes, of the sort described in input–output and linear programming analysis. For such purposes the economist's now conventional conceptualization of technological innovation as a change of a neoclassical production function – an alteration of relationships between inputs and output across the entire array of known techniques – has turned out to be less helpful than one might wish. On more than one occasion, regrettably, it has led historical discussions of invention and diffusion into paradox and confusion.

But rather than eschewing any reliance on the neoclassical construct of a smooth, continuous production function, the subsequent studies restrict themselves to employing it circumspectly, at low levels of aggregation. At the level of specificity and detail thereby assayed, one may at least hope to exchange breadth of 'coverage' for the insights that occasionally come from perceiving, in concrete and particular circumstances, the operation of some general and rather abstract principles.

Thus the empirical findings of the second essay[1] regarding the historical importance of learning by doing as a source of endogenous improvements in industrial efficiency – and the implications this carries for a reappraisal of America's early tariff policies – relate not to ante-bellum industry in the large, but instead to the circumstances of the manufacture of coarse cotton sheetings in water-powered, integrated spinning and weaving mills of the so-called 'Massachusetts' type. Still more restricted is the scope and character of the evidence examined by the third essay,[2] which discloses, at an early phase of America's industrial development, the existence of learning-by-doing in the context of *fixed* production facilities. The case of the number 2 mill of the Lawrence cotton textile company of Lowell, Massachusetts, in the years between 1834 and 1856, probably represents the earliest well-documented instance of short-run learning effects. This would make its story the true precursor to the experience of the Swedish steel-mill built at Horndal a century later, and of the still more widely known phenomenon encountered in modern airframe production.

In similar fashion, the two following studies probe for some deeper understanding of diffusion phenomena along a very narrow line of inquiry.

[1] Chapter 2 was published in substantially its present form under the same title in the *Journal of Economic History*, Vol. XXX, 3 (September 1970). The footnotes indicate passages in which new material has been added.

[2] Chapter 3 was published under the same title in *Explorations in Economic History*, Vol. 10, 2 (Winter 1973).

The fourth essay[1] describes how during the decades preceding the Civil War, farmers in the American Midwest were led to abandon their hitherto ubiquitous reliance on the scythe and the grain cradle, and to take up the horse-drawn reaping machines of Hussey and McCormick. It draws particular attention to the dual influence of changing relative factor-prices and scale considerations in bringing about the widespread adoption of this novel embodiment of a labor-saving but comparatively fixed capital-intensive technique of harvesting wheat and other small grains.

The fifth essay[2] seeks an explanation for the rather different reception accorded the same innovation by contemporary grain farmers in the British Isles. In the ensuing analysis of the economic obstacles to harvest mechanization that arose in Britain, particular attention is directed to the consequences of the close 'technical interrelatedness' of these machines with the existing terrain configuration and field layouts. The latter features peculiar to Britain's farm landscape were largely man-made, the durable legacy of agrarian improvement efforts reaching back to the sixteenth century, and even earlier. Here we find a new concrete meaning for the old Veblenian generalizations about the 'disadvantages of an early start' in the process of economic modernization.

Finally, the sixth essay[3] deals with the interplay between micro- and macrocosmic aspects of the history of technology. It addresses itself to the immensely complicated problems raised by an attempt to quantify comprehensively the 'contribution' that a single, distinct technological innovation may make to aggregate economic growth. But rather than conducting the discussion on a general methodological plane, it adheres to the specific concentration of Robert W. Fogel's and Albert Fishlow's path-breaking research on the impact of the steam railroad in America.

Three broad strands of thought, themselves closely interwoven, may be traced through the following pages. Their emergence amidst the intricate details of many different, highly specific historical investigations is made more curious, and perhaps on that score alone deserving of notice, by the fact that none of these studies was originally conceived as an element of the larger design which they now form – at least not within the consciousness of the writer.

When stated simply and separately, these three substantive themes are utterly familiar. Indeed, it may be felt that they verge on the bromidic.

[1] Chapter 4 was originally published under the same title in *Industrialization in Two Systems: Essays in Honor of Alexander Gerschenkron*, ed. Henry Rosovsky (New York: John Wiley and Sons, 1966), Ch. 1, and has since been reprinted elsewhere in slightly different forms.

[2] Chapter 5 was published under the same title in *Essays on a Mature Economy: Britain after 1840*, ed. D. N. McCloskey (London: Methuen, 1971), Ch. 5.

[3] Chapter 6 was originally published under the same title in the *Economic History Review* Second Series, Vol. XXII, 3 (August 1969), and has since been reprinted elsewhere.

Yet I shall risk stating them explicitly, in the belief that together they form an important, insufficiently appreciated, and too infrequently articulated point of view about the role technology plays in the history of economic growth, and the bearing this has upon the role of history in technological progress and economic development.

THE CRITICAL ROLE OF THE CHOICE OF TECHNIQUE IN TECHNOLOGICAL HISTORY

The macrocosmic character of a society's technological progress may be conceived in terms of the direction and rate of changes in the global constraints on resource use formed by available processes of production. In the last part of the opening essay these changes are shown as being 'guided,' or directed in an *ex post* manner by previous myopic decisions – decisions having their objective in the minimization of current (as distinct from future) private costs of production. Control over the future course of technological progress in this global sense is an unintended, essentially historical consequence which flows from the localization (at one place or another in the spectrum of techniques) of improvements that result from learning by doing. Because technological 'learning' depends upon the accumulation of actual production experience, short-sighted choices about what to produce, and especially about how to produce it using presently known methods, also in effect govern what subsequently comes to be learned. Choices of technique become the link through which prevailing economic conditions may influence the future dimensions of technological knowledge. This is not the only link imaginable. But it may be far more important historically than the rational, forward-looking responses of optimizing inventors and innovators, which economists have been inclined to depict as responsible for the appearance of market- or demand-induced changes in the state of technology.

To the foregoing conception, Chapters 4 and 5 add the argument that the gradual adoption, or diffusion, of innovations is not simply a temporary disequilibrium phenomenon reflecting differences in the alacrity with which different entrepreneurs respond to a uniform economic stimulus, the opportunity for each to make a profit by cutting costs.[1] Rather, diffusion is portrayed as the reflection of a changing (equilibrium) distribution of production among the different techniques, each one chosen rationally by the members of a *heterogenous* population of firms, a population for which it could not be said that the latest method that has become available at any moment *ipso facto* constituted the dominant, best-practice

[1] The theoretical argument underlying these studies is explicitly elaborated in my lengthy paper, 'A Contribution to the Theory of Diffusion,' Stanford Research Center in Economic Growth *Memorandum* No. 71, June 1969.

technique. Thus, it is suggested, a deeper understanding of the conditions affecting the speed and ultimate extent of an innovation's diffusion is to be obtained only by explicitly analyzing the specific choice of technique problem which its advent would have presented to objectively dissimilar members of the relevant (historical) population of potential adopters.

On both grounds one is asked to recognize the longer-run, dynamic and macrocosmic implications of essentially static choice of technique decisions made at the micro-economic level. Doing so, however, calls sharply into question the penchant historians of technology have displayed for separating their task into two distinct undertakings: writing the 'history of common practices' and the 'history of inventions,' and supposing that it will prove useful to continue to pursue each enterprise rather independently of the other.

As a consequence, the first sort of history too often emerges as a placid reconstruction of the technological banalities. A succession of static idealizations is offered in answer to the question: 'What method typified productive activity in this or that epoch of the society's development?' Whereas the second sort too frequently degenerates into a disjointed chronicling of 'the march of invention.' The designing of novel devices – of those 'waves of gadgets' which swept across eighteenth-century England, in the imagery of a schoolboy's famous script – comes to be viewed by these accounts less as an instrumentality and more as an activity valued for its own sake, a signal manifestation of man's comparative success in refashioning his environment through *creative* achievements.

Free reign thus has been given to interpreting the advent and subsequent elaboration of inventions as material reflections of the less mundane attainments of the society under consideration – including its supposed capabilities for tolerating, if not actually engendering change. And many subtle interrelationships between the science, the plastic art and the 'practical discoveries' of different cultures have been elucidated as a result.

It does not denigrate such accomplishments of historical craft to point out the error of their oft-implied suggestion that there is a line of *uni-directional* historical causation flowing from the loftier sphere of concerns to those mundane realms where producers are forced to choose among alternative methods. Moreover, it will not be a sufficient remedy to notice that the scope of fruitful scientific inquiry has been limited significantly at various points in time by the prevailing state of the technical arts. That much has been said often enough. Yet a satisfactory understanding of why and how technology changes cannot be expected to emerge so long as historians of technology ignore the reciprocal influence exerted by producer's *choices* of technique – first upon the actual extent to which specific technical innovations come into use, and thereby upon the future locus of 'learning' activities.

THE PERVASIVE INFLUENCE OF INCREASING RETURNS TO SCALE

In the essays devoted to the diffusion of the reaping machine, economies of scale internal to the operations of the establishment (the farm) are shown to have played a key part in the choice between mechanized and hand techniques. The critical observation here was one made long ago, and amplified by Allyn Young in his famous Presidential Address to Section F of the British Association in 1928:[1]

It would be wasteful to make a hammer to drive a single nail; it would be better to use whatever awkward implement lies conveniently at hand. It would be wasteful to furnish a factory with an elaborate equipment of specially constructed jigs, lathes, drills, presses and conveyors to build a hundred automobiles; it would be better to rely most upon tools and machines of standard types, so as to make a relative larger use of directly-applied and a relatively smaller use of indirectly-applied labour. Mr. Ford's methods would be absurdly uneconomical if his output were very small, and would be unprofitable even if his output were what many other manufacturers of automobiles would call large.

Increased roundaboutness, and greater capital-intensity, are thus important characteristics of processes that can be most advantageously employed when 'the market,' i.e., the scale of production, becomes enlarged.

This is an empirical proposition, having to do with the 'lumpiness' of the capital goods whose services are enlisted as replacements for the more readily divisible input of labor; it carries immediate implications for the mode of production where either economic or legal and institutional conditions prevent the services of lumpy instruments of production from being utilized on a more divisible basis (such as has been achieved by time-sharing on large computers). In the cases discussed in Chapters 4 and 5, it meant that under any prevailing regime of relative input prices there were some farms whose grain acreage remained below a break-even scale, or 'threshold size,' beyond which it would become profitable to *switch* from hand- to machine-methods of harvesting. Given the initial distribution of farm sizes, changes in relative factor-prices, and in the real costs of adjusting the individual enterprise scale, therefore altered the extent to which the population of grain farmers could profitably adopt mechanization.

Another kind of scale effect – to which Adam Smith himself gave more emphasis – is brought to the fore by the studies of endogenous technical progress and the sources of productivity growth among the leading cotton textile firms of New England. Learning by doing is an irreversible form of increasing returns, deriving from the temporal accumulation of production experience relevant for devising cost-reducing innovations. It

[1] Allyn Young, 'Increasing Returns and Economic Progress,' *Economic Journal*, Vol. 38 (December 1928), p. 530.

directly corresponds to Alfred Marshall's notion that a competitive firm can have an *historical downward-sloped* supply schedule (the 'particular expenses curve,' as he called it) along which it achieves permanent reductions in its supply price *pari passu* with increases of the cumulative quantity produced.

Irreversible scale effects of this kind are designated as being 'neutral' (or 'quasi-neutral') if their realization proceeds in ways that do not significantly disturb the relative input proportions characteristic of the production activity, the specific learning environment wherein they occur. In the scheme envisaged by Chapter 1, where this terminology is introduced, localized learning effects possessing the neutrality property become the mediating process through which past choices among productive techniques can indirectly influence the future *global* drift of technological change. Yet, this important property assigned to irreversible scale economies most probably arises ultimately from the existence of conventional scale effects at the microcosmic engineering level where new production methods are designed.

At least, Chapter 1 shows that the generation of quasi-neutral sequences of 'localized' (or activity-specific) innovations through learning by doing may be traced, ultimately, to the presence of two conditions in the relevant technological microcosm: *indivisibilities* within, and strong *complementaries* (technical interrelatedness) among the constituent sub-activities, elements or components of the complete production activity. The former give rise to potential regions of sharply increasing returns as alternations are made in the capacities of individual sub-activities within the overall engineering design. In conjunction, the latter create bottlenecks (or regions of sharply diminishing returns), and thwart the substitution of sub-activities where capacity is more divisible for those sub-processes in which it is not.

Perhaps not surprisingly, both kinds of scale economies reappear in important roles in the sixth and concluding essay. The appropriate measurement of the direct cost-reducing impact of an innovation such as the steam railroad depends in considerable degree on the answer to an empirical question: Was or was not the technique of railroad freight-haulage, *by comparison* with the other, pre-existing modes of overland wagon carriage and waterborne transportation, subject to substantially decreasing long-run marginal costs? Furthermore, measurement of the indirect impact upon productive capacity (entailed by the direct effect of the railroad in *differentially* reducing the prices of transported goods) turns out to be no simple affair when it cannot be assumed that the activities constituting the rest of the economy operate with constant returns to scale technologies. The alteration of the structure of *delivered* input and output prices throughout the economy must have occasioned further cost reduction. These would come via the realization of conventional and irreversible scale economies in branches of production that depended upon significant amounts of transport services to assemble their inputs and reach their ultimate customers. Failure to take them into account leads to an improper minimization of the measured 'contribution to economic growth' assigned to innovations in transport technology, and, more generally, to an inadequate appreciation of the interlocking and mutually supportive aspects of technical advances.

There are still weightier methodological implications remaining to be noticed. The ubiquity of production technologies characterized by indivisibilities, and the emergence of new indivisibilities with the advance of technical knowledge, strikes a damaging blow at the body of theory which represents a decentralized market economy as attaining, or always tending to attain a general equilibrium configuration in the prices and quantities of a pre-specified array of inputs and outputs. Indivisibilities mean that abrupt, qualitative changes may accompany innocent continuous variations in a quantitative dimension like size: make a drum 'bigger' and its timbre deepens; reduce a symphony orchestra to a handful of players, or shorten a sonata to a few bars, and one gets a quite different musical evening.

The metaphysics of this need not engage us at the moment, for the significance of indivisibilities here is that they cause conventional economies of scale. Increasing returns to scale, in turn, leads to the existence of *gaps* in the price response of suppliers (and purchasers); instead of having continuous supply and demand schedules over whose range each point is a permissible response, there will exist lacunae within which dreaded inequalities between supply and demand may establish themselves. Commodities may not be produced at all, because the size of the market is too small to make it possible for the factors of production to earn their opportunity costs. Such awkward situations, which are seen here as ubiquitous, simply destroy the conditions under which it has been possible to mathematically prove the *existence* of a competitive general equilibrium.[1]

Were this not enough, it is not all. The presence of scale economies, including those of the irreversible sort which derive from 'learning'

[1] Cf., e.g., K. J. Arrow and F. H. Hahn, *General Competitive Analysis* (Edinburgh: Oliver & Boyd, 1971), pp. 52–62, 169–73, for discussion of the relationship between the assumption of 'divisibility' and the axioms of the convexity of production possibility sets and preference sets. Chapter 7 of Arrow and Hahn's work is entitled 'Markets With Non-Convex Preferences and Production,' and takes as its epigraph the following lines from Milton, *Paradise Lost*:

> 'A gulf profound as that Serbonion Bog
> Betwixt Damiata and mount Casius old,
> Where Armies whole have sunk.'

phenomena, also removes the mathematical basis for supposing that market configurations we might still wish (albeit with shaken confidence) to treat as equilibrium solutions, will be globally *stable* equilibria. Yet, upon just that supposition rest many of the economist's casual empirical applications of the methods of comparative statics and comparative dynamics.

Equilibrium concepts were imported into economics (from zoology, it seems) well before the mechanical paradigm began to establish its present hegemony. Nevertheless, the preoccupation with general equilibrium behavior, and the interpretation of observed economic phenomena as 'positions of rest' has undoubtedly received enormous impetus from the systematic application of mathematical methods – the introduction of which was suggested to Walras by his vision of 'the pure theory of economics or the theory of exchange and value in exchange' as 'a physico-mathematical science like mechanics or hydrodynamics.'[1]

In the age before electronic computers made the numerical investigation of large equation-systems a practical proposition, the reliance upon mathematics to reveal their salient properties by extracting autonomously determined equilibrium solutions was irresistibly attractive. Neoclassical economics, in its rigorous development of general equilibrium models was thus instructed by the example of classical mechanics axiomatically to exclude from consideration those conditions of economic reality (such as the presence of indivisibilities) that would jeopardize the equilibrium-seeking analytical enterprise.

In attempting to demonstrate how a decentralized system would work, it considered a prespecified array of commodities for which consumers had autonomous, non-economically determined (and convex) preferences; suppliers of these commodities had access to convex production possibility sets, autonomously dictated for them by the state of technological knowledge. To be sure, the neoclassical theory of induced technological change, which Chapter 1 critically examines, ventured to relax the assumption of a 'given' technology. But it did so, in effect, by treating innovations as perfectly divisible and additive commodities that could be generated by 'suppliers' who were furnished with an autonomously determined innovation possibility set. The latter 'technology,' like all other technologies envisaged in this view of the world, is of course assumed to be (convex) free from any taint of increasing returns.

One may judge the foregoing to have been an eminently successful strategy. It is important, however, not to imagine that it has been pursued completely without cost: the discipline that adopted it thereby was led away from conceptions of economic processes which integrated

[1] Léon Walras, *Elements of Pure Economics or the Theory of Social Wealth*, William Jaffé, transl. (London: George Allen and Unwin, 1954), p. 71.

observations from the real world, and within which history was of some consequence.

THE SIGNIFICANCE OF HISTORY IN ECONOMIC PROGRESS

One may come by many routes to appreciate the uses of history. A thorough indoctrination in the modes of thought formalized by the now-dominant neoclassical tradition of economic analysis, however, does not happen to be among them. Consequently, to look backward in time for the well-springs of the current flow of economic events it is necessary for an economist nowadays to start with another, an explictly *historical* vision. But because such a vision finds no natural place within the most fully articulated corpus of economic theory, any attempt to sustain it must force one continually to seek out *the limits* of that body of thought. Here at least, the critical impulse is just the other side of the positive intention to expose the significance of history in technological progress, and more generally in economic growth. The latter intent, then, is a third theme uniting the studies of technology represented by the present volume.

As a unifying purpose it will only appear as perverse as the need for it seems paradoxical. Surely by now everyone has heard that the study of economic history lately has been enjoying a revival – some would say renaissance – from which there has emerged a literature that increasingly looks like applied neoclassical economics.[1] It has come to be known as 'the new economic history.' Some sensations of being confronted by a paradox, and possibly some reproaches for ingratitude, may be elicited by my expressing even a few grave reservations about these exciting developments. Yet such reactions only can become bothersomely acute if the proximate origins of the recent movement in economic history have been forgotten entirely.

The 'old' economic history, that supposedly plodding and oft-benighted descendant of the German Historical School, had turned inward and sought no further enlightenment from neoclassical economic theories. Quite the reverse. One should recall how a generation of economists put aside the short-run preoccupations of Keynesian analysis *and turned to history* for deeper insight into the great social transformation entailed in modern economic growth; how expectantly they sought there the

[1] For a description couched in just these terms, cf. Peter Temin, ed., *New Economic History* (Harmondsworth, Middlesex: Penguin Books, 1973), Introduction, esp. pp. 7–8. The profession's seemingly satisfied eagerness to emphasize this aspect of its recent works is ever so gracefully held up for (well-deserved) ridicule by William N. Parker, 'From Old to New to Old Economic History,' *Journal of Economic History*, Vol. XXXI (March 1971), esp. pp. 6–10.

cumulative causes behind the glaring international disparities of economic life in the era following World War II. Many soon moved on, perhaps disillusioned by the intractability of economic development as a subject about which new general laws of economic behavior (theorems, no less) might be deduced from abstract premises.

The smaller band of interdisciplinary migrants who stayed – the 'colonizers' as they could today be more sinisterly described – struggled on, mainly using the pre-Keynesian tools they had brought with them. By perseverance and good fortune they managed to consolidate a curious, and rather remarkable intellectual revolution. To the study of economic history they extended the greater logical coherence, deductive power, and capacity for precise quantitative implementation of modern economic analysis – a system of thought which in its pure form happens to be fundamentally *ahistorical*, if not actually anti-historical.

There persists, as a result, an unresolved tension between the economic historian's newly augmented ability to discern the workings of market forces in historical events, and his dependence for that power upon a mechanistic world view that allows to the past at best a transient role in shaping the future. This deep conflict, far more than economists' evident difficulties in reconciling the claims of semantic precision with the achievement of an engaging prose style, explains why so many conventional historians continue to look upon the works of the new economic history with mingled feelings of awe and dismay.

Regarded from this vantage point, the methodological polemics to which we recently have been treated, concerning the legitimacy and utility of employing mathematics and statistical techniques in historical investigations, unfortunately have been occupied with superficialities, the mere symptoms of a deeper distress. The real issue is not the one Robert W. Fogel has sought to define by (accurately) describing the new economic history as being methodologically committed to a 'scientific' approach to history, to the 'attempt to cast all explanations of past economic development in the form of valid hypothetico-deductive models.'[1] Rather, the crux of the matter is the peculiar character of the formal models which neoclassical theory suggests we should use in such undertakings.

The faithful adherence of economists during the last one hundred years to the conception of the market system as a *mechanical analogue* which rapidly settles down to maintain a 'steady' circular flow of production and consumption – a general equilibrium constellation of the exchange values (prices) and quantities of all commodities – is surely one of the great anomalies of modern intellectual history. The recurring suggestion of Biology as a more suitable paradigm for economics and the other *social*

[1] R. W. Fogel, 'The New Economic History: Its Findings and Methods,' *Economic History Review*, Vol. 19 (December 1966), pp. 642–56.

sciences has been easily shrugged off, like Alfred Marshall's use of organic analogies, as 'vague,' if not 'mystical.' Physics remains the model. But it is as though physics had stopped with classical mechanics; as if the thermodynamic revolution had never occurred, and outside the life sciences phenomena involving irreversible processes and qualitative change had remained beyond the reach of scientific explanation. As Georgescu-Roegen recently has remarked:[1]

Once, it is true, physicists, mathematicians, and philosophers were one in singing the apotheosis of mechanics as the highest triumph of human reason. But by the time Jevons and Walras began laying the cornerstones of modern economics, a spectacular revolution in physics had already brought the downfall of the mechanistic dogma both in the natural sciences and in philosophy. And the curious fact is that none of the architects of 'the mechanics of utility and self-interest' and even none of the latter-day model builders seem to have been aware at any time of this downfall. Otherwise, one could not understand why they have clung to the mechanistic framework with the fervor with which they have. Even an economist of Frank H. Knight's philosophical finesse not long ago [1935] referred to mechanics as 'the sister science' of economics.

Economics repeatedly is attacked for embracing the fiction of man as a mechanism, a rational utility-maximizing instrument whose behavior with regard to economic affairs (if not to all the intriguing aspects of life) is free of all habituating influences, and thus remains devoid of all cultural and social propensities. Although less explicitly decried, the implied *timelessness* of this fiction must be troubling to every historian, even those who do not proclaim themselves to be Humanists and therefore offended by modeling men as machines. Like the mechanical conception of the physical universe, this construction of man bespeaks a longing for permanence in human affairs, for the certainty of laws governing a natural order based on uniform matter. *Homo economicus* thus enters upon the neoclassical stage fully grown, not as the child of an ascendant rationalistic and materialistic culture, not as a creature adapting to specific epochs in social and economic evolution. He is, for those who fully accept the faith, a hypothetical man for all seasons. In representing the presumption of an unvarying, ever-present element in human actions relating to the allocation of resources, the fiction of 'economic man' embodies the aspiration of the neoclassical school to imitate the great success which Newtonian mechanics achieved by confining its attention exclusively to locomotion, a uniform property of matter.

The concept of uniformity of matter often has been defined by reference to behavior that is determined only by the *present* state of matter, strictly,

[1] N. Georgescu-Roegen, *The Entropy Law and the Economic Process* (Cambridge, Mass.: Harvard University Press, 1971), pp. 2–3.

by conditions ascertainable without possession of any information regarding its past states. Such behavior, being independent of the position 'the present' occupies on the axis of Time, leads to the analysis of static and dynamic systems without history. Classical mechanics taught that in an account of a cannon ball's flight we should record the moments from firing. But in the equations for the missile's trajectory, Time will not appear explicitly; that the occasion was 21 October 1805 in the English Channel, instead of the morning of 12 April 1861 in Charleston harbor, should make no sensible difference. Such 'history' as there is would appear only in the initial data of an incomplete causal system, the otherwise unexplained touching of match to powder.[1] Analogously, modern economics' account of the upward path of a commodity's relative price from the moment an excess demand for it has materialized, logically allows no *essential* way for the past history of prices to enter the story. Only if the traders are envisaged as subscribing to the very fallacy which all price theory sets out to explode, that is to say, only if they believe that it is not prevailing market forces but the past trend which somehow forms future prices, could price-*history* influence the actual course of adjustments – by shaping (unfounded) expectations about the future.

Long before Newton's triumph, the belief in the uniformity of matter – as distinct from the existence of some uniform properties – had proved antithetical to the serious study of history. Once Descartes had answered Montaigne's skepticism as Plato had answered Socrates's, by anchoring down the concepts of science on the unchanging foundations of geometry, he could allot no serious importance to historical change and development. Thus, it may be said that for the economists who fully embraced the theoretical orientation deriving from Jevons and Walras, as for Descartes, and Plato before him,[2]

> the temporal flux of historical events lost all fundamental significance, acquiring a theoretical interest only by providing illustrations of the unchanging geometrical laws – the events in question losing, in the process, both their individualities and their dates. 'Histories and fables' were admirable for the same reasons as travel: they were a way of broadening the mind. But serious philosophy did not begin until one saw past the surface flux of events to the geometrical skeleton beneath.

There is much about history to be studied, and there is ample room for divergent traditions to contribute to knowledge of the past. It would be a tragic error (once again) to seek to exclude ahistorical approaches from

[1] This is precisely the role allotted to 'history' in the economic dynamics of P. A. Samuelson, *Foundations of Economic Analysis* (Cambridge, Mass.: Harvard University Press, 1947), Ch. XI, esp. p. 315.

[2] Stephen Toulmin and June Goodfield, *The Discovery of Time* (London: Pelican Books, 1967), p. 97.

the enterprise. Nonetheless the differences of approach are worth recognizing.[1]

The applied economist may draw upon historical data to test a hypothesis, but he will not be terribly interested in the dates attached to the observations – simply because he is not worried how the partial model he happens to be testing, if it is not falsified, will fit in with other things known or conjectured *about that period.* In doing econometric work as a rule he will regard with annoyance any indications of significant structural changes within the time span covered by his observations – where the term 'structure' here subsumes the proposed account of the relationships among the relevant variables of the economic system, or sub-system currently under consideration. He is trying to find – or so it often seems – a set of observations for which his descriptive structure or model will 'work,' and he does not bother to label tested models according to when or where they do *not* work. An equation that fits the data well for half the available run of time-series observations and not for the rest is, for the ordinary applied economist, a failure; he will have to resist the impulse to discard the recalcitrant data in presenting his results.

By contrast, the demonstration that a particular model fails to rationalize the data for certain societies, or periods in the history of one society, must be viewed by econometric *historians* as of no less interest than the finding that there is after all some body of data to which the model is appropriate. The economic historian may hail the half-failed regression equation as nothing less than triumph – in the sense that by uncovering the occurrence of a change in economic structure, it signals him to set to work to learn what happened in history.

I think it is instructive to envisage the economic historian's world as a succession of working models or applicable theories, each appropriate to a particular social, temporal and technological setting. His ultimate goal is to explain the sequence of different models. Why do they change as they do? What determines the rate at which one supplants another? It could be said that really the historian shares the neoclassical economist's faith in the quest for some stable set of general relations governing the succession of particular representations of states of economic life. In practice, however, such a system is too complex, and involves too many non-economic variables affecting tastes, technology, and the institutional settings within which exchange can take place; it is simply not expeditious even to try writing the whole thing down in closed form. And so, rather

[1] I have sought to delineate some of the differences of approach noted by the following paragraphs in a paper on 'The Future of Econometric History,' which appears as a 'Comment' on Gavin Wright's admirable survey of econometric studies of history in *Frontiers of Quantitative Economics*, M. D. Intriligator, ed. (Amsterdam: North-Holland Publishing Co., 1971), pp. 459–67.

than obscure the developments in which he is interested, the economic historian may be quite content to try to accumulate a sequence of partial models, or incomplete causal systems, each of which has been found to be consistent with as much of the evidence as is available for some historical settings, and rather less appropriate as a rationalization for empirical observations relating to other settings.

Much of the econometric historian's efforts necessarily will be spent in doing the same sort of thing that ordinary applied economists do – looking for a simple, comprehensive structure consistent with a given set of data. It does not follow that the ultimate purposes of the two are identical, or that it is desirable for the new economic (and econometric) history to assimilate majority-attitudes within the larger profession of economics concerning undertakings of this kind. If they spend their time piecing together an appropriate neoclassical (or Marxian or Veblenian) model, historians ought not to be encouraged to suppress their ultimate interest in discovering the limits of its applicability.

There is no doubt that non-economists will find many passages of this book discouragingly hard going. All of the following studies look to some aspect of neoclassical theory as a particularly fruitful source of hypotheses with which to approach economic life in nineteenth-century America and Britain; they make free use, for example, of the model of the cost-minimizing firm as a means of organizing and interrogating the historical evidence. But it is equally true that the historian who manages to penetrate beneath the surface of these chapters should find them in many respects congenial to his way of looking at the world – indeed, more so than will the pure neoclassical economist.

I have already intimated as much in considering the broader implications of the second unifying theme. The pervasiveness of economies of scale opens up the prospect that past market configurations, which neoclassical theory tempts one to interpret as globally stable equilibria, were in reality unstable positions away from which the system moved when disturbed by shifts in demand. The added presence of 'learning' effects in production (and the implied suggestion that they may have also been present in consumption, in the form of habituation or endogenous taste formation) introduces a degree of irreversibility in the ensuing market adjustments of relative costs and prices. As a result of this, previous economic configurations become *irrevocably* lost, and in trying to work backwards by entertaining counterfactual variations on the present, one cannot hope to exhibit the workings of historical processes.

In these essays, therefore, one may glimpse a view of the past that is very different from the Newtonian conception embraced by Walras and now brought to its full realization in the work of Debreu and Arrow. They project instead, Adam Smith's and Allyn Young's vision of secular change in the economic order as an unstable, cumulative process to which decentralized market economies are particularly subject.[1] When the dynamic paths followed by industries and economies are not stable, when neither the individual economic agents nor the ensemble of actors that we refer to as the economic system can be presumed to be found always in the near neighborhood of an autonomously determined equilibrium, the relevant notion of equilibrium becomes one of an endogenously positioned moving target – an elusive, moving quarry whose track is altered by the very actions of the pursuing pack.

Under such conditions, marked divergences between ultimate outcomes may flow from seemingly negligible differences in remote beginnings. There is no reason to suppose that dynamic processes are *ergodic*, in the sense of ultimately shaking free of hysteresis effects and converging from dispersed initial positions towards a pre-destined steady state. To understand the process of modern economic growth and technological development in such an untidy world necessarily calls for the study of history. For, change itself ceases to be mere locomotion. Economic growth takes on an essentially *historical* character, and the shape of the future may be presumed to bear the heavy impress of the past.

[1] Cf. Nicholas Kaldor, 'The Irrelevance of Equilibrium Economics,' *Economic Journal*, Vol. 82 (December 1972), pp. 1237–52, for a recent attempt to articulate this vision.

Concepts and preconceptions

1 Labor scarcity and the problem of technological practice and progress in nineteenth-century America

A new view of Anglo-American differences in technology during the nineteenth century is set forth by this essay. It offers fresh arguments for the reassertion of historical propositions that have become both familiar and increasingly suspect in the train of discussions ignited by the publication of H. J. Habakkuk's extended essay on the subject more than a decade ago.[1] On this new view, as of old, America's comparative natural resource abundance (expressed in a land–labor ratio higher than that prevailing in Britain) appears as a powerful influence promoting substitutions of capital for labor – not only in industry but in agricultural production as well. The resulting choice of more mechanized, capital-intensive techniques is further seen to have led, in America, to a flow of new production methods whose bias was toward augmentation of the country's labor resources rather than its capital-endowment. In the context of this view, moreover, it could be said that a high rate of innovation biased toward labor-augmentation was attributable to the comparative scarcity of labor (*vis-à-vis* land) that characterized the United States economy of the pre-Civil War era.

The positive, constructive aspect of the present essay is, like Habakkuk's contribution, essentially an exercise in analytical interpretation. By that I do not mean to suggest that either this work or that which may be thought to have inspired it is purely theoretical, in the sense of transcending empirical issues of particular relevance to the history of production methods in the nineteenth century. Quite the opposite. My ultimate purpose is to indicate a conceptual framework within which new empirical questions emerge, and still-unresolved issues of fact acquire a significance that justifies some redirection of research efforts on the part of historians interested in the evolution of modern technology. But no new descriptive account comparing American and British technologies is to be found herein. Instead, an alternative *explicandum* is proposed for the set of historical generalizations concerning America's propensity to 'invent and adopt mechanical methods more rapidly' than Britain, which form the *explanas* of the so-called Rothbarth–Habakkuk thesis.

[1] H. J. Habakkuk, *American and British Technology in the Nineteenth Century* (Cambridge: Cambridge University Press, 1962).

I

In devoting a major portion of the present essay to this explanatory enterprise I have accepted implicitly those familiar generalizations about the comparative character of nineteenth-century American technology as deserving of some coherent rationalization, if only on account of the frequency with which they have been asserted in respectable places and the consequent credibility that still adheres to them. Yet it seems preferable here to continue to avoid referring to these generalizations as 'the facts,' or even as 'the stylized facts.' For, not the least of the difficulties with the entire subject is that its empirical outlines at present are so hazy, and so inadequately corroborated even in their grossest imprecisions.

To motivate what might otherwise appear to the uninitiated as the exceedingly curious exercise of explaining yet uncorroborated 'facts,' and bothering to propose a novel view of nineteenth-century technological developments for this purpose, some preliminary historiographic excursions are needed. These occupy the remainder of this first part of the essay. They will serve also to indicate the way in which the approach of the recent literature on the subject has been extended in the part immediately following, wherein I present a neoclassical critique of the less-discussed of the two principal aspects of the Rothbarth–Habakkuk thesis – that concerning the character of technological *innovation* in America.

Together, Parts I and II lay a critical basis for the somewhat radical departures proposed by the final section. But for the cognoscenti, and those who start from a position of skepticism about the neoclassical view of technological progress, this much preparation is not mandatory. Even readers who merely dislike suspense will find it quite possible to turn now to Part III and work their way back to this point.

FACT, THEORY AND THE ROTHBARTH–HABAKKUK THESIS

The comparative discussion of American and British technology during the nineteenth century has not been conducted in a total factual vacuum, although much less is known about the subject than might be supposed from the recent interest it has managed to capture from students of economic history. There does exist, of course, the rich legacy of pertinent and oft-cited technical commentary, left to us by competent foreign observers who traveled in the United States before (and after) the Civil War. It is the comparative remarks of the British contingent among them,[1]

[1] E. Hoon Cawley, ed., *The American Diaries of Richard Cobden* (Princeton, N.J.: Princeton University Press, 1952); James Montgomery, *A Practical Detail of the Cotton Manufacture of the United States of America Contrasted and Compared with that of Great Britain* (Glasgow: John Niven, 1840); [Great Britain] *Parliamentary Accounts and Papers*, 1854, XXXVI

men the likes of Richard Cobden, James Montgomery, George Wallis and Joseph Whitworth, which form the historical point of departure for Habakkuk's essentially speculative essay. Unfortunately, as Habakkuk himself was compeled to acknowledge, these sources provide none of the systematic data which might give precision to their impressions and furnish some 'reasonably unambiguous measure of what it is, in economic terms, that needs to be explained.' 'But,' he persisted, 'that something needs to be explained seems clear.'[1]

The existence of this body of contemporary comment does indeed warrant a careful inquiry, still to be conducted, into just what it was about technological practices in American industry and agriculture that visitors from Victoria's Britain might have found so remarkable. A conclusion that 'something needs to be explained' is not terribly helpful after all, unless one has more than a glimmer of what the nature of that 'something' was. And here, alas, the trouble is that we are provided with too many broad hints rather than too few.

On reflection, it would be most surprising were the travelers' accounts to have delineated between three assertions which today appear to us as conceptually quite distinct: (1) that Americans characteristically used more capital equipment than their counterparts in Britain; (2) that the Americans were outstripping their British cousins in the *development* of labor-saving, mechanical devices; and (3) that Americans simply were exceling in the invention and introduction of what, to the eyes of visitors, were novel production methods of all sorts. The difference separating the first assertion from the latter two, for example, corresponds to a distinction between (1) movements to *disparate positions along a common unit isoquant* (or its factor-price frontier dual) reflecting the state of technology known to both countries, and (2 and 3) some particular pattern of *inter-country divergence between the frontiers* describing their respective technologies.[2] Palpably this taxonomic principle was not one that commended itself

[1] Habakkuk, *American and British Technology*, p. 5.

[2] This interpretation was proposed by Peter Temin, 'Labor Scarcity and the Problem of American Industrial Efficiency in the 1850's,' *Journal of Economic History*, Vol. XXVI (September 1966). In a well-behaved two-factor model of aggregate production subject to constant returns to scale, both the unit isoquant and the factor-price frontier dual will slope downward to the right, and be convex. If the isoquants (frontiers) for two different economies do not coincide because knowledge is not shared completely, the dominant, technologically superior isoquant (factor-price frontier) is the one lying everywhere closer to (farther from) the origin.

'Special Reports of Mr. George Wallis and Mr. Joseph Whitworth on the New York Industrial Exhibition,' and 1854–5, L, 'Report of the Committee on the Machinery of the United States.' The Reports of the British Commissioners Wallis and Whitworth are reprinted, with comments, in Nathan Rosenberg, ed., *The American System of Manufactures* (Edinburgh: Edinburgh University Press, 1969).

to nineteenth-century observers, who were not conjuring up aggregate production functions but were seeking some generalizations from technological comparisons made at a micro-economic level. Moreover, uninformed by such conceptual distinctions, their testimony and the quantitative fragments preserved therein were not structured so to inform us whether the relevant problem for the historian is or is not simply one of accounting for a recurring pattern of divergences in the choice of techniques made in many branches of industry conducted on opposite sides of the Atlantic. Perhaps it is necessary instead, or in addition, to explain why there occurred some systematic bifurcation of the course of technological progress in these two societies whose cultural and scientific heritages held so much in common.

Burdened with the indiscriminate support which contemporary visitors' commentaries might afford for each of the three foregoing comparative statements about America's technology, Habakkuk – and Erwin Rothbarth[1] before him – proceeded by seeking in the influence of American land abundance, or equivalently in 'labor scarcity,' a common explanation for them all. In no small part, the sheer complexity of the resulting argument, and the manifest difficulty many subsequent critics and commentators encountered in trying to grasp it in its entirety, can be traced to this underlying empirical ambiguity.[2] Habakkuk seemingly was obliged at each stage of the discussion to examine the implications that every new consideration would carry for (1) the factor proportions chosen, specifically the degree of mechanization; (2) the direction toward which inventive efforts at factor-saving tended to be skewed; and (3) the characteristic pace of technological innovation actually achieved.

Still further complexities were entailed when questions about the influences exercised by 'labor scarcity' were pursued all the way up to the plane of macro-economic speculation regarding the American economy's growth in comparison with that of the British economy, rather than being treated exclusively at the level of the behavior of micro-economic units considered to be statistically representative of those found in each country. It is, of course, the latter abstraction rather than the former to which the commentaries of contemporary observers and the ancilliary illustrations Habakkuk draws from technological history may be related – however inconclusively. Those sources provide no information about entire economies, or even about entire industrial sectors.

[1] Cf. E. Rothbarth, 'Causes of the Superior Efficiency of U.S.A. Industry as Compared with British Industry,' *Economic Journal*, Vol. 56 (September 1946).

[2] In addition to Temin, 'Labor Scarcity' (1966), cf. David S. Landes, 'Factor Costs and Demand: Determinants of Economic Growth,' *Business History*, Vol. VII, 1 (January 1965); S. B. Saul, ed., *Technological Change: the United States and Britain in the Nineteenth Century* (London: Methuen, 1970), 'Editor's Introduction.'

Why then does Habakkuk's essay struggle so to ascertain the likely macro-economic consequences for the rate of investment and the pace of industrial capital formation, indirectly linking these too to America's comparatively abundant natural resource endowment? Economic reasoning here again is being asked to do duty in place of absent facts about the comparative rate of capital accumulation, which Habakkuk recognizes as having had a hand in shaping the peculiar cost-price environment in which America's industrial (and agricultural) producers found themselves.[1]

At the resulting macro-economic level of discourse, partial equilibrium analysis is left behind, and the several distinguishable strands of Habakkuk's discussion of the evolving character of American technology are drawn together in the effort to infer what consequences for capital formation could (did?) flow from the existence of effective incentives to replace labor with capital, and to search out new and more capital-intensive techniques. The end-result is an extraordinarily involuted form of argument, aptly described by David Landes[2] as history written in the subjunctive and conditional modes:[3]

> The dearness of American labour and the inelasticity of its supply provide an adequate explanation of why, from a given range of techniques, the choice of the American manufacturer should have been biased towards those which were more productive per unit of labour because they were more expensive in capital per unit of output. The same circumstances might also have exerted favourable influences on the rate of investment by providing an incentive to devise new labour-saving methods, and because capital-intensity of investment increased the ability to devise such methods and also increased the prepensity to save out of profits. The main point is the favourable effect of labour-scarcity on technical progress.

Initiated in the concrete and microcosmic testimony of nineteenth-century travelers, relating events and practices observed in particular branches of industry and regions of the United States and Great Britain, Habakkuk's explanatory argument emerges as abstract and macrocosmic, not only in its ultimate concerns but in the aggregative mode of theorizing invoked to resolve essentially empirical questions.

Little wonder that critics and proponents alike have sought to break up the Rothbarth–Habakkuk thesis into separate, disconnected propositions about the peculiar influence factor-endowment differences might exercise upon various facets of nations' respective technological development, each of which might then be reconsidered on its own merits. Little wonder, too, that such piecemeal reconsiderations – not being firmly

[1] Cf. *American and British Technology*, esp. pp. 43–63 on 'Labour-Scarcity and the Rate of Investment.' The significance of this difficult section of Habakkuk's essay has not been adequately appreciated, and it will be more fully reconsidered below.

[2] Cf. Landes, 'Factor Costs and Demand' (1965), p. 17.

[3] Habakkuk, *American and British Technology*, pp. 62–3.

anchored to a specific, well-documented phenomenon which demanded of the historian some explanation – should have rapidly drifted toward a preoccupation with the pure economic logic of proferred 'causes.' After all, if there was any part of the thesis that could be restated as a very general theoretical proposition, a strong theorem perhaps, some reliable inference might be made from the undisputed premise of comparative land-abundance to the existence of an (otherwise unconfirmed) 'effect' in the character of American technology. Conversely, general theoretical considerations which impeached the logic of one or another partial rendition of the argument, increasingly appeared to throw into doubt the very existence of the phenomenon for which an explanation was ostensibly being sought.

Proceeding thus on a piecemeal basis, it became easier effectively to set aside Habakkuk's conviction that there was *something* which required an explanation. Some agnosticism on this very point might justifiably have been voiced more than a decade ago, founded on nothing more (or less) sophisticated than a skeptical discounting of the worth of traveler's reports in this connexion.[1] But curiously enough, such expressions of doubt nowadays derive from the difficulties encountered in framing a rigorously neoclassical, theoretical case in support of any of the several alleged technological consequences of the high land–labor ratio in early nineteenth-century America as it is for the historian.

THE 'BASIC THEOREM' AND THE NEOCLASSICAL CRITIQUE

The logical dismantling of the Rothbarth–Habakkuk thesis at the hands of economic historians has not yet been completed, but the very considerable headway made thus far is epitomized by Peter Temin's most recent (1971) reformulation of his original (1966) critique.[2] Widely misunderstood at the time, Temin's initial critical foray can now be seen to have launched a damaging attack upon the supposed connexion between the greater

[1] William N. Parker's 'Review' of *American and British Technology* [*Business History Review*, Vol. 37 (Spring/Summer 1963), pp. 121–2] expressed some early doubt about the basic factual judgements underlying the book, but there has been no serious consideration on the part of economic historians of the frailties of the 'traveler's account' as a source for generalizations about an entire economy. More attention might be paid, in particular, to the part played by British and Continental European preconceptions about America. If foreign observers found essentially what they were prepared to understand, it is relevant to ask what they had read and heard prior to looking for themselves, and how much their preconceptions reflected conditions which had long ceased to obtain – if indeed they ever did obtain. Viewed from this angle, the persistence of the 'labor scarcity' theme in foreign visitors' accounts of America is as much a problem for intellectual and literary historians as it is for the historian of the American economy.

[2] Cf. Peter Temin, 'Labor Scarcity in America,' *Journal of Interdisciplinary History*, Vol. I, 2 (Winter 1971).

availability of land in the United States and the putative establishment there of higher wage rates *vis-à-vis* the rental price of capital equipment.

The essence of the Rothbarth–Habakkuk thesis is restated by Temin in the form(s) of a 'Basic Theorem of labor scarcity':[1]

> If one country has a higher ratio of land to labor than another, all other things being equal, then this country will use more – or perhaps better – machinery for each worker in manufacturing than the other country.

Either more machinery, or better machinery, would suffice to account for the superior efficiency of American workers with which Rothbarth was mainly concerned, and the prevalent use of either would serve to explain at least one of the possible interpretations Habakkuk gave to the commentaries of British visitors on nineteenth-century American technological practices. Temin then insists on a clear separation between the version of the Basic Theorem referring to 'more machinery' and that referring to 'better machinery': the former has to do with international differences in the choice of technique, whereas the latter version alludes to the influence of labor scarcity on invention and innovation. The assumptions and logic underlying these two 'theorems' are quite different, and Temin maintains they should be separately considered, and independently evaluated.[2]

Since neoclassical production theory seemed to offer firmer guidance for a discussion of the 'more machinery' version, Temin originally set aside the problem of the rate and bias of innovative activity. Even in subsequent reformulations the central question continues to be the possibility of establishing a strong, general inference from the premise of an economy endowed with a comparatively high land–labor ratio (and no higher overall ratio of capital to labor), to the conclusion that in its industrial pursuits the degree of mechanization, indicated in the amount of capital used per worker, must be comparatively high as well.

Working within the confines of a simple characterization of the American (and the British) economy as comprised of an agricultural and an industrial sector, Temin can show that the foregoing version of the Basic Theorem is generally valid *only* when land and capital can be treated as inputs specific to agricultural and industrial production, respectively, and labor

[1] *Ibid.*, p. 252.

[2] This separation was proposed by Temin in his first critique, 'Labor Scarcity' (1966), but the discussion was made more difficult to follow by his seeming acceptance – at some points – of Habakkuk's notion that at least one version of the thesis must be valid. Thus, the logical objections Temin raised against the 'more machinery' version threatened to reduce the Rothbarth–Habakkuk thesis to a set of propositions which had to be accepted as pertaining to Anglo-American differences in innovation. In 'Labor Scarcity in America' (1971), however, the discussions of the two versions are 'entirely independent, as are the conclusions.' (cf. *ibid.*, p. 254.)

is taken to be a homogeneous mobile factor employed in each sector under conditions of constant returns to scale. Under these conditions an increase of the amount of land (natural resources) in relation to the total labor force would, by raising the marginal physical product of labor exclusively in agriculture, draw thither some additional workers from elsewhere in the economy. In the sectoral redistribution of the labor force thus induced, the per worker endowment of capital left for those who remain engaged in industrial pursuits necessarily would be increased – inasmuch as the overall capital–labor endowment of the economy is, by assumption, unchanged. Hence the marginal physical product of *capital* must be reduced. The meaning of the expression 'labor scarcity' is utterly unambiguous in this specific context: a higher land–labor ratio in the economy, and therefore specifically in agriculture, corresponds with the existence of a higher capital–labor ratio and – if factors are paid in strict proportion to their respective marginal products – a higher wage–rental price ratio in the industrial sector.

The trouble with the Rothbarth–Habakkuk thesis, according to Temin, is precisely that the Basic Theorem which assures this absence of ambiguity about the meaning of 'labor scarcity' can validly be asserted only under the *improbable* conditions just reviewed. Were land not a factor specific to agriculture, were it taken, more plausibly, to enter the production of industrial goods (along with capital and labor, of course) directly, or indirectly in the form of raw material inputs, then an enlarged land endowment would not obviously favor the envisaged incremental absorption of labor by agrarian pursuits. Alternatively, were capital equipment employed along with land and labor in agriculture, an increase in the total natural resource endowment would lead to the greater absorption of both labor and capital in agriculture. It would then be necessary closely to specify the production technology available to each sector before one could predict whether or not the capital–labor ratio in the industrial sector would be raised thereby. But, says Temin, 'no one maintains that the truth of the Basic Theorem depends on such precise – and obscure – knowledge.'[1]

Thus the only compellingly general theoretical connexion between the comparative abundance of natural resources *vis-à-vis* labor in America and the persistence of a comparatively high wage–rental price ratio in her manufacturing industries is found to turn on an unacceptable characterization of the economy. Rather than debating the inherent implausibility of taking labor to be the only one of the three factors employable

[1] Temin, 'Labor Scarcity in America' (1971), p. 256, n. 7. It is not clear whether or not Temin approves of this tacit agreement to decide the issue without benefit of further empirical research on nineteenth-century technology, but the article in question certainly abides by the rules of that game.

in more than a single sphere of production, a point which subsequently was pressed by Robert Fogel[1] and by Edward Ames and Nathan Rosenberg,[2] Temin perceived another difficulty. If capital and labor alone were used by industries possessing the same technical knowledge in both America and Britain, and the scale of production made no difference to the efficiency with which these inputs were utilized, a higher marginal productivity of capital could arise only where the (industrial) capital–labor ratio was lower. Accepting the tenets of neoclassical distribution theory, the observation that the American rate of return on capital exceeded the rate prevailing in England would require, in the context of the preceding model, the paradoxical inference that the degree of mechanization in American industry was comparatively low.[3] Alternatively, abandoning the supposition of a shared technology, it might be inferred that American industrial technology was superior – hardly the historical condition that is contemplated in the 'more machinery' version of the Rothbarth–Habakkuk argument.

The extreme consternation with which these last announcements were received by the profession, and the vigorous rebuttals directed against this aspect of Temin's original critique,[4] reflected nothing so much as a general confusion regarding the position the point occupies in the logical structure of his attack on the Rothbarth–Habakkuk thesis. It is best seen as a 'proof by contradiction': the only general theoretical specification of a static two-sector model which will validate the 'more machinery' form of the Basic Theorem, itself cannot be squared with the gross factual observation that prevailing interest rates in mid-nineteenth-century America were higher than those in England. The main thrust of Temin's critique is thus entirely consistent with the theoretical point brought out first by Fogel's seemingly antagonistic article on 'The Specification Problem in Economic History' and later proved with still greater generality by Ronald Jones.[5] Under the postulate of a common

[1] Cf. Robert W. Fogel, 'The Specification Problem in Economic History,' *Journal of Economic History*, XXVII (September 1967).

[2] Cf. Edward Ames and Nathan Rosenberg, 'The Enfield Arsenal in Theory and History,' *The Economic Journal*, Vol. LXXVIII, 312 (December 1968).

[3] Cf. Temin, 'Labor Scarcity' (1966), p. 291 for the argument based on comparison of British and American interest rates on low-risk securities during the first half of the nineteenth century.

[4] Cf. Fogel, 'The Specification Problem' (1967), pp. 299–308; Ian M. Drummond, 'Labor Scarcity and the Problem of American Industrial Efficiency in the 1850's: A Comment,' *Journal of Economic History*, Vol. XXVII, 3 (September 1967), esp. pp. 387, 390; Ames and Rosenberg, 'Enfield Arsenal' (1968), esp. pp. 827–31.

[5] Cf. Ronald W. Jones, 'A Three-Factor Model in Theory, Trade and History,' in *Trade, Balance of Payments, and Growth: Papers in International Economics in Honor of C. P. Kindleberger*, ed. Jagdish Bhagwati, Ronald W. Jones, Robert A. Mundell, and Jaroslav Vanek (Amsterdam: North-Holland Publishing Company, 1971). Drummond, 'Labor

technology and constant returns to scale, the greater availability of natural resources for use by industry, as well as by agriculture, would tend to raise the marginal productivity of both labor and capital; the comparatively high rate of return does not imply lower capital-intensity of American industrial production in the context of the sort of two-sector model where, as we have seen, the Basic Theorem need not hold.

A still more pointed theoretical demurrer from the paradoxical inference might have been made by first acknowledging that it is not essential to postulate (as did Temin) the existence of a continuously differentiable two-factor aggregate production function for the manufacturing sector in order to derive a unique, downward-sloping and convex factor-price frontier relating the wage rate to the rental rate of capital. Samuelson[1] has shown how a smooth, neoclassical-looking factor-price frontier for an industry or an economy may be generated when there are many (an infinity of) distinct, fixed-coefficient methods of combining manpower and machinery to produce a specified output. But it is now equally well known that the very real possibility of the 'reswitching of techniques' means that in general the rate of return will not order techniques uniquely according to their respective capital-intensities.[2] This, of course, is a two-edged argument: it cuts no less lethally against the suppositions of Rothbarth and Habakkuk that any mechanism which established a higher wage–rental ratio in the United States economy must have led to the choice there of more highly mechanized, more capital-intensive methods.

To bury the 'more machinery' version of the Rothbarth–Habakkuk thesis at this point, or even to reject its empirical presuppositions on the strength of the foregoing analytical objections, would be premature. First, from Temin we have learned only that *very general* neoclassical models do not suffice to establish the suggested connexion between land abundance and a high price for labor relative to the rental rate of capital. But then there may have been specific characteristics of nineteenth-century technology which served to link the adoption of more mechanized, capital-intensive methods with the comparative cheapness of natural resources.

[1] Cf. Paul A. Samuelson, 'Parable and Realism in Capital Theory: The Surrogate Production Function,' *Review of Economic Studies*, Vol. 29 (3), No. 80 (June 1962).

[2] Cf., e.g., Luigi Spaventa, 'Realism Without Parables in Capital Theory,' *Recherches Récentes sur la Fonction de Production* (Namur: Centre D'Etudes et des Recherches Universitaires de Namur, 1968), pp. 15–45; Luigi Pasinetti, 'Switches of Technique and the "Rate of Return" in Capital Theory,' *Economic Journal*, Vol. 79 (September 1969).

Scarcity: A Comment' (1967), also contested the inference Temin drew from the comparison of American and British rates of return, but suggested discarding the whole neoclassical apparatus of continuously differentiable production functions and the assumption of long-run equilibrium as the way to avoid being forced to conclude that the capital–labor ratio was lower in America than in Britain.

And such technologically determined links need not have turned so exclusively upon the responses of producers to variations of the wage–rental ratio envisaged by the familiar exercises in neoclassic production theory.[1]

In the second place, it is not at all clear that the faulty line of connexion exposed by Temin's analysis is really the one contemplated in Habakkuk's book. As the neoclassical critique has been developed, it is more directly apposite to the long-standing contention voiced by America's early nineteenth-century proponents of tariff protection,[2] and reiterated by Rothbarth: the existence of land freely available for independent cultivation saddled industrial firms with higher relative wage rates, no matter what the size of the labor force they might collectively seek to employ. Notice, however, that in this there appears no trace of the dynamic considerations which so complicate Habakkuk's formulation.[3]

Whereas the question of the rate of investment and the expansion of manufacturing is quite irrelevant to the static mechanism envisaged by writers from Hamilton and Gallatin to Rothbarth, it is vital in Habakkuk's account of the connexion between land abundance and the comparative dearness of labor (*vis-à-vis* capital) in industrial pursuits. In the context of the peculiar labor supply conditions which are supposed to have been characteristic of the United States as a region of recent settlement, the *expansion* of non-agricultural activities is portrayed as having kept the relative price of industrial labor high and tending to rise. Unfortunately, what is missing is a more detailed account of the supply side of the putative difference between the dynamic labor market adjustment process in Britain and America. Without this, the specific influences exercised by the existence of the frontier and other concomitants of natural resource abundance remain not only historically undocumented but dissatisfyingly vague. Instead, however, Habakkuk's discussion focuses attention almost wholly upon the demand side of the industrial labor market.

It was in just this connexion that it appeared necessary to explain why in America the comparative absence of a reserve army of industrial labor did not so depress the profit rate as to have curtailed expansion (capital-widening) of that sector, thereby removing the upward pressure on the

[1] This rather heretical line of consideration is pursued a bit further in Part III, below.

[2] Cf., e.g., Alexander Hamilton's 'Report on Manufactures' (1791), *American State Papers, Finances* (Washington: Gales and Seaton, 1832), Vol. I, pp. 123–7; and the following passage in Albert Gallatin's 'Report on Manufactures' (1810), *American State Papers, Finances* (Washington: Gales and Seaton, 1832), Vol. II, p. 430: 'The most prominent of those causes (which impede the introduction and retard the progress of manufacturers in the United States) are the abundance of land compared with the population, the high price of labor and the want of a sufficient capital.'

[3] The following paragraphs are derived from *American and British Technology*, pp. 8–9, 43–63.

level of real wage rates which is said to have underlaid the widespread adoption of capital-intensive techniques. And here, Habakkuk maintained, the response of invention and innovation to the latent and manifest conditions of labor scarcity had to be reckoned with, entering the argument through the favorable bearing they would have exerted upon the profitability of continued industrial investment.

The one point that should be clear from all this is that references to the theses of Rothbarth and Habakkuk within a single, hyphenated breath should allude only – as they do here – to these writers' shared view that America's land abundance powerfully shaped the special characteristics of her technological practice and progress during the nineteenth century, and did so by raising the price of labor relative to capital equipment. The way this is held to have come about, however, is quite different in the two accounts.

A second point to be carried away is that the total separation and independence which Temin insists on maintaining between assessments of the version of the Basic Theorem having to do with the choice of technique on the one hand, and with technological progress on the other, is more faithful to the spirit of Rothbarth's static argument than it is to Habakkuk's.[1]

In Habakkuk's account the achievement of a (differentially) rapid rate of technological progress, and especially a form of technical advance biased toward augmenting the productive capacities of the industrial labor force, plays an integral part in the dynamic process whereby America's high land–labor ratio 'caused' the wage–rental ratio in manufacturing to be set at a comparatively high level. Hence the logical critique of the Rothbarth–Habakkuk thesis cannot be regarded as complete without a close look at the 'better machinery' version of the Basic Theorem – not only its own account, but in view of the supporting role it occupies in the 'more machinery' argument set forth by Habakkuk.

[1] Perhaps this is why Temin's brief gesture to deal with the 'better machinery' version of the Basic Theorem has naught to dc with the range of considerations broached by Habakkuk. Cf. 'Labor Scarcity' (1966), pp. 292–3; 'Labor Scarcity in America' (1971), pp. 260–1. Instead, rather revealingly, Temin's discussion under this head is focused on the very policy question that had occupied those who were first to advance the static argument ultimately embraced by Rothbarth: so long as tariffs can be imposed it is not necessary for a country to develop technological superiority in order to prevent its comparative land abundance from dictating complete specialization in non-industrial pursuits. It is true that Alexander Hamilton advocated patent protection and other subsidies to invention, but he was even more forceful in support of tariffs on manufactured imports. Hence, says Temin, it is easy to reject the logic of the proposition that the appearance of 'better machines' was in any way a necessary condition of the establishment of manufacturing in a comparatively land abundant region. This line of argumentation says nothing about the economic or non-economic determinants of actual technical invention and innovation during the nineteenth century, and therefore it is not pursued in the following pages.

II

The facts of nineteenth-century America's comparative technological progress *vis-à-vis* Great Britain remain just as obscure, and are if anything rather less immediately accessible than those which bear on the suspected contrast between these countries in regard to the prevailing degree of mechanization, or, distinctly, the capital-intensity of their average-practice methods of production. Yet, it happens that the first steps toward a systematic determination of the facts of the matter, ostensibly an empirical undertaking not relying upon theoretical inferences from the existence of prior causes, have been directed toward verifying whether or not there is some substance to the 'better machinery' version of the Rothbarth–Habakkuk thesis.[1] To date the results are still too limited in their scope to be conclusive, but they have done nothing to dispel doubts that cost-reducing innovation in America during the first three-quarters of the nineteenth century actually occurred more rapidly than in contemporary Britain, and/or manifested a more pronounced labor-saving bias.

Such doubts have been harbored for some time, fostered by suspicions that the analytical props available to support this wing of the Rothbarth–Habakkuk edifice really were too shaky to be trusted. Indeed, from an early point in the current discussions economic historians have steered away from serious re-evaluation of the propositions about the rate and bias of innovation, precisely because standard economic analysis was thought to offer less reliable guidance there than on questions of the choice among alternative known techniques of production.[2] As has been pointed out, however, in the more celebrated rendering of the Rothbarth–Habakkuk thesis the problems of technological progress and technological practice cannot be so casually divorced. A belief in the differential success Americans enjoyed in devising new, and more labor-saving mechanical methods, forms an integral element in Habakkuk's (although not Rothbarth's) account of the way land abundance encouraged the more pervasive adoption of capital-intensive techniques in the United States. And, as I shall try to show, the proposition with regard to the comparative labor-saving bias of American innovation is a central tenet of the faith.

The attractions of seeking empirical rather than theoretical support

[1] The principal reference here is to the recent work of Ephraim Asher, 'Industrial Efficiency and Biased Technical Change in American and British Manufacturing: The Case of Textiles in the Nineteenth Century,' *Journal of Economic History*, Vol. 32, 2 (June 1972). Asher makes use of the econometric method of ascertaining the existence and direction of (Hicksian) factor-saving bias that was first applied in a study of the character of aggregate technological change in the U.S. during the twentieth century: Paul A. David and Th. van de Klundert, 'Biased Efficiency Growth and Capital–Labor Substitution in the U.S., 1899–1960,' *American Economic Review*, Vol. 55, 3 (June 1965).

[2] Cf. Parker, 'Review of *American and British Technology*' (1963), p. 122; Temin, 'Labor Scarcity' (1966), esp. pp. 281, 293.

for that judgement about historical facts will become more fully apparent in the following section, which reviews the difficulties economic historians have had in establishing a compeling rationale to connect causally the condition of natural resource abundance with the supposed peculiarities of the pace and bias of American innovation.

Yet, recourse to measurement in this field cannot be expected to get far without the aid of theory; the very definition of an operational concept of technological progress at any aggregative level of empirical analysis requires a number of strong, and highly abstract assumptions. From the ensuing discussion it will be seen, moreover, that in order to interpret and compare such measures of the dimensions of technological change as the data available permit the historian to construct on well-chosen assumptions, an operational distinction must be made between differences in the bias of technological progress and differences in its rate. This can be done within the framework of a model of factor-augmenting efficiency changes by making yet a further theoretical commitment: one must embrace the concept of a concave, downward-sloping 'innovation-possibility frontier,' thereby developing a logic for the Rothbarth–Habakkuk thesis along the lines of the neoclassical theory of induced technical progress due to Kennedy, Weizsäcker and Samuelson.[1] The resulting dynamic distinction, between different positions on a common innovation-possibility frontier and the existence of different frontiers in different countries, formally resembles the static dichotomy Peter Temin employed – separating differences in the choice of techniques along a common unit isoquant (or factor-price frontier) from international divergences between the relevant technology sets from which such choices might be made. In that aspect, as well as in others, my recasting of the 'better machinery' version in this neoclassical mold should be read as a sequel to, and the logical completion of, the analytical re-evaluation of the Rothbarth–Habakkuk thesis initiated in Temin's work.

There is some parallelism in the outcome as well. For, it will be shown that the necessary aggregative model of induced technological biases, upon which we would need to rely – if only in verifying the facts as to the comparative bias and pace of factor-augmenting technical change in America and Britain – cannot readily be squared with certain gross features of the United States' record of economic growth in the nineteenth century. The empirical difficulties here are twofold. First, inasmuch as the concept of an innovation-possibility frontier envisages a strictly

[1] Cf. Charles Kennedy, 'Induced Bias in Innovation and the Theory of Distribution,' *Economic Journal*, Vol. LXXIV (September 1964); Paul A. Samuelson, 'A Theory of Induced Innovation Along Kennedy–Weisacker [sic] Lines,' *The Review of Economics and Statistics*, Vol. 47, 4 (November 1965).

uniform global alteration of factor efficiency across the entire spectrum of techniques, it must have difficulty assimilating the finding that nineteenth-century technological progress in the American economy exhibited a pronounced capital-*deepening* bias.[1] Second, it turns out that the particular pattern of changes in macro-production relationships observed for the United States cannot be rationalized within the framework of a *stable* innovation-possibility frontier. While shifts of the innovation-possibility frontier are entirely conceivable, the necessity of accepting their occurrence in this context signifies a practical failure of the underlying theoretical construct. For the latter treats the position of the frontier as established autonomously for each economy, and has no explanation to offer for it.

The parallelism to which I have referred, however, is not total. Where Temin's readers were left an impression that incompatibility with standard neoclassical analysis was sufficient cause for discarding the Rothbarth–Habakkuk thesis and turning historical research toward other, more promising problems, the suggestion here is that it is the attempt to provide neoclassical interpretations of that thesis which ultimately must be abandoned. Anticipation of more widespread concurrence in this view,[2] has motivated my effort in Part III of the present essay to reconstruct the entire argument along rather different lines. Although no more than a tentative reformulation can be offered at this stage, without some logically coherent set of hypotheses being thus set at risk, it is unlikely that the much needed systematic historical investigations will be undertaken.

LAND ABUNDANCE AND INDUCED INNOVATION: A MOST DIFFICULT CONNEXION

In a famous, fatefully vague passage of his *Theory of Wages* (1932), J. R. Hicks suggested that a factor-saving bias in technological change might somehow be induced by the prevailing structure of factor-prices, and more specifically that labor-saving inventions historically have been elicited in response to 'the dearness of labor':[3]

[1] This part of the argument draws upon the quantitative findings briefly reported by Moses Abramovitz and Paul A. David, 'Reinterpreting Economic Growth: Parables and Realities of the American Experience,' *American Economic Review*, Vol. 58, 2 (May 1973), and discussed by the same authors at greater length in 'Economic Growth in America: Historical Parables and Realities,' *De Economist*, Vol. 121, 3 (May/June 1973).

[2] Cf., e.g., the opening remarks, expressing dissatisfaction with the outcome of 'standard' theorizing on the subject, in Nathan Rosenberg, 'The Direction of Technological Change: Inducement Mechanisms and Focusing Devices,' *Economic Development and Cultural Change*, Vol. 18, No. 1, part 1 (October 1969).

[3] J. R. Hicks, *The Theory of Wages* (London: Macmillan Co., 1932), pp. 124–5.

The real reason for the predominance of labour-saving inventions is surely that which was hinted at in our discussion of substitution. A change in the relative prices of the factors of production is itself a spur to invention, and to invention of a particular kind – directed to economizing the use of a factor which has become relatively expensive. The general tendency to a more rapid increase of capital than labour which has marked European history during the last few centuries has naturally provided a stimulus to labour-saving inventions.

Contrary to the impression sometimes conveyed, Habakkuk – along with W. E. G. Salter, William Fellner, and many others more recently – recognized the flaw in attempts to interpret this as a static resource-allocation theory of invention and innovation, if the latter are to be regarded as activities essentially distinct from the process of factor-substitutions.[1]

For a firm in a competitive market setting which has arrived at a minimum-cost equilibrium with regard to the disposition of its factors of production, all inputs are equally 'dear' and 'productive' at the margin. Why, then, should a wage–rental ratio that was high in America *vis-à-vis* Britain – even supposing that situation may be linked to the comparative land-abundance of the former country along the lines of Habakkuk's argument – stimulate Americans to search for methods that saved *labor* as opposed to any of the other costly inputs? Salter made the point forcefully in commenting critically on Hicks's suggestion:[2]

If... the theory implies that dearer labour stimulates the search for new knowledge aimed specifically at saving labour, then it is open to serious objections. The entrepreneur is interested in reducing costs in total, not particular costs such as labour costs or capital costs. When labour costs rise any advance that reduces total cost is welcome, and whether this is achieved by saving labour or capital is irrelevant.

This critique of the Hicksian theory of induced invention has persuaded economic historians, without prompting them to inquire more closely into what it is that they mean when speaking of 'inventions' and 'changes

[1] Cf. Habakkuk, *American and British Technology*, pp. 9, 53; W. E. G. Salter, *Productivity and Technical Change* (Cambridge: Cambridge University Press, 1960), esp. Ch. III: William Fellner, 'Does the Market Direct the Relative Factor-Saving Effects of Technological Progress?' in Universities-National Bureau Committee for Economic Research, *The Rate and Direction of Inventive Activity* (Princeton, N.J.: Princeton University Press, 1962); *idem*, 'Two Propositions in the Theory of Induced Innovations,' *Economic Journal*, Vol. LXXI (June 1961). Samuelson ['A Theory of Induced Innovation' (1965), p. 355] unjustly groups Habakkuk among those 'who have attempted to use this theory [the 'vague notion that high wages somehow induce "labor saving" inventions'] to interpret the history of American prosperity.' Rosenberg, 'Direction of Technological Change' (1969), pp. 2–3 reviews the objections to Hicks's theory by Salter, Fellner and Samuelson.

[2] Salter, *Productivity*, pp. 43–4.

in the state of the arts.'[1] Consequently, they have not taken up the idea that in some respects the concrete development of new techniques – at least the translation of fundamental principles of mechanical, chemical or electrical engineering into new specific designs – 'blueprints' for machines and processes – usefully could be treated as involving factor-substitution choices guided by alterations in the constellation of input prices.[2] Certainly this essentially Hicksian conception was not pursued in Habakkuk's work.

Nor was use made of Fellner's point that the theoretical position is much altered by dropping the assumption of perfect competition in all markets.[3] A firm operating in *differentially* imperfect factor-markets, possessing greater monopsony power in dealing with some sets of suppliers than with others, might well anticipate that any attempt to expand output would entail alteration of the relative marginal costs of their inputs, and thus devote resources to developing techniques having a countervailing factor-saving bias. The lack of attention paid to this point is curious, since, in quite another connexion, Habakkuk was prepared to assert that during the early nineteenth century *labor* market imperfections were more pronounced in the United States than in England.[4] The latter historical proposition, although eminently testable, has remained unsubstantiated – perhaps because its significance in the present connexion has not been duly appreciated. To stress it would be to make a critical part of the argument turn, once again, on a view of the difference between the workings of labor markets in the two countries, rather than on anything about the nature of technology and inventive activity in the nineteenth century. But in the broader context of Habakkuk's thesis, there would remain the difficulty that the existence of more serious labor market imperfections was a condition both distinct from the *comparative* dearness of labor in

[1] Cf., e.g., Nathan Rosenberg, *Technology and American Economic Growth* (New York: Harper and Row, 1972), pp. 23–4, for recognition that 'the distinction between innovation and adaption (or between a shift in a production function and movement along an existing production function) is frequently much overstated.' Yet, later in the same work Rosenberg (pp. 55–7) accepts the Salter–Fellner critique of Hicks which is predicated on just this distinction.

[2] Cf. Salter, *Productivity*, pp. 13–16, for an extension of the approach taken by H. B. Chenery, 'Process and Production Functions from Engineering Data,' in W. W. Leontief and others, *Studies in the Structure of the American Economy* (New York: Oxford University Press, 1953). These ideas are given a more formal neoclassical development – and robbed, in the process, of some of their operational content – by S. Ahmad, 'On the Theory of Induced Innovation,' *Economic Journal*, Vol. LXXVI (June 1966). The bearing of this conception of 'innovations' upon questions of interest to economic historians is further considered in Part III, below.

[3] Cf. Fellner, 'Two Propositions' (1961), p. 307.

[4] Cf. Habakkuk, *American and British Technology*, p. 63.

America, and not connected in any obvious way to the comparative abundance of natural resources there.

Much the same objections also may be lodged against Nathan Rosenberg's recent effort to resuscitate the seemingly moribund 'better machinery' version of the Rothbath–Habakkuk thesis by applying another proposition due to Fellner – this one regarding the influence of expectations about the future course of macro-economic scarcity relations:[1]

Firmly-held expectations about the future rise in the relative price of *any* input will be expected to induce exploratory activity to economize on the use of that particular input.... It was not so much the high level of wages [sic] but rather the persistent experience of pressures on the labor market, the numerous opportunities elsewhere in a resource-abundant environment, and the high degree of labor mobility which conditioned people to expect further future increases in the cost of labor relative to other inputs. Hence a strong bias toward the development of labor-saving techniques.

The difficulty here lies not so much in the internal logic of Fellner's proposition,[2] as it does in Rosenberg's presupposition that American resource abundance had somehow expectations there of a *differentially faster* rise in the price of labor relative to capital. Both countries, however, were experiencing a secular rise of the overall capital–labor ratio, and a rational basis for the requisite *divergence* of anticipated trends in the factor–price ratio simply has not been established. Indeed, if Habakkuk's characterization of the labor-supply situations prevailing during the first half of the nineteenth century is accepted, American industrial employers should have been looking forward to a gradual filling-up of the country, to mounting barriers against entry into independent cultivation, and to the eventual 'closing of the frontier'; whereas, across the Atlantic, their counterparts faced a bleaker future of exhausted agricultural labor-reserves, more pervasive and aggressive trade unionism, and still other pressures raising real wages.

If it has been difficult to forge durable theoretical connexions along

[1] Rosenberg, *Technology*, p. 57.

[2] Fellner, 'Two Propositions' (1961), pp. 306, 307, notes that for firms rationally to devote resources to develop a relatively labor-saving technique (as a speculation on the expected future price of labor relative to capital in the economy at large) it must be anticipated that 'no further invention and innovation will become available' during the relevant investment planning horizon. This *seems* to acknowledge the very real difficulty which arises in the Fellner model if the externalities created by investment in invention are taken into account. When externalities are large in relation to the total privately appropriable benefits anticipated, firms will have a powerful incentive to become technological 'free-riders,' especially when the supposed signals guiding the direction of inventive activity are not only relevant to firms in general, but are plainly there for all to see. On the problem of externalities in this connexion, cf. the seminal paper by Kenneth J. Arrow, 'Economic Welfare and the Allocation of Resources for Invention,' in *The Rate and Direction of Inventive Activity, op. cit.*

these lines between macro-scarcity relationships and the directions toward which inventive activities in America and Britain were bent, finding a general way to account in these same terms for international differences of the *pace* of innovation appears to pose still more vexing difficulties. This is not so surprising, for when attention is focused exclusively on the determination of the realized rate of technological progress, a peculiarity of the whole theoretical approach becomes more obvious. Economists have spilt ink copiously over the demand side of the ledger, but surely the pace of innovation (and, on further reflection, even its factor-saving bias) must in some way have reflected influences to be found on the supply side.[1]

William Parker voiced his sense of discomfort with the general thrust of the Rothbarth–Habakkuk thesis on this score fully a decade ago. Yet, in doing so he took up the polar opposing position that the course of technological advances might not be only supply-determined but essentially exogenous except in some very long-run sense:[2]

> ... we need to know much more about the sheer technical and intellectual constraints on invention before we can be wholly at ease with an economic explanation of its course. It is after all an international process, whose course in the largest sense must surely depend upon the general growth of scientific and technical knowledge rather than the factor endowment of any specific national state.

The best empirical defense that presently could be mounted for the prevailing contrary assumption of an infinitely elastic supply of 'innovations,' would have to rest in large part on Jacob Schmookler's[3] subsequently published investigations of historical and contemporary patent statistics relating to hundreds of inventions in four branches of industry in the United States. Schmookler concluded that the potential commercial profitability of invention, signalled by the recent state of the market for the class of products in which they would be embodied, played the dominant role in determining the volume of patenting (inventive) activity. Moreover,

[1] Rosenberg (*Technology*, p. 59) deftly characterizes the deficiency of a theory of the *demand* for inventions by recalling this bit of Elizabethan badinage from act III, scene I of *Henry IV, Part I*:

> 'Glendower: I can call spirits from the vasty deep.
> Hotspur: Why, so can I, or so can any man; but will they come when you do call for them?'

Economists have not been free of qualms about the inadequate attention their discipline has been able to give to the supply side of the question. Cf., e.g., K. J. Arrow, 'Classificatory Notes on the Production and Transmission of Technological Knowledge,' *American Economic Review*, LIX, 2 (May 1969), esp. pp. 34–5.

[2] Parker, 'Review of *American and British Technology*' (1963), p. 122.

[3] J. Schmookler, *Invention and Economic Growth* (Cambridge, Mass.: Harvard University Press, 1966).

in most of the instances of 'important' inventions studied by Schmookler, the necessary fundamental scientific knowledge was found to have been available long before the actual appearance of the invention.

Clearly either of the extreme positions in this debate is less tenable than the intervening ground, where both economic and non-market supply constraints and demand conditions can be admitted to have governed the historical course of inventive activity. The obvious rejoinder to Schmookler's argument (leaving aside the problems raised by the elision between 'patenting' and inventive activity) is that the sample studied is not only small but selective, and rather arbitrarily confines empirical analysis to fields in which there was a great deal of patenting activity. Is it demand deficiencies or supply constraints which explain the fact that in each historical period there were other, less fully developed branches of industrial endeavor, where the general level of patenting activity remained minimal?

Pursued a little further, this query leads one toward the attractively plausible notion that in a given state of basic scientific understanding about the world, whether or not technological innovations can be generated in rapid succession is likely to depend upon the particular directions in which the search for new, cost-reducing methods happens to be extended. The strong focus upon the development of a *machine* technology often is remarked on as a ubiquitous feature of industrialization during the nineteenth century, in contrast with the more recent experience of modern economic growth. Further, to follow the phrasing of the observation by Rosenberg,[1]

> the invention of new machines or machine-made products . . . involved the solution of problems which required mechanical skill, ingenuity and versatility but not, typically, a recourse to scientific knowledge or elaborate experimental methods.

Presumably this was the range of considerations prompting Habakkuk to suggest, at several junctures of his argument, that early in the nineteenth century 'the more capital-intensive of the existing range of techniques had the greatest possibilities of technical progress.' At one point, a hundred pages later in the book, he comes right out with the obverse:[2]

> It is also probable that the technical knowledge of the nineteenth century was more capable of providing solutions to problems of labour-scarcity than to those of natural-resource scarcity. . . . [T]he main reason is that most solutions of natural-resource scarcity problems have been obtained from chemistry and electrical

[1] Rosenberg, *Technology*, p. 54.

[2] Habakkuk, *American and British Technology*, p. 161. The previously cited allusions appear on pp. 50 and 61. The interpretation offered here of the way they fit into Habakkuk's larger design differs from that of Temin, 'Labor Scarcity' (1966), p. 293.

engineering, and in the nineteenth century these lagged behind mechanical engineering, simply because they are more difficult subjects.

Can we not surmise from such assertions, in which the state of fundamental scientific knowledge *and* the scale of the commitment of resources to invention are implicitly being held constant, that the essential Habakkukian view regards the pace of American technological progress as having been indirectly governed at any moment in time by the very same forces which would determine the factor-saving bias of the search for innovations? On such a reading, the question of the pace and that of the bias of technological progress are seen to be really no more independent of each other than the argument about the choice of technique in America is separable from that concerning the character of technological progress.[1] Moreover, it then appears that the hinge upon which swings the logic of Habakkuk's entire thesis is the factual allegation that comparatively land-abundant America experienced a relatively labor-saving bias in technical change. But, as we have just shown, this is precisely the coupling that economic historians have found it most difficult to effect on purely theoretical grounds. What stronger motivation could be provided for considering the means available for resolving the issue empirically?

TOWARD AN EMPIRICAL RESOLUTION OF THE QUESTION OF BIAS

Ephraim Asher[2] recently has undertaken to test the factual basis of the Rothbarth–Habakkuk thesis on invention and innovation by measuring and comparing the factor-saving bias of technical change in manufacturing industries located in the United States and Britain. His published findings relate only to the cotton textile and woolens industries during the latter half of the nineteenth century. But the proposal is more general, and quite intriguing as a way out of the apparent *cul de sac* into which economic historians interested in the character of American technology have been

[1] For the benefit of the reader who has lost his way momentarily in this thicket, the point at which the exposition of Habakkuk's thesis has now arrived may be rapidly summarized: (1) Land abundance *and* continuing rapid industrial expansion allegedly lead to a high relative wage–rental ratio and thus to adoption of capital-intensive practices. (2) But, for comparatively rapid industrial expansion to be *sustained* in America, rapid technological progress, or innovation especially biased toward labor-saving, supposedly was required to mitigate the adverse cost effects of pressing hard against the labor-supply constraint. (3) More rapid technological advance would be readily forthcoming from a search for new mechanical devices, but not so otherwise. Hence the historian's hang-up: if a labor-saving bias of inventive activity cannot theoretically be deduced for the United States from the fact of the country's comparative land abundance, how can the argument be supported?

[2] Cf. Asher, 'Biased Technical Change' (1972).

led by the pursuit of theories of induced technical change. It is a proposal built upon the possibility of econometrically estimating the absolute difference between the respective growth rates of labor-efficiency and capital-efficiency in the context of a CES production model in which technological changes take a purely 'factor-augmenting' form.[1]

To grasp the main advantages and drawbacks of this approach, perhaps it will prove easiest to start with the idea of a production function whose arguments are the inputs (L') labor measured in *efficiency* units, and (K') capital services measured in *efficiency* units. For the M-th industry we write the production function

$$Y_M = P(K'_M, L'_M), \tag{1}$$

where $P(\cdot)$ is taken to be (a) twice differentiable, (b) homogeneous of the first degree and (c) of the CES form, i.e., the elasticity of substitution between the inputs is constant. The dated inputs each may be conveniently represented as the product of a 'natural units' measure (L measured in man-hours, for example) and a corresponding (man-hours) efficiency index: $L'(t) = L(t) \cdot E_L(t)$, and $K'(t) = K(t) \cdot E_K(t)$.

Further, it is assumed that the factors of production receive remuneration equal to the value of their respective marginal products in the M-th industry, letting w and r represent real wage rate and the real rental rate, respectively. Then, denoting the marginal physical products of labor and capital – each measured for this purpose in *natural* units – by $(\partial Y/\partial L)$ and $(\partial Y/\partial K)$, respectively, the foregoing equilibrium conditions are expressed as follows:

$$\left.\begin{aligned} \partial Y/\partial L &= E_L(\partial Y/\partial L') = w \\ \partial Y/\partial K &= E_K(\partial Y/\partial K') = r \end{aligned}\right\}. \tag{2}$$

Using the familiar definition of the (constant) elasticity of substitution σ, we may immediately write:

$$\sigma \equiv \frac{d\{\ln(K'/L')\}}{d\{\ln(\partial Y/\partial L')/(\partial Y/\partial K')\}}. \tag{3}$$

Then, adopting the general notation in which proportional rates of change are indicated by asterisks directly above the variable (e.g., $dK'/K' \equiv \overset{*}{K'} = \overset{*}{K} + \overset{*}{E}_K$), and identifying the rates of change of the input efficiency-indexes by

$$\lambda_L \equiv \overset{*}{E}_L \quad \text{and} \quad \lambda_K \equiv \overset{*}{E}_K. \tag{4}$$

[1] That is to say technological progress is conceived of as simply reducing the per unit requirements of the inputs measured in 'natural' dimensions (man-hours, constant quality machine hours, etc.), without altering the ease of substitution among the inputs. For a more formal, but possibly a less heuristically appealing exposition of the model than the one offered below, cf. David and van de Klundert, 'Biased Efficiency Growth' (1965).

Equation (2) may be invoked to justify rewriting (3) in the form:

$$\sigma = \frac{(\overset{*}{K}+\lambda_K)-(\overset{*}{L}+\lambda_L)}{(\overset{*}{w}-\lambda_L)-(\overset{*}{r}-\lambda_K)}. \tag{5}$$

It is then simple enough first to rearrange equation (5) as

$$(\overset{*}{K}-\overset{*}{L}) = \sigma(\overset{*}{w}-\overset{*}{r})+(1-\sigma)(\lambda_L-\lambda_K), \tag{5a}$$

next, to add $\{-\sigma(\overset{*}{K}-\overset{*}{L})\}$ to both sides, and thus to arrive at the basic relationship from which Asher's approach may be regarded as having derived:

$$(\overset{*}{K}-\overset{*}{L}) = \left\{\frac{\sigma}{1-\sigma}\right\}(\overset{*}{\theta}_L-\overset{*}{\theta}_K)+(\lambda_L-\lambda_K). \tag{6}$$

In this last expression θ_L and θ_K represent the shares of (output elasticities with respect to) the inputs of labor and capital, respectively; hence

$$\left.\begin{aligned}\overset{*}{\theta}_L &= \overset{*}{w}+(\overset{*}{L}-\overset{*}{Y})\\ \overset{*}{\theta}_K &= \overset{*}{r}+(\overset{*}{K}-\overset{*}{Y})\end{aligned}\right\}. \tag{7}$$

Now, let it be specified that the difference between the respective rates of factor-augmenting efficiency change is a *constant*:

$$\beta = (\lambda_L-\lambda_K), \tag{8}$$

and immediately we see that equation (6) furnishes the basis for a linear regression model from which (joint) estimates of the parameters σ and β can be obtained.[1] The interpretation of these estimates is quite straightforward. Upon a closer inspection of equations (5a) and (2) it will be found

[1] An 'impossibility theorem,' due to P. Diamond and D. McFadden but still unpublished, holds that the constancy of β is required to estimate the parameter σ in this model, and vice versa. Cf. the account of this theorem by M. Nerlove, 'Empirical Studies of the CES and Related Production Function,' in *The Theory and Empirical Analysis of Production*, ed. M. Brown (New York: Columbia University Press for National Bureau of Economic Research, 1967). Abramovitz and David, 'Reinterpreting Economic Growth' (1973) makes use of equation (6) in the text above as a least squares regression model. Alternatively, holding σ and β constant, equation (6) may be integrated as a linear, first-order differential equation, yielding

$$\ln(K/L)_t = A_0+\left\{\frac{\sigma}{1-\sigma}\right\}\ln(\theta_L/\theta_K)_t+\{\beta\}t,$$

where t (time) is the index of integration and A_0 is a constant. The latter is the form underlying the regression models employed in David and van de Klundert, 'Biased Efficiency Growth' (1965), and Asher, 'Biased Technical Change' (1972).

that when $\sigma < 1$, the condition of efficiency-changes having a labor-augmenting bias ($\beta > 0$) is equivalent to the Hicksian definition of a *labor-saving* bias in technological progress, inasmuch as it will tend to depress the ratio between the marginal productivity of labor and the marginal productivity of capital. Analogously, the condition ($\beta < 0$) is equivalent to a Hicksian *capital*-saving bias when $\sigma < 1$.[1] Thus the estimation of the two parameters of the model yields an unambiguous test for the presence (and the nature) of biases in technological change – or however one chooses to label the process through which alterations in factor-efficiency are brought about.

Some years ago, in an econometric study based on this approach, and employing annual observations for the United States Private Domestic Economy in the period 1899–1960, Theo van de Klundert and I found that the elasticity of substitution was well below unity and technological change during the twentieth century had indeed exhibited a significant (Hicksian) bias toward labor-saving. Much more recently, albeit from an analysis of a less ample and less reliable body of aggregate statistics, Moses Abramovitz and I have concluded that the same statements broadly hold true for the United States Domestic Economy of the nineteenth century.[2] Asher's published results likewise attest to the existence of a relatively rapid rate of labor-augmenting technical change (again equivalent to a Hicksian labor-saving bias) in the cases of the American cotton textiles industry and the woolens industry during the period 1850–1900.

Was this bias toward labor augmentation peculiar to American experience? The corresponding aggregate production function studies for Britain have yet to be attempted, but the results of Asher's parallel econometric investigation of the production of cotton goods and woolens would suggest otherwise. Moreover, there is no econometric support from either branch of the textile industry for the view that the parameter β, measuring the absolute bias in favor of labor-augmentation, was *greater* in America than in the United Kingdom. Quite the contrary.

[1] $\beta = 0$ obviously represents the case of Hicks neutrality. When $\sigma > 1$, $\beta > 0$ corresponds to Hicks capital-saving, and $\beta < 0$ to Hicks labor-saving technical progress. Cf. David and van de Klundert, 'Biased Efficiency Growth' (1965), for an alternative demonstration of these correspondences.

[2] Cf. *ibid.*, p. 377, for the estimates: $\hat{\sigma} = 0{\cdot}316$ and $\hat{\beta} = 0{\cdot}007$ per annum, relating to the period 1899–1960. Using the same definitions of capital (net reproducible and non-reproducible stock, in constant dollars) and labor (man-hours), Abramovitz and David, 'Reinterpreting Economic Growth' (1973) reports the following estimates for the period 1800–1903/7: $\hat{\sigma} = 0{\cdot}20$ and $\hat{\beta} = 0{\cdot}017$ per annum. Abramovitz and David, 'Economic Growth in America' (1973) reports the following 95 per cent confidence interval for the difference between the growth rates of labor efficiency and *reproducible* capital efficiency over the period 1800–1903/7: $0{\cdot}0086 < \beta' < 0{\cdot}0215$.

For Asher, at least, 'the resulting estimates appear to be in contrast with what one might have expected from the Rothbarth–Habakkuk implications.'[1]

There is an antecedent question, however, which must not be glossed over. What precisely does the Rothbarth–Habakkuk thesis have to say about the *comparative* magnitude of the measure (β) upon which Asher's work has focused? Putting the issue more concretely, let us suppose it is more generally found that with $\sigma < 1$ in both countries, $\beta^{GB} > \beta^{US} > 0$. Would this imply rejection of the empirical premises of the 'better machinery' version of the Rothbarth–Habakkuk thesis (1) because the rate of labor-augmenting technical progress was higher in Britain than in America, or (2) because the mix of innovations in Britain was skewed more toward labor-augmentation than was the case in America, or (3) because both the foregoing inferences are warranted? The conclusions to be extracted from such cross-country comparisons of β (for the same industry or sector) are not immediately clear, since the absolute difference between the rates of factor-efficiency change confounds in a single measure two distinguishable conditions. One has had to do with the comparative *rate*, the other with the comparative *direction* of innovative activities. If Asher's proposal is to prove fruitful, it will be necessary first to remove this ambiguity.

A convenient and intuitively appealing measure of the *relative* bias, or 'mix,' of factor-augmenting technical change in the M-th sector of the i-th country is provided by B_M^i in the expression

$$(\lambda_K^i)_M = (1-B_M^i)\,(\lambda_L^i)_M. \tag{9}$$

For the condition of a (Hicks) *labor*-saving bias when $\sigma < 1$ we simply require $B > 0$, whereas for a (Hicks) *capital*-saving bias we require that $B < 0$. The absolute measure of bias employed by Asher – defined in equation (8) – can be then decomposed into two elements by writing it as

$$\beta_M^i = B_M\,(\lambda_L^i)_M. \tag{10}$$

From this it is obvious that $\beta_M^i = B^i{}_M = 0$ whenever technical change is Hicks neutral. Equally obviously, when that condition does not obtain in either the i-th or the j-th country, merely comparing β_M^i with β_M^j provides no basis from which to reach conclusive inferences about the separate inequality-relations in which we really are interested. For example, from the finding that $(\beta^i > \beta^j > 0)$ we may confidently infer

[1] Asher, 'Biased Technical Change' (1972), p. 441. These econometric findings are accepted for the purposes of the present discussion, where concern lies mainly with the problem of how such results should be interpreted.

that all the elements – B^i, B^j and λ_L^i, λ_L^j – entering the comparison are positive. But since

$$(\beta^i - \beta^j) \gtrless 0 \Rightarrow (B^i/B^j) \gtrless (\lambda_L^j/\lambda_L^i), \tag{11}$$

$(\beta^i > \beta^j)$ alone does not imply that $(B^i > B^j)$ and/or that $(\lambda_L^i > \lambda_L^j)$.

Now, identifying Britain as the *i*-th country and America as the *j*-th, the preceding formalities should make it clear that Asher's 'unexpected' findings ($\beta^{GB} > \beta^{US} > 0$, for the textile industries) do serve to rule out the *joint* assertion: $(\lambda_L^{US} > \lambda_L^{GB})$ *and* $(B^{US} > B^{GB})$. This can be construed as a direct contradiction of Habakkuk's position only insofar as he is thought to have also subscribed to a theory of the *rate* of technological innovation in which $(\lambda_L^j \gtrless \lambda_L^i) \Leftrightarrow (B^j \gtrless B^i)$. For, if the *joint* assertion is false, *both* of its parts must be false. We need, then, to reconsider the problem of the relationship between the rate and the direction of inventive activity.

AN INTERPRETATION ON KENNEDY–WEIZSÄCKER–SAMUELSON LINES

The notion that the rate of labor-augmenting technical change and the relative bias of innovation are constrained to bear some stable relationship to each other is hardly a peculiarity of Habakkuk's views about nineteenth-century technology. Precisely that implication is embedded in the concept of an 'innovation-possibility frontier' (IPF), describing the locus along which trade-offs must be made between achieving a higher rate of labor-augmentation and a lower rate of capital-augmentation, or vice versa.[1] The IPF is nothing but a transformation function defined for the respective factor-efficiency growth rates yielded by innovative activities, and, by analogy with familiar transformation loci in the sphere of production theory, it is usually posited as being downward-sloping and concave. (See Figure 1.)

Characteristically, for economic theorizing has yet fully to come to grips with the supply side of technological change, the position and precise shape of the IPF are not explicitly explained. Presumably they are in some manner determined for any economy or branch of industrial endeavor by the state of the relevant body of fundamental scientific knowledge, and the amounts and qualities of the various resources devoted to generating technological innovations.[2] The simplest way of proceeding is to neglect any national differences which might have a bearing upon

[1] Cf. Kennedy, 'Induced Bias' (1964), for the original exposition of the concept.

[2] The significant property of the usual characterization of the IPF is that neither its position nor shape depends upon the prevailing relative factor proportions or the relative factor shares. Cf. Samuelson, 'A Theory of Induced Innovation' (1965), pp. 351–2. A very different theory results if one discards the assumption that firms have special knowledge of the shape

the position of the IPF, and suppose that innovation in America and Britain during the nineteenth century was constrained by a common frontier – CC' in Figure 1. Now, in a comparison between two positions along *any* negatively sloping frontier, where λ_L is higher, λ_K, and *a fortiori* λ_K/λ_L ($\equiv 1-B$) must be lower. Consequently, confining the discussion to the case of a common, internationally shared IPF would be sufficient to establish that $B^{US} \gtreqless B^{GB} \Leftrightarrow \lambda_L^{US} \gtreqless \lambda_L^{GB}$. Under these conditions Asher's proposal would constitute a conclusive test of the factual premises of the 'better machinery' version of the Rothbarth–Habakkuk thesis, inasmuch as either of the statements ($\lambda_L^{US} > \lambda_L^{GB}$) or ($B^{US} > B^{GB}$) implies the other, and hence implies ($\beta^{US} > \beta^{GB}$).

The foregoing is, however, a terribly strict construction, neglecting Habakkuk's many allusions to cultural, institutional and economic divergences which could have given rise to a difference between the positions of the relevant innovation-possibility frontiers for America and Britain.[1] But to admit the existence of different national IPF's threatens to return us to our original quandary; as we really do not know *a priori* which country enjoyed the superior set of innovation-possibilities, some additional piece of information must now be provided before we can usefully interpret the comparison between β^{US} and β^{GB}.

To be able to make *independently* justifiable assertions regarding the comparative magnitude of the measures of relative bias, B, therefore would be quite helpful.[2] Consider Figure 1: the slope of the ray OA extending from the origin is $1/(1-B^{US})$. Were it established that $B^{US} > B^{GB}$, Britain would have to be located somewhere to the right of the ray OA. If in addition $\beta^{US} < \beta^{GB}$, we would also know that the British position lay above the locus PP' paralleling the 45° line. Thus we could place it unambiguously within the shaded wedge *above* the American

[1] Cf. Habakkuk, *American and British Technology*, pp. 111–12, 115–16, on social attitudes toward innovation, patent laws, and 'the possibility that Americans were, from the circumstances of their life, better at inventing than the English.' Arguing that, if true, the latter could have been only a question of the mobilization of ingenuity (*Ibid.*, pp. 116–17, 216–17) Habakkuk suggests finally – as is noted above – that 'the principle difference... was not that the Americans were more inventive or better able to develop their inventions; it lay in the objects to which these abilities were devoted.' (*Ibid.*, p. 118.)

[2] The following may be derived from equation (11) without aid of geometry.

---

of the trade-off locus, and replaces it by the assumption that in forming a view as to the probability of achieving the same percentage reduction in production costs via labor-augmenting improvements as by capital-augmenting improvements (given a level of expenditure on invention), the prevailing relative weights of these factors in total production costs will be considered. The implicit theorizing about the processes of technological change contained in the construct of a stable 'efficient possible innovations' locus, such as the IPF introduced by Kennedy, is exposed to a more thorough-going criticism in the concluding section of Part II, below.

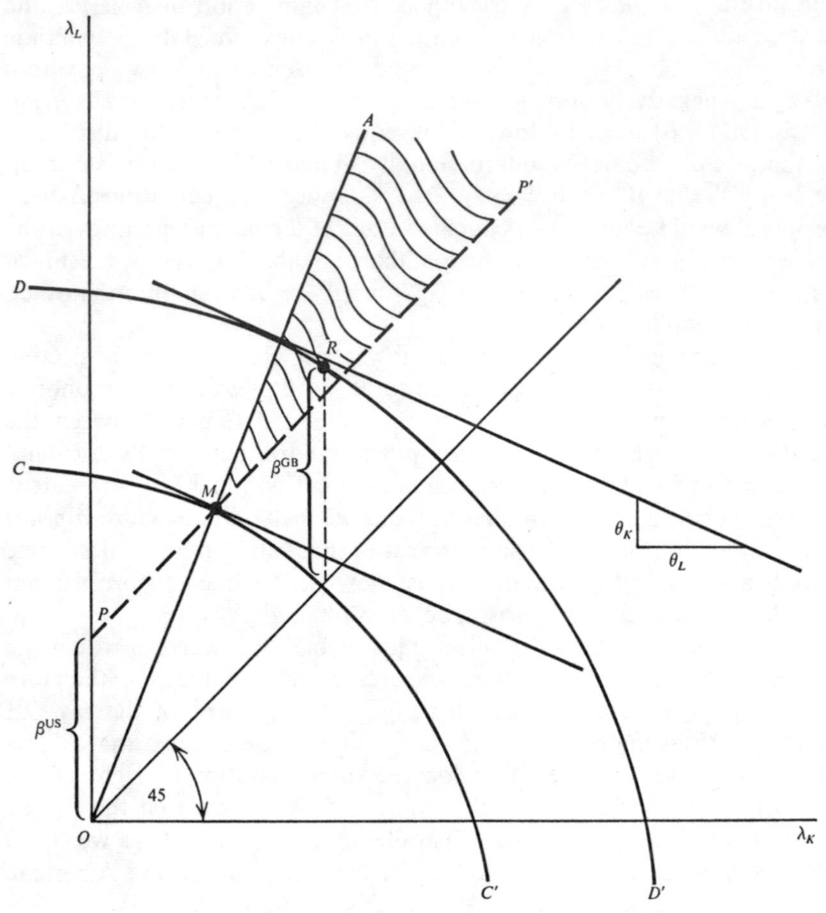

Figure 1 Locating America and Britain on their respective innovation-possibility frontiers

position at M, the intersection of OA with PP'. The point R on the frontier DD', for example, satisfies these restrictions. On the other hand, if $\beta^{US} > \beta^{GB}$ (again given $B^{US} > B^{GB}$) there would be no basis for inferring that the IPF relevant to Britain's innovative activities must have dominated the American IPF since the British position might lie anywhere between OMP' and the abscissa.[1]

Here it seems some empirical use might be found for the theory of induced innovation due to Kennedy, Weizsäcker and Samuelson, which

[1] Were it established that $B^{US} < B^{GB}$, Britain's position in Figure 1 could be placed to the left of OA, and the finding that $\beta^{US} > \beta^{GB}$ would imply a location on an inferior IPF, within the wedge PMO. The opposite finding, $\beta^{US} < \beta^{GB}$, however, would be inconclusive as to the relative position of the British IPF.

tells us that when resources are being allocated optimally among the alternative innovation-possibilities described by an IPF, the measure of relative bias defined here (B) would be greater where the relative share of labor (θ_L/θ_K) was larger.

To see why this should be so is not difficult.[1] In a constant returns to scale production activity which uses labor and capital inputs, the overall rate of cost-reduction per unit of output resulting from any mix of factor-augmenting innovations is

$$\lambda = \theta_L\lambda_L + \theta_K\lambda_K. \tag{12}$$

For a competitive industry or enterprise facing given factor-prices, w and r, and also for the economy as a whole *in the short run*, the maximum λ_L achievable within the constraint imposed by the IPF

$$\lambda_L = H(\lambda_K), \qquad H' < 0, \qquad H'' > 0 \tag{13}$$

will obviously be reached where

$$(\mathrm{d}\{\lambda_L\})\theta_L = -(\mathrm{d}\{\lambda_K\})\theta_K. \tag{14}$$

Since the slope of the frontier is $H'(\lambda_K)$, a function of λ_K, let us represent it as the specific (invertible) function $F(\lambda_K/[H(\lambda_K)])$:

$$\frac{-\mathrm{d}\{\lambda_L\}}{\mathrm{d}\{\lambda_K\}} = F(\lambda_K/\lambda_L), \tag{15}$$

and observe that the concavity property, $H''(\lambda_K) > 0$, implies $F'(\cdot) > 0$. Now consider the inverse function $F^{-1}(\cdot) \equiv G(\cdot)$, and use the first-order condition given by equation (14) to express $G(\cdot)$'s argument in terms of the factor-share ratio, thus:

$$\lambda_K/\lambda_L = G(-[\mathrm{d}\lambda_L/\mathrm{d}\lambda_K]) = G(\theta_K/\theta_L). \tag{16}$$

Applying the inverse function rule, and noting that $F'(\cdot) > 0$, serves to establish immediately the sign of the first derivative: $G'(\cdot) = [F'(\cdot)]^{-1} > 0$. Therefore, recalling the definition of B from equation (9), we arrive at the relationship already indicated:

$$\frac{\mathrm{d}B}{\mathrm{d}(\theta_K/\theta_L)} = \frac{\mathrm{d}B}{\mathrm{d}(1-B)}\{G'(\theta_K/\theta_L)\} < 0. \tag{17}$$

The slope of the line(s) tangent to the IPF(s) in Figure 1 is equal to the ratio θ_K/θ_L. From this representation of the first-order condition for max λ, it is evident – for those who find geometry easier – that the higher is labor's share, θ_L, the flatter will be that slope. Hence, the greater will be the *optimal* ratio (λ_L/λ_K) and the corresponding measure of relative bias, B.

[1] The exposition through to equation (14), below, has been adapted from Samuelson, 'A Theory of Induced Innovation (1965), pp. 343–4.

For the branches of textile manufacture studied by Asher, the share of wages in value added was persistently higher in the United States than in the United Kingdom.[1] The same condition appears to have been obtained with respect to the aggregate share of wages in national income, at least during the latter half of the nineteenth century.[2] In the context of the model we have just considered, this added information (i.e., that $\theta_L^{US} > \theta_L^{GB}$) implies three things about the state of affairs with regard to textile innovations. (1) The inference that $B^{US} > B^{GB}$ would follow immediately if Britain and America were on the same IPF, but the situation thereby envisaged could not be consistent with Asher's determination that $\beta^{US} < \beta^{GB}$. Hence, the latter must imply that the two countries were *not* constrained by a common IPF. (2) If the only inter-country differences in the positions of innovation-possibility frontiers were homothetic displacements of the sort depicted in Figure 1, along any given ray from the origin the slopes of alternative IPF's would be identical and it could be quite generally asserted that $B^{US} > B^{GB}$. That being the case, the condition $\beta^{GB} > \beta^{US}$ must imply – as we have indicated – that the IPF relevant to textile innovations in Britain was superior to, i.e., dominated the existing frontier which constrained such innovations in America. (3) To say this – placing Britain at a point above the *PP'* locus in Figure 1, yet below the ray *OA* – is tantamount to concluding that the British textile industries enjoyed both a comparatively rapid rate of general efficiency improvement (λ) *vis-à-vis* the same industries in the United States, *and* a higher rate of labor-augmentation (λ_L) in particular.[3]

The preceding line of analysis cannot be immediately generalized

[1] Cf. the sources cited by Asher, 'Biased Technical Change' (1973), p. 438, n. 12. The published article does not present the underlying labor share data.

[2] Phyllis Deane and W. A. Cole, *British Economic Growth, 1688–1955* (Cambridge: Cambridge University Press, 1962), Table 65, p. 247, give figures for the period 1860–1914 indicating wages and salaries represented 47–50 per cent of Great Britain's net national income at factor-cost. For the United States, Robert Gallman's estimates put the share of labor in N.N.P. just below 70 per cent, on the average. in the period 1840–1900. Cf. Lance E. Davis, *et al.*, *American Economic Growth* (New York: Harper & Row, 1972), Ch. 2. An alternative set of labor share estimates presented by Abramovitz and David, 'Reinterpreting Economic Growth' (1973), Table 2, suggest somewhat lower figures than Gallman's: 55 per cent of G.N.P. and 60–5 per cent of N.N.P. during the period 1855–1905.

[3] From the latter inferences it should be obvious that an alternative route must exist to extract the same information. The rate of labor-efficiency growth (λ_L) could be estimated simultaneously with (σ), making use of the marginal-productivity relationship which is often employed indirectly to estimate the parameters of CES functions. Cf., e.g., the discussion in David and van de Klundert, 'Biased Efficiency Growth' (1965), pp. 366–70. Note, however, that as (β) is also jointly estimated with (σ), it would be necessary to develop a regression procedure constraining the estimates of (β) and (λ_L) to be consistent with the same value for (σ). The estimation of (λ_L) by regression would also require an explicit specification of the time-path of labor efficiency, and observations on real output, for each country considered.

and applied to the problem of the comparative characteristics of technological progress in the American economy, or even in its industrial sector. The difficulty has nothing to do with the 'representative' or 'unrepresentative' nature of the textile industries, but rather stems from the limitations of the analysis itself. It is a short-run argument in which optimization of the factor-saving mix of innovations is carried out on the basis of fixed relative factor-prices (and factor-shares). Clearly, this assumption is more appropriate to discussion of individual branches of industry than entire economies; but even within an industry that behaved as a price-taker in the factor markets, the induced changes in relative factor-efficiencies would affect the factor-shares, and impinge upon the subsequent bias of innovations.

At the macro-economic level, therefore, the question becomes one of whether or not there is some process driving the system toward a long-run equilibrium mix of λ_L and λ_K, and on what that equilibrium depends. Samuelson[1] has dealt with this question explicitly, and his treatment of it is sufficiently accessible to obviate the need for a full recapitulation here. The essential point is simple enough. Starting with the secular tendency for capital to accumulate at a rate more rapid than that at which the labor force was growing, there would be (as Hicks observed) an incipient tendency for the wage–rental price ratio to rise. Given that the elasticity of substitution (σ) was less than unity, θ_L, the share of the factor that was tending to become less abundant should have tended to rise. But, by the mechanism Kennedy and Weizsäcker envisage, this rise would raise B, increasing the relative labor-saving bias of induced innovation. At some point the faster rate of labor-efficiency improvement relative to capital-efficiency improvement would become sufficient to offset the incipient effect upon factor-shares of the rise of the aggregate capital–labor ratio. The level θ_L^* at which the labor share is stabilized, and the equilibrium relative bias $B(\theta_L^*)$, therefore must depend upon the magnitude of σ, the shape of the IPF, and the rate of capital accumulation in relation to the growth of the labor force. In the context of the foregoing *comparative* analysis, only the last-mentioned of these appears as a potential source of international divergence in the character of technological progress.[2]

Habakkuk stressed the effect of a more rapid pace of non-agricultural capital accumulation (capital-widening) in a region of recent settlement, which, by forcing up the relative price of labor, promoted Americans' choices of more capital-intensive industrial methods. From the neoclassical

[1] Cf. Samuelson, 'A Theory of Induced Innovation' (1965), pp. 346–50.

[2] The first two are ruled out by the assumption that (a) the production functions, equation (1), for each country are identical except for differences in the levels of factor-efficiencies, and (b) the members of the family of innovation-possibility frontiers are homothetic, i.e., have the same shape and orientation with references to the origin.

model just considered it would appear that if the situation envisaged by Habakkuk also entailed a faster rise of the capital–labor ratio for the American economy as a whole, it would have served also to induce there a more marked labor-saving bias in technological progress. But this need not have resulted in either a rate of labor-augmentation, or an overall pace of efficiency growth surpassing that enjoyed in Britain, and in the proposed reconstruction of the argument on Kennedy–Weizsäcker–Samuelson lines, the connexion with America's comparative land-abundance could at best be said to be very tenuous. In this last respect, as in others, there is more than a superficial resemblance between the long-run Kennedy–Weizsäcker–Samuelson model and Fellner's proposition, considered earlier, about the effects of (anticipated) trends in macro-scarcity relationships upon the bias of innovative activities.

RENEWED DOUBTS AND RESERVATIONS: WAS THIS NEOCLASSICAL TRIP NECESSARY?

Proponents of the Rothbarth–Habakkuk thesis will have drawn little comfort, if any, from the conclusions indicated by our efforts at recasting the 'better machinery' version of that argument. Yet, regarding the preceding neoclassical analysis of induced variations of the national characteristics of technological change in a wider, non-polemical perspective, the real trouble is that one simply hesitates to take it at all seriously. Two sorts of objections must be registered now that the nature of the analysis has been exposed. Both styles of criticism seem serious enough to militate against acceptance of the whole effort here to develop some satisfyingly unambiguous interpretations of the econometric tests proposed by Asher.

The first are essentially empirical objections to postulating the existence of an innovation-possibility frontier whose position, for the industry, sector or economy considered, remains more or less stationary within the period of the analysis. To the extent that the specification of a constant difference between λ_L and λ_K in equation (6) is correct, and the resulting regression estimates of β are accepted, the latter are in conflict with the short-run theory of induced innovation which has been enlisted to clarify their meaning. For the purpose of the regression analysis, the absolute bias of technological change (β) is treated as an exogenously determined constant; changes in capital–labor proportions are taken to depend upon that influence and the (parametric) alterations of factor-prices, or equivalently of factor-shares. Where there are no variations of the latter, statistically significant estimates of σ and β could not be secured by this approach. Yet, on the assumption that the factor-augmenting mix of technical change is also a subject of choices made under the constraint of a *stable* IPF, β could not be a constant. Like the capital–labor proportions used in production, β also should tend to vary positively with (θ_L/θ_K).[1] Thus, if there exists an IPF and a Kennedy–Weizsäcker–Samuelson optimization process, the empirical constancy of β implies that frontier shifts about over time, so that the locus of tangency points corresponding to different values of (θ_L/θ_K) is a straight line having a positive unit slope – like *PP'* in Figure 1.[2]

So much for the attempted application of the short-run version of the theory. Now it must be noticed that the record of American economic growth during the nineteenth century exhibits a number of features which, taken together, compel us also to abandon any long-run interpretations of the character of aggregate technological progress that are based on the existence of a stationary IPF. Over the course of the century, and particularly over the interval from the early 1830s to the late 1880s, the rapid secular rise of the aggregate reproducible capital–labor ratio was accompanied by an increase in the total (reproducible and non-reproducible) property share, and a corresponding decline in the share going to labor.[3] This would not appear anomalous were it thought that the aggregate elasticity of substitution was greater than unity. But, as Moses Abramovitz and I have recently argued,[4] within the framework of the sort of aggregate production model described by equation (1) the available econometric indications are more generally consistent with the alternative conclusion: $\sigma < 1$, and the United States experienced an era of relatively labor-saving efficiency growth ($\beta > 0$) during which labor measured in efficiency units (L') was becoming more abundant in relation to reproducible capital measured in efficiency units (K') – the rise of the capital–labor ratio measured in natural units notwithstanding.

The mere appearance of a secular drift in the ratio of factor-shares obviously is inconsistent with the view that this historical record would reflect the dynamic equilibrium outcome indicated by the long-run version of the Kennedy–Weizsäcker–Samuelson theory of induced innovation. Even were one prepared, as seems more reasonable, to take the historical observations as representing the movement of the system

[1] From equation (17) we know that B, and hence λ_L/λ_K, increases with θ_L. Along $H(\cdot)$ we also know that λ_L rises with the measure of relative bias, B. Hence the product $(B)(\lambda_L) = \beta$ must increase with θ_L.

[2] There is a further implication: so long as θ_L was becoming larger through time, the required temporal shifts of the IPF could not be homothetic, but rather would have to skew higher frontiers more toward higher λ_L than toward higher λ_K. A secularly declining θ_L, however, could induce adjustments consistent with the constancy of β and a homothetically shifting IPF.

[3] Cf. Abramovitz and David, 'Reinterpreting Economic Growth' (1973), Table 2.

[4] Cf. *Ibid.* The argument is made at greater length in Abramovitz and David, 'Economic Growth in America' (1973).

toward a long-run equilibrium mix of λ_L and λ_K, the theory and the facts just cannot be squared under the assumption of a stationary IPF. Given the growth of the ratio K/L, and the finding that $\sigma < 1$, the incipient tendency would have been for the labor share to rise rather than fall. The *induction* of a higher (λ_L/λ_K) – and, if the IPF were stationary, a higher β as well – would have been capable of preventing that tendency toward a higher θ_L from actually materializing. But, precisely because the continuity and concavity restrictions placed on the IPF guarantee that when $\sigma < 1$ the resulting equilibrium position is stable, there is no way for an induced bias in technological change to have lowered θ_L in these circumstances. Hence, the observed reduction of θ_L must have stemmed from an *autonomous* bias, which is to say from some temporal shift in the position of the innovation-possibility frontier.

Now, such temporal displacements are not inconceivable. Quite the opposite: it would be a palpably implausible theory of technological progress which left absolutely no room for the position of the IPF to undergo some alterations. The point is, rather, that our being forced by the data to conjure up these shifts signals an essential failure of the existing theory as an explanatory device, and its consequent degeneration (in this historical context) into a mere taxonomic principle. Since the shape and position of the IPF are not really accounted for by the theory of induced innovation, its application here has succeeded only in taking an old, familiar historical problem and restating it in new, more precise, and considerably more esoteric terms. Like the shifts of a neoclassical production function inferred within the framework of a static comparison of macro-economic production relationships, the inferred temporal – and international – displacements of the IPF are truly 'measures of our ignorance.'

To delimit more precisely the areas of one's ignorance would not be a negligible contribution were there adequate reason to believe that the dimensions in which such measurements are made will prove helpful in the ensuing search for understanding. It is necessary, however, to draw attention to a second, and a more fundamental set of objections which call into question the usefulness of the neoclassical view of technological progress, and the terms in which it has led economic theorists to formulate the notion of investment in invention and innovation as a constrained optimization problem.

The neoclassical conception that has gained currency during the past two decades strives to distinguish technological progress from the substitution of one factor of production for another within the limits set by a pre-existing state of knowledge. And the kind of 'knowledge' – the alteration of which is thought to be the substance of technological progress – referred to in this scheme is very fundamental and rather abstract. Technological progress means major modifications in the body of established scientific and engineering principles, which permit improvements of the efficiencies of some or all of the factors employed in all the specific methods available for producing stipulated commodities.

This seems to be the only way to make sense of the textbook representation of 'technological change' as the upward-shifting of a smooth concave locus describing the relationship between labor productivity and the capital–labor ratio, or, alternatively as the displacement toward the origin of a continuous unit isoquant summarizing an entire (constant returns to scale) production function. Only some truly 'fundamental' advance of knowledge would be capable of altering input requirements per unit of output across the whole of the spectrum of specific production techniques, each of which here is associated with a different characteristic set of input proportions. Surely this is the sense, if there is any, behind the notion of technological progress taking a purely input-augmenting form, and thereby leaving unaffected the ease of substitution between, say, labor and capital, while changing the unit input requirements of each by a standard proportional amount ($-\lambda_L$ and $-\lambda_K$) for every possible combination in which the two might be employed.

It is thus thoroughly in keeping with this neoclassical conception for factor-augmenting advances (in basic technical and organizational knowledge) to be depicted as the goal of purposive inventive activities – as they are in Samuelson's model along Kennedy–Weizsäcker lines, or in the similar model of Drandakis and Phelps.[1] Yet, it scarcely seems plausible to regard firms and independent inventors as capable of perceiving the global factor-efficiency implications of alternative fundamental innovations, much less as having the information required to assess the expected trade-offs involved in reallocating resources among different fundamental innovation-strategies. Would such additions to the set of general 'principles' really be the object of the firm's interests and concerns? Is it sensible to think their perceptions of trade-offs between alternative lines of research would be so utterly independent of information regarding the prevailing patterns of factor-use and the consequent distributions of production costs – as the Kennedy–Weizsäcker formulation of the innovation-possibility frontier would have us believe?[2]

[1] Cf. E. M. Drandakis and E. S. Phelps, 'A Model of Induced Invention, Growth, and Distribution,' *Economic Journal*, Vol. LXXVI (December 1966).

[2] Samuelson, in 'A Theory of Induced Innovation' (1965), p. 352, voiced his own doubts on this score, and proposed a radical reformulation of the Kennedy–Weizsäcker model aimed at discarding 'the implicit theorizing by which one assumes that the entrepreneur can have knowledge that the [IPF] transformation function remains invariant over time and has naught to do with the costs and shares of the factors themselves.' The common sense behind the last clause seems to me to derive in the last analysis from an implicit – but alas only momentary – abandonment by Samuelson of the neoclassical conception of techno-

A further aspect of the implausibility of the 'implicit theorizing' represented by the latter construct emerges when it is observed that the fundamental sorts of advances in knowledge contemplated therein imply the maximum conceivable degree of 'technological spill-over.' Such spill-overs, by and large, would represent *externalities* from the viewpoint of a particular firm, or even that of a collection of cost-minimizing competitors operating in a homogeneous factor-market environment, for the productivities of the whole array of known production techniques would be affected, not only those techniques being used or likely to be used by the enterprise or industry. Indeed, the more fundamental the advances in knowledge, the less 'patentable' they will be – which is to say, the harder the innovators would find it to enforce proprietary rights in the benefits they convey. When externalities are large in relation to the privately appropriable gains anticipated, however, firms have an incentive to behave as technological 'free-riders,' and discrepancies arise between the socially optimal and the privately profitable patterns of investment in technological change.[1] Unless we are simply to shut our eyes to this range of issues it is hard to understand why different firms – not being identically situated with respect to the opportunities for internalizing the benefits of any given innovation – would pay attention to the shape and position of the *unique*, technologically determined and socially relevant IPF, even if they were capable of discerning its outlines.

Beyond these questions of *a priori* plausibility, the economic historian should certainly be entitled to ask whether it can really be thought helpful conceptually to limit the scope of 'invention' and 'innovation' so that the words apply only to technological developments of general consequence, 'improvements' which would have equally multiplied the productive capacity of the capital represented by the wagon and team, the canal barge, *and* the steam locomotive. Is it useful thus to exclude implicitly from the theoretical analysis of technical change an innovation like the Atkins cam-driven grain-raking device of the latter 1850s, which raised the productivity of labor engaged in harvesting wheat with horse-powered reaping machines,[2] but implied nothing for the productivity of the men and women who gathered in the grain with scythe, sickle and rake?

The alternative to the neoclassical conception which these rhetorical

[1] Cf. the discussion and references in footnote 2 on p. 36 above.

[2] Cf. P. A. David, 'The Mechanization of Reaping in the Ante-Bellum Midwest,' in *Industrialization in Two Systems*, ed. H. Rosovsky (New York: Wiley and Sons, 1966), pp. 22, 34, for discussion and further references on the 'self-rake' reaping machines. This study appears as Chapter 4 of the present volume.

---

logical changes as global phenomena which shift entire production functions, rather than altering the productivity of specific techniques. The latter, alternative conception is discussed in the text below.

questions pose is a view of technological changes as being 'localized' rather than global, to use the term recently introduced by Atkinson and Stiglitz.[1] That is to say, the innovations under consideration are taken to consist of concrete changes in specific production methods: new machine designs, blueprints and process specifications. Being more-or-less focused on the reduction of production costs at a particular point in the spectrum of techniques, the inventions and innovations thus conceived will as a rule have little technical 'spill-over effect' upon the efficiency of input usage in other techniques.

This is a view commending itself naturally to historians and others who seek guidance for research on micro-economic questions concerning choice among discrete production processes, or studying the diffusion of specific techniques embodied in new materials and machines, or, again, examining the circumstances of the generation and subsequent refinement of new production methods. Yet, significantly, it is also a view that receives support from the outcome of attempting to apply the neo-classical model of factor-augmenting technological change in discussions of historical experience which are pitched at a much higher level of aggregation.

As has already been pointed out, econometric exercises applying variants of the CES production relationship in equation (6) to the data for the United States Domestic Economy in the period 1800–1905 indicate a low elasticity of substitution, coupled with a pronounced absolute bias toward labor-augmentation. Indeed, the estimates of β thus obtained are so large in relation to the concurrent (residual) rate of total factor productivity growth (λ) that the two magnitudes can be reconciled only by inferring that while λ_L was positive, λ_K was *negative*.[2] The latter condition, operating to raise the desired capital–output ratio corresponding to any given level of the real rate of return, is referred to as a capital-*deepening* bias in technological progress.[3] Figure 2, adapted from my work with Moses Abramovitz,[4] depicts the full set of these findings for nineteenth-century America in terms of the global displacements of the unit isoquants

[1] Cf. Anthony B. Atkinson and Joseph E. Stiglitz, 'A New View of Technological Change,' *Economic Journal*, Vol. LXXIX (September 1969).

[2] Cf. Abramovitz and David, 'Reinterpreting Economic Growth' (1973); 'Economic Growth in America' (1973).

[3] Note that as the IPF crosses the λ_L axis in Figure 1, there is no *formal* difficulty with the idea of an innovation mix in which $\lambda_K < 0$. It is argued, below, however, that that condition is difficult to square with the literal interpretation of an aggregate production function subject to factor-augmenting shifts.

[4] Cf. M. Abramovitz and P. A. David, 'Towards Historically Relevant Parables of Growth,' Stanford Center for Research in Economic Growth *Memorandum* 122 (Presented at the Southern Economic Association Conference Miami Beach, 4–6 November 1971), pp. 21–2.

relating the reproducible capital and man-hours input requirements for real gross domestic product. The literal interpretation of the condition ($\lambda_K < 0$) is that the new isoquant (F^tF^t) *crosses* the old (F^oF^o) and lies farther away from the origin for those techniques in the region below $\alpha(k^o)$, but closer to the origin for those techniques characterized by capital-intensities above k^o.

Yet, to accept this literal interpretation is most difficult. For, it implies that 'innovations' entailing technological retrogression somehow could

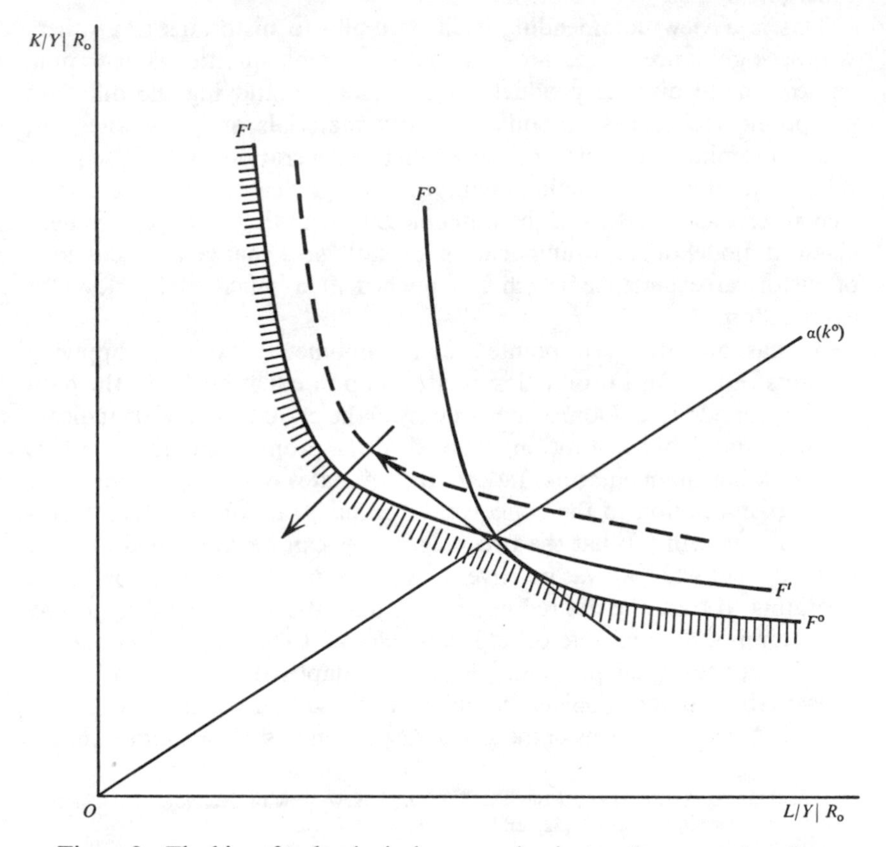

Figure 2 The bias of technological progress in nineteenth-century America
Labor (L)-augmenting and reproducible capital (K)-deepening efficiency changes displace the separable production function's unit isoquant leftward toward greater capital-intensity, while land (R)-augmentation displaces it homothetically toward the origin. The specification of that part of the generalized factor-augmented production function which describes the substitution possibilities (CES) between capital and labor – as given by equation (1) – makes the shape and displacements of the unit isoquants invariant to the level of unimproved land (non-reproducible capital) inputs, R_o.

have been forced upon those producers who sought to use methods less capital-intensive than the $\alpha(k^o)$ technique. The way to make more immediate sense of the econometric evidence is to suppose, correctly, that (1) the American economy moved away from the exploitation of the more labor-intensive techniques during the course of the nineteenth century, and (2) technological progress was actually *localized* at the opposite, more capital-intensive end of the spectrum. But in admitting that labor efficiency in some, high labor-intensity techniques remained unaffected, and in showing the relevant new unit isoquant to be formed as the convex hull of the original and the displaced contours, Figure 2 effectively abandons the concept of purely factor-augmenting technological change upon which the innovation-possibility frontiers of the recent mathematical theories of induced innovation are founded.

In the remaining part of this essay, I have sought to make good this escape from the confines of the neoclassical conception of technological changes as global in nature, and therefore also logically distinct from the phenomenon of factor substitution. On adopting the alternative view of localized productivity improvements propounded by Atkinson and Stiglitz, key propositions of the Rothbarth–Habakkuk thesis suddenly begin to re-emerge as logically coherent, and historically interesting assertions regarding the comparative characteristics of technological progress in nineteenth-century America.

III

The way now lies open for an attempt at a radical reformulation of the Rothbarth–Habakkuk thesis. Our immediate aim must be to improve the logical coherence and *a priori* plausibility of the explanation being offered for the putative facts of Anglo-American technological experience during the nineteenth century. The ultimate purpose, of course, is to provide a theoretical framework within which empirical inquiry can be pursued more fruitfully, and thereby to enhance the likelihood of the true nature of the relevant facts becoming more firmly established.

As soon as one is ready to discard the neoclassical conception of technological progress which insists that innovation and factor-substitution be viewed as logically distinct phenomena, there is no longer any great difficulty in taking an important first step toward this proximate objective. Specifically, it becomes possible to indicate how the realized factor-saving bias of 'changes in the state of the technical arts' may come under the influence of factor-prices – directly, as well as indirectly through the medium of choice-of-technique decisions. In regard to the latter, we may for the present purposes eschew less orthodox, 'behavioral' approaches to the decision-making activities of firms; the prevailing structure of input prices will therefore continue to be cast in the governing role assigned to them by the traditional theory of the rational, cost-minimizing firm.[1] This will furnish us with a set of plausible, empirically testable (refutable) propositions causally linking the existence of a high relative wage–rental–price ratio with the emergence in that economic environment of a bias toward the development of comparatively capital-intensive techniques, and also (under conditions to be specified) the achievement of a comparatively high rate of total input productivity improvement. The foregoing are precisely those 'observations' which, within the neoclassical framework of analysis, have been interpreted as indicative of a 'bias' toward labor-saving innovations and a more rapid pace of 'technological progress' in America. But it should be stressed that the reformulation advanced by the first stage of the argument here is not another try for a theory of optimizing adjustments in the rate and direction of *global* technological progress. It is, rather, an alternative rationalization of phenomena which the reigning neoclassical conception of technical change has seduced us to think of in those terms.

To complete the job, we must take the further step and reconnect the condition of comparative natural resource abundance with the choice of more capital-intensive methods, both directly and indirectly via the influence land abundance exercised in establishing a relatively high wage–rental–price ratio in America. American comparative resource abundance during the nineteenth century then may be held to have been linked with the differentiating characteristics of the country's technological development, which I take to be the essential proposition common to the works of Habakkuk and Rothbarth.

THE ESSENTIALS OF THE FIRST STEP

The hitherto so elusive connexion of the conditions of comparative factor-scarcity with the characteristics of technological change can be made in two ways. At the micro-economic level, where it is relevant to discuss the existing state of technology in terms of distinct available processes of production, at least some innovations may be regarded as having been designed to effect factor-substitutions. In the subsequent elaboration on this original Hicksian notion, it will be most convenient to draw upon the concepts of a 'Fundamental Production Function' (FPF) and an 'Available Process Frontier' (APF), which Chenery and Salter have introduced to clarify the relationship between the neoclassical

[1] Cf. R. R. Nelson and S. G. Winter, 'Towards an Evolutionary Theory of Economic Capabilities,' *American Economic Review*, LXIII, 2 (May 1973), pp. 440–9, for a brief account of a more thorough-going reconstruction of economic theories dealing with technological choice and change.

classical construct of a production function based on fundamental knowledge, and the notion of a discrete 'process,' or production technique, which is familiar from input–output and linear programming analysis.[1]

Proceeding from that base, it is then quite straightforward to develop the second, and rather less familiar, line of connexion, by taking up Atkinson and Stiglitz's[2] recent suggestion that learning by doing could give rise to *localized* technical change. In this case, however, the viewpoint adopted is very Olympian, and the evolution of the technology of an industry or sector appears, over time, to have been 'guided' by the same set of conditions which impinge upon the choice of technique. This turns out to be an accurate surmise from outward appearances, because it is the decisions about which techniques to employ in production which in reality are controlling the rate and direction of technological change. And, in the argument advanced here, the peculiar externalities and uncertainties surrounding investments in innovation may well result in a justifiable myopic disregard for the possibilities of generating new techniques from experience with the old. This leaves the operative objectives of choice-of-technique decision confined largely to the minimization of current and extremely near-term costs of production.

Despite a degree of novelty in the design of this latter connexion, in the main it is forged from elements already introduced in Parts I and II, and in any event, its central propositions are not at all difficult to grasp without the help of formal analytical preparations.

A succession of technological advances, each of which happens to be localized at some point in the spectrum of techniques, will be most likely to have a Hicks neutral (or possibly a quasi-neutral) appearance when viewed in the small – for reasons still to be considered. Yet, if that is so, the cumulative effect of their persisting localization must impart a distinct factor-saving bias to the manner in which the global pattern of technological constraints becomes modified with the passing of time. A glance at Figure 4, about which more will be said shortly, may be helpful here.

The very fact that technical innovations do have a local character may be thought to come about from men's preoccupation with problems that are close at hand. Atkinson and Stiglitz do not spell out this deeper cause, but Nathan Rosenberg[3] recently has called attention to the important, and insufficiently recognized, degree to which the signals directing attention to opportunities for improvement in production techniques emanate from actual experience in trying to operate with the original version of one or another particular method of production. This constitutes

[1] Cf. footnote 2 on p. 35 above, for references to the work of Chenery and Salter in this connexion.

[2] Cf. Atkinson and Stiglitz, 'New View of Technological Change' (1969), pp. 574–5.

[3] Cf. Rosenberg, 'The Direction of Technological Change' (1969).

a long-run form of learning by doing,[1] a process of interplay between men and machines or between groups and organizations, rather than the passive conformity of the ordinary agents of production to rules and systems laid down by heroic inventor–entrepreneurs.

Viewed in a broader perspective, the actual innovations thus generated are bound to reflect the prior decisions of firms to operate at some particular place within the spectrum of techniques. Moreover, in the course of myriad specific improvements made on the processes that remain in use some new and rather more generally applicable knowledge may be gained. Thus, even refinements of an extremely parochial sort may carry some potential 'spill-over,' as the experience thereby acquired modifies and makes more efficient the application of basic engineering or other scientific principles relevant for the redesign of kindred processes.

In the dynamic and essentially historical process envisaged, the squeaking axle will get greased, and the successful repetitious application of grease to cart-axles may even be generalized to the immersion of propellor- and drive-shaft bearings in lubricant baths. But all this will follow with greater certainty if one had decided on using the cart, and not a sled in the first instance. *Historical* choices among techniques thus rule the future. Now, if economic historians have emerged with any conclusion from their recent micro-economic studies of industrial and agricultural production practices, it is that relative input prices *can* be taken to have played a powerful role in determining the choices made among alternative available techniques, even when they did not exercise exclusive control. This generalization seems not less valid for the British economy – accusations of mismanagement by late Victorian entrepreneurs notwithstanding, than it is for the American economy of the nineteenth century.[2]

As one may observe within the foregoing scheme, marked changes in relative factor-prices would tend to lead to the extension of the technical

[1] For historical evidence on the existence of both long-run and short-run learning effects (albeit for only one branch of early American industry), and for citations to the relevant theoretical and more recent empirical work in this subject, cf. Paul A. David, 'Learning By Doing and Tariff Protection: A Reconsideration of the Case of the Ante-Bellum United States Cotton Textile Industry,' *Journal of Economic History*, Vol. XXX, 3 (September 1970); *idem*, 'The Use and Abuse of Prior Information in Econometric History: A Rejoinder to Professor Williamson on the Antebellum Cotton Textile Industry,' *Journal of Economic History*, Vol. XXXII, 3 (September 1972); *idem*, 'The "Horndal Effect" in Lowell, 1834–1856: A Short-Run Learning Curve for Integrated Cotton Textile Mills,' *Explorations in Economic History*, Vol. 10, 2 (Winter 1973). Chapters 2 and 3 of the present volume present the studies originally reported in the first and the third of these articles respectively.

[2] Cf., e.g., the survey of recent work in nineteenth-century American economic history given by William N. Parker, 'From Old to New to Old in Economic History,' *Journal of Economic History*, Vol. XXXI, 1 (March 1971), esp. pp. 5–6. Much the same impression also emerges as a dominant theme of the papers gathered in the recent conference volume, *Essays on a Mature Economy: Britain after 1840*, ed. D. N. McCloskey (London: Methuen, 1971).

arts in the direction where proportionately less use was made of the dearer factor. By the same token, in a region marked by the *comparative* scarcity of labor (relative to capital), where producers displayed an understandable preference for the more capital-intensive among the techniques made available to them, the recorded course of technical improvements may be expected to exhibit a bias toward conserving on labor rather than on capital – when compared with the drift of technological progress that is taking place elsewhere in the world.

Having arrived at this point it is immediately possible to pick up the thread of the earlier discussion, in Part II, regarding the determinants of the rate of technological progress. For now we have a suitable conceptual framework in which to adapt, and so to adopt Habakkuk's interesting hypothesis that in the nineteenth century 'the more capital-intensive of the existing range of techniques had the greatest possibilities of technical progress.'[1] In a given state of basic scientific understanding about the structure of materials and the transmission and conversion of energy, whether or not (quasi-neutral) technological innovations turn out to be generated in rapid succession is indeed likely to depend upon the particular portion of the technical spectrum where most of the long-run 'learning' happens to be going on.

But here again, before leaping to make this a basis for a normative model of historical investment in learning by doing,[2] the realities of the nineteenth-century setting should be recalled. The less well understood are the scientific foundations of the technologies currently available for use, and the less able firms are to internalize the benefits arising from such improvements as the workers, foremen, engineers and managers are able to effect by responding to perceived production bottlenecks, the less likely it is that the choices made among available methods of production would have passed beyond simple cost-minimization to consider the differential potentialities for their rapid future development.

The most noteworthy aspect of the line of connexion thus re-established between factor-prices, the choice of technique and the rate and direction of global technological changes, is its fundamentally evolutionary character. The drift of technological developments generated over time within a fairly stable economic environment needs to be viewed, first and foremost, as a distinctively *historical* phenomenon, inasmuch as it may arise through the myopic selections *past* producers made from among the different species of techniques with which they originally had to work. The element of 'guidance' present in the long-run process is attached in

[1] Cf. Habakkuk, *American and British Technology*, p. 61, and the related citations appearing in footnote 2, p. 38, above.

[2] This theoretically promising terrain has been briefly surveyed by Atkinson and Stiglitz, 'New View of Technological Change' (1969).

large measure to the circumstances surrounding those choices, and involves no forward-looking, induced innovational responses to current market signals or future portents. Only on a flagrantly homocentric misconstruction could the historical course of technological progress in a given society be read as the direct expression of such rational, optimizing intentions.

SOME DETAILS OF THE FIRST STEP

A number of the points touched on by the argument just sketched may well bear a more formal explication and rigorous development. The three following sections, however, should be read as preliminary notes aimed primarily at clarifying less familiar details of the two lines of connexion – in order of their preceding appearance rather than of an axiomatic presentation.

(*a*) *Technological change in the small and innovation as substitution*

Salter has suggested that there are two ranges of alternative techniques which are relevant in discussing the concept of a production function for an industry. The first is the relatively narrow range of techniques that actually have been developed, and therefore are immediately available to producers. They are described in machinery catalogues, in engineers' and architects' blueprints and specifications, and by the formal systems and rules of thumb which managers, supervisors and craftsmen have worked out for operating existing facilities.

In Figure 3 the elements of this narrow range of techniques are described by reference to their respective characteristic sets of labor- and capital-requirements per unit of output. For graphical convenience, constant returns to scale are assumed throughout, and each linear, fixed-coefficient 'process' (or, alternatively, 'technique') is represented by an L-shaped boundary like TT'. We may imagine that at a given point of time only two such processes have been developed: let them have the capital-intensities indicated by the slopes of the rays α and γ drawn from the origin. Together, they present the industry with the convex 'available process frontier' labelled APF^{o} – since by employing them in varying linear combinations it will be possible to achieve (aggregate) capital-intensity levels in the range between α and γ.[1]

[1] The assumption that the individual processes are of the Leontief-type could be relaxed, admitting substitution possibilities within the definition of 'a technique,' without altering the essential character of the argument – so long as the positive elasticity of substitution along a continuous version of the TT'-boundary was substantially smaller than the elasticity of substitution along FPF^{o} in Figure 3.

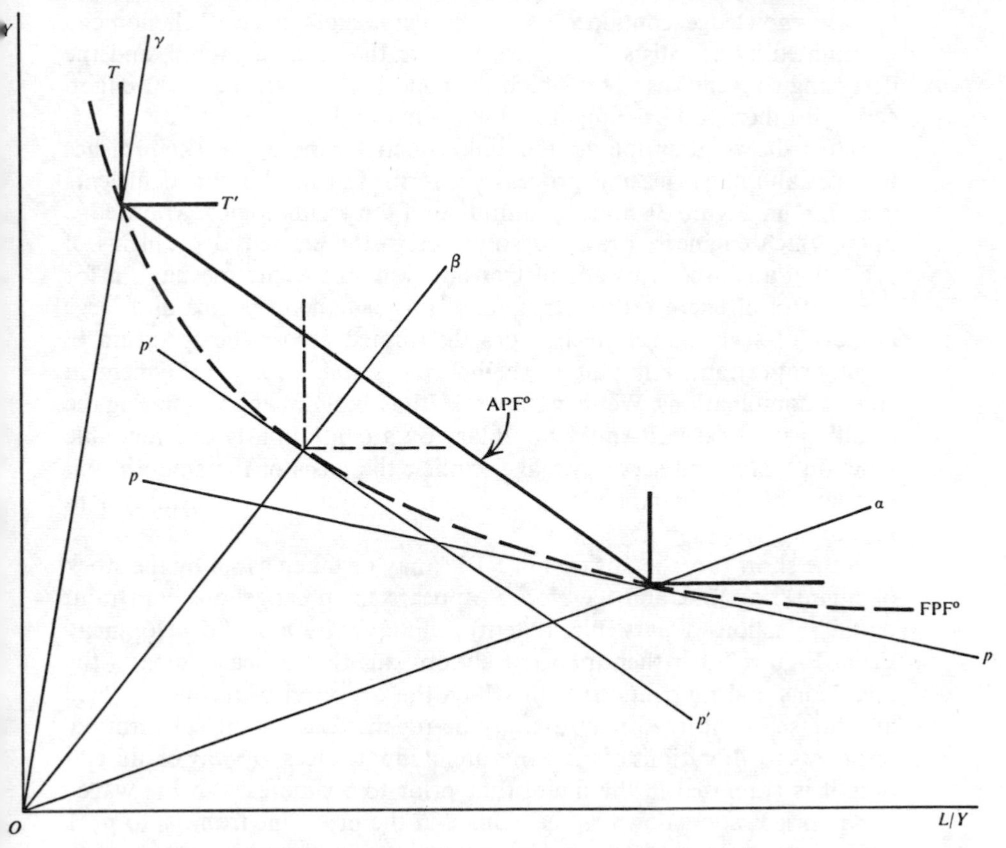

Figure 3 Technical innovation viewed as substitution

Underlying this first, restricted range, is the broader array of potential processes that could be designed within the currently existing state of knowledge – with some high probability of *technical* success. The critical economic hypothesis here is that the difference between the richness of the underlying range of techniques, and the sparseness of the set of available processes, is not a matter of accident or of inevitability, but of choice. For, as Salter urges us to recognize, other techniques contained in the latent set would have been developed had it seemed commercially promising to make the effort.

The immediately operative body of knowledge defining the set of latent techniques is that which spans the gap between basic science and its practical applications to production problems. These 'principles of practice,' if we may so describe them – in physical and chemical engineering, in aeronautics, metallurgy and agronomy – derive not only from the deeper sort of knowledge contained in the principles of physical phenomena formulated by scientists. They also include the rules of thumb, and the designing conventions upon which the collective experience of the engineering brotherhood has taught its members to rely.

Salter draws attention to the link which Chenery's work furnishes between alternative feasible processes described by fixed input coefficients (e.g., TT' in Figure 3), and the minutiae of the technological knowledge upon which engineers draw for solutions to the elemental problems of providing a source, a means of transmission, and some mechanism for the control of energy. The next link is to posit the existence of a large number of such discrete techniques distributed across the spectrum of input proportions, any pair of them being suitable for employment in convex combinations. We then may take the liberty of approximating the resulting minimal unit–input boundary by a continuously differentiable function.[1] This will serve to make explicit the basis of the smooth unit isoquant of the fundamental production function (FPF°) exhibited in Figure 3.

In the short run, the position of FPF° may be taken as set by the stock of engineering lore, and therefore it appears as an exogenous constraint on the selections of particular (latent) techniques for actual development. From Figure 3 it is then immediately obvious that at least some of the inventions and innovations with which the historian of technology will find himself concerned may usefully be regarded as 'factor-substitution' responses to alterations of the structure of input prices. By way of illustration, it is supposed in the figure that prior to an increase in the wage–rental–price ratio (shown as the change in the price-line from pp to $p'p'$), producers were content to minimize production costs by working with the available α-technique. An immediate consequence of the indicated rise in the relative price of labor would be to render them indifferent, on the same considerations, between staying with α or switching to γ. And a further marginal change of relative wage rates in the same direction would decide the question in favor of the more capital-intensive available process. But an alternative response could also be elicited from some producers by this factor-price change: the allocation of resources to the actual development of a β-technique. For the latter now would offer a

[1] For references to the works of Salter and Chenery, cf. footnote 2 on p. 35, above. From a knowledge of the frequency density of the distribution of fixed-coefficient techniques, a specific form for the limiting approximation may be obtained. For example, the Cobb–Douglas form of (neoclassical) production function is implied when the underlying fixed-coefficient techniques follow a Pareto frequency distribution. Cf. Hendrik S. Houthakker, 'The Pareto Distribution and the Cobb–Douglas Production Function in Activity Analysis,' *Review of Economic Studies*, XXIII (1), No. 60 (1955–6), pp. 27–31.

margin of cost-savings over linear combinations of the immediately available processes along APF^o. Whether or not that potentiality for gain appears large enough and certain enough to justify the necessary commitment of resources, of course, is quite another question.

The foregoing will do well enough to account for factor-price-induced 'biases,' when technological change is viewed in microcosm; and within a time span short enough to justify treating the fundamental production function as 'given,' and thus positioned independently of the influence of prices upon producers' choices. But for a longer-run and truly historical view, it becomes necessary to depart from this (Salter's) assumption. We must recognize that the price-guided choices made along the APF will be altering the shape and position of the latter frontier, and so also that of the FPF via the effect such decisions must have upon the acquisition of engineering knowledge.

(*b*) *Learning, and the local neutrality of endogenous innovations*

Although the localization of technological change has been linked with the existence of learning phenomena in the work of Atkinson and Stiglitz, the factor-saving characteristics of the emerging stream of improvements have yet to be explicitly discussed. From Figure 4, however, it can be seen that the supposition of localized changes which take the form of a persisting advance toward the origin *along a specific process-ray*, plays a quite crucial role in explaining how factor-prices may govern the long-run bias of technological progress. For it is this property which translates an essentially static, price-induced switch to a more capital-intensive available technique (now β), into a dynamic skewing of the available process frontier in the relatively labor-saving direction – the secular transformation of APF^o into APF^t shown by Figure 4.

In detailing the salient consequences of endogenous technical changes which raise productivity but do not alter the input proportions associated with the original version of the technique, it will be convenient to refer to this condition as one of 'locally neutral' technological progress.[1] Three aspects of the story told by Figure 4 deserve immediate comment, so that the full implications of that condition will be appreciated.

[1] It should be noted from the following discussion that we can no longer make precise use of the (Salter) definition of 'neutral' inventions as those which would leave the factor-proportions in use undisturbed so long as factor-prices remained unchanged. Similarly, the strict Hicksian definition of neutral inventions (as those which would leave relative factor-prices unchanged along any given K/L-ray) is no longer appropriate. This break with established usage of the 'neutrality' concept is caused by the stipulation that technical change is *localized*; it has nothing to do with the smoothness, or 'kinkedness,' of the unit isoquant.

The first concerns the presence of some element of dynamic irreversibility in the global bias that develops when the progress of innovation carries producers down an initially selected process-ray. A relative rise in the price of labor (shown by the change in the price-line from *pp* to *p′p′*) is clearly sufficient to launch an incidental, myopic exploration of the β-ray. But as this learning process proceeds, the power of price varia-

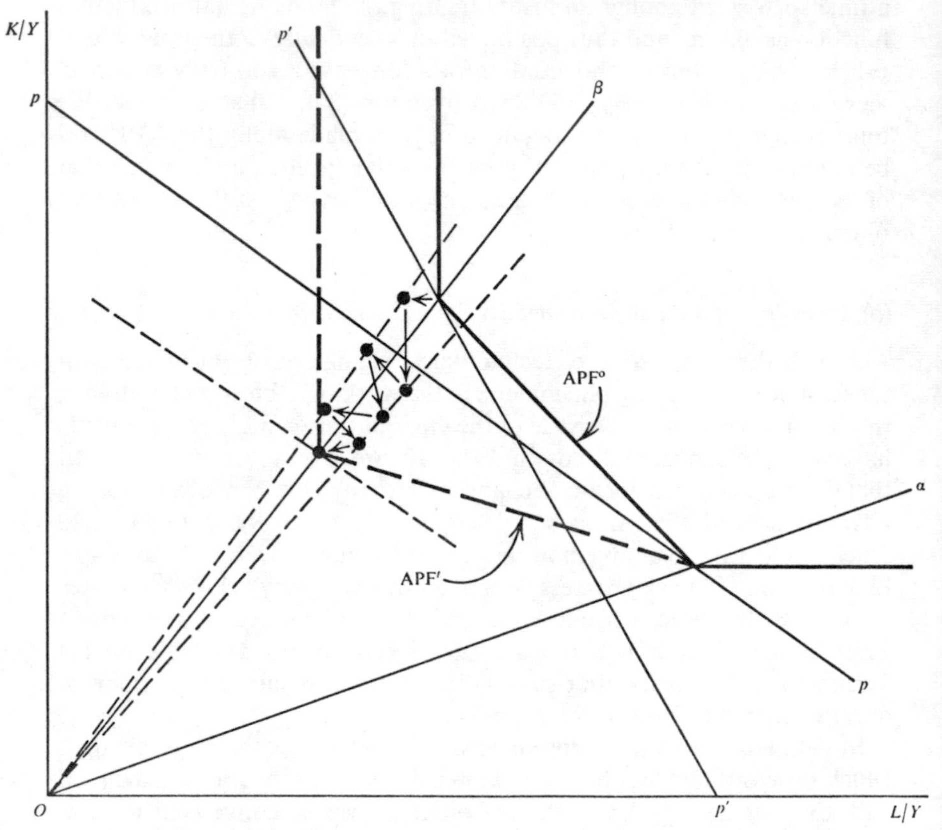

Figure 4 Localized learning and the global bias of technological progress

tions to halt the emerging labor-saving drift of the APF begins to diminish. Once the point of the β-ray's intersection with *pp* is reached, a mere restoration of the *status quo ante* in the factor markets would surely not draw even the most myopic producer back to the α-technique. Eventually, even with progress down the β-ray occurring at a retarded rate – as we generally expect to happen on a learning curve – the APF would become approximately L-shaped. With the β-technique's emergence as *the* dominant method, the prior condition of local neutrality becomes one of globally neutral advance. There is thus some theoretical basis for seeking the origins of the modern configuration of a society's technology in the accidents of its *remote* factor-price history.

To be sure, the historical process just envisaged could be interrupted by a dramatic alteration of input prices, returning producers to the use of the α-technique. This brings us to the second aspect: if the potential time-rate of advance down all rays were the same,[1] then the more pronounced was the secular rise in the relative price of labor, the less likely it would be that factor-price fluctuations of any given amplitude would interrupt the global labor-saving drift of technological progress. Note, however, that in the present myopic formulation of the choice of technique problem, the influence of the trend in relative wages does not operate via *expectations*, as Fellner's propositions about induced innovation would have us conclude.[2]

On the other hand, if, as Habakkuk suggests, during the nineteenth century the possibilities for rapid locally neutral progress were greater in the region surrounding a more mechanized, capital-intensive technique such as β, than they were in the region around α, the global labor-saving bias would tend to emerge – even were there no trend toward higher real wages and cyclical relative input-price fluctuations caused the price-line to swing back and forth between pp and $p'p'$. From this it should be clear that the question of the global bias of technological change still cannot be disentangled from the question of the realized rate of (localized) technical progress, but that in this formulation the former is made to depend on the latter, and both are governed by the past history of choices of technique.[3]

The third point of interest concerns the relationship between movements in the available process frontier (APF) and the position of the fundamental unit isoquant (FPF) of Figure 3. Under the conditions just considered, we might imagine producers in a high real wage region (call it America) moving down ray β, while those in another region (call it Britain) are proceeding along ray α – albeit at a slower pace. Given this differential rate of advance, the accumulation of experience with the design and operation of the β-process would have fundamental spill-over effects drawing the FPF toward the origin while slowly skewing it in a more labor-saving direction. We may suppose, not unrealistically, that British

[1] This might be alternatively expressed as the (testable) proposition that the coefficients of the long-run learning curves associated with high- and low-capital-intensity techniques were much the same. On estimating long-run learning coefficients from historical data, cf. David, 'Learning By Doing' (1970).

[2] Cf. Fellner, 'Two Propositions' (1961).

[3] In the neoclassical models of induced innovation discussed in Part II, above, it will be recalled, the realized rate of technical progress depended upon the factor-saving bias (given the position of *the* innovation-possibility frontier), and neither were immediately connected with the choice of techniques.

and American engineers had more or less equal access to the basic principles underlying the new FPF. Therefore, before the point had been reached at which British producers would find it attractive on narrow cost-minimization considerations to switch from α to the developed β-technique (copying American machines or importing them), even if British firms continue to face an unaltered relative factor-price environment indicated by the slope of *pp* they might well make considerable efforts to develop some (latent) process in the range between α and β. Here we envisage that the faster locally neutral learning going on in America (with processes of a more capital-intensive nature) leads ultimately to a labor-saving displacement of the common (internationally shared) FPF. This truly neoclassical labor-saving bias in the movement of the underlying spectrum of latent techniques, without requiring changes in relative factor-prices within either America or Britain, would elicit a corresponding drift within the (manifest) set of processes made available to producers in each region.

In the foregoing scenario, there would appear no lack of seemingly spontaneous (non-price-induced) inventive activity, and much evidence of engineering sophistication among the British. Indeed, viewed in microcosmic detail their technical innovations would (confusingly) display a *local* labor-saving 'bias' with a frequency matching that observed among the local improvements being made by their American cousins. But each time a traveling-party returned with information about processes available on the other side of the Atlantic, the outcome of Britain's technical efforts to date would convey the impression of a reluctance fully to embrace the principles of capital-intensive design, and her technological development would appear to be bobbing along in the American's wake.

All this is rather satisfyingly plausible, and helps us to make more sense of an oft-told tale. But as it all hangs on the notion that localized *learning* carries producers down their respective process-rays toward the origin in Figure 4, the latter supposition cannot lightly be left unsupported.

(c) *Local neutrality, random walks and compulsive sequences*

The most straightforward defense to be offered for assuming the local neutrality of technological changes which arise through learning by doing is an unabashed appeal to the Principle of Insufficient Reason. Why should we assume otherwise? True, it has been seen how a change in factor-prices (or a foreign-generated shift of the FPF) could induce an effort to develop a new, hitherto latent, process with input proportions *different* from these currently available. But that is a phenomenon quite distinct from the one referred to as 'learning by doing.'

Notice the strong symmetry between this argument and the position that was originally taken – in regard to the global neutrality or biases of 'inventions' – by Hicks in *The Theory of Wages* (1932): individual 'autonomous' inventions, he suggested, might be labor- or capital-saving, but on average they could only be expected to be neutral.[1] Within the scheme of things, it then became necessary to hypothesize the existence (indeed, the preponderance) of 'induced' inventions – in order to account for the historical appearance of a global bias in the course of technological progress. Although the modern treatment of the problem by economists unfortunately went astray at precisely that juncture, essentially the same argument about 'neutrality' can be applied to the present discussion of localized improvements effected within a stable 'learning environment.'

In the absence of any indications to the contrary,[2] it is most plausible to suppose that in Figure 4 a mapping of the 'new-variant' points generated by experience with the β-technique under practical production conditions would show a frequency distribution whose density was greatest in the region immediately surrounding the β-ray. This is visually suggested by the dotted lines fanning out from the origin around that ray, indicating the range of capital-intensities diverging by a given *multiple* from the one characterizing the original version of the β-technique. Within this region we should expect to find a high, constant percentage of the variants ('mutations,' if one approves of the evolutionary analogy) generated by experience with the β-technique.[3] The local technical innovations

[1] From the vantage point that hindsight affords, it appears that in this argument Hicks embraced either a popular fallacy regarding the operation of the so-called 'law of averages,' or a fallacious conception of 'the representative invention,' or possibly both. These fallacies are discussed below, pp. 77–78.

[2] Although this empirical question emerges here as one of considerable importance, it has to date received scant attention. I posed the question in my study of 'The "Horndal Effect" in Lowell' (1973), without indicating its larger significance. It is pointed out there that the readily available historical evidence – confined to the case of the ante-bellum American cotton textile industry – is consistent with endogenous productivity improvements having had no systematic effect on the capital–labor ratio over the period 1839–55. This, it is suggested, could imply that whereas short-run learning effects (on fixed facilities) had a purely labor-augmenting bias, the endogenous improvements incorporated in the re-design of plants had a relatively capital-augmenting bias. Note that the objections raised to the neoclassical factor-augmenting model of technological change, in Part II, above, carry considerably less force when – as is the case under consideration – the activity studied is very narrowly defined.

[3] Of course it is quite arbitrary to assume (as in Figure 4) that the distribution of variants is such that the 'confidence region' around a ray fans *out* as one moves *away* from the origin. The opposite assumption might be made, suggesting a more rapid convergence of the technical improvements that would be generated from initially quite disparate techniques. Naturally, at time 0 only the portion of the indicated region that lies below the APF° would be of interest to producers. See the following discussion for conditions under which the variance of the distribution of variants ('mutation') will *increase* as one moves downward from the initial point causing the confidence region around the ray to fan out as one moves *toward* the origin, on contradiction of Figure 4.

belonging to the sequence which is thus distributed can be described as possessing the property of 'quasi-neutrality,' for, none need be strictly neutral in its effect upon input proportions.

Formulating our assumption about local neutrality in these weaker, probabilistic terms, and defending it by a straightforward appeal to the Principle of Insufficient Reason, suffices to render the foregoing argument generally plausible. From a methodological standpoint it is perfectly respectable thus to leave a major premise of the argument hostage to future empirical investigations – to be preserved or sacrificed in the light of evidence still to be uncovered. Nevertheless, it would be obviously far more satisfying (from a forensic standpoint) to be able to mount a more compelling *a priori* case on behalf of this vital assumption. But equally obviously, to do so requires a deeper treatment of the character of the search for beneficial modifications of established production methods. In place of high-level implicit theorizing excused by ignorance, ultimately we must aim to offer some detailed specification of the sort of stochastic leaning processes that would generate the hypothesized sequences of quasi-neutral innovations.

Perhaps the most important point to stress in this connexion is that economists and economic historians ought not feel obliged by considerations of professional honor to furnish explanations which are purely 'economic.' There may be potent technological conditions controling localized technical change in ways that render it quasi-neutral. Indeed, if such innovation-sequences are each conceived of as emerging from the production experiences of a perfectly competitive firm, it appears possible to explain their generation *only* by reference to fundamental properties or characteristics of the relevant production technology. They must arise in some basic sense from the brute facts of nature. More precisely, they must stem from restrictions that regularities in the natural order place upon the range of improvements which are readily 'discoverable' through an incremental search initiated in the neighborhood of a utilized industrial process.

A useful way to come to subscribe to this proposition is to start by contradicting it, by trying to build a probabilistic model on the supposition that there were no such 'physical' restrictions upon the results flowing from endogenous innovation. It has been already pointed out that in the absence of inducements created by external changes (in relative input prices, or in the state of fundamental scientific and technical knowledge), a perfectly competitive enterprise would have no incentive to modify its existing production process(es) so as to save on the use of one input rather than another input. We might imagine, therefore, that the course of localized technological progress due purely to learning by doing would resemble the outcome of a *random walk* in the space defined by the input coefficients, and more specifically an unimpeded random walk whose locus was free to drift.[1] The question is: Can we get predictable quasi-neutrality with such a model?

For the sake of utter simplicity consider Figure 5(a), wherein individual technical 'improvements' or 'useful innovations' are represented as discrete

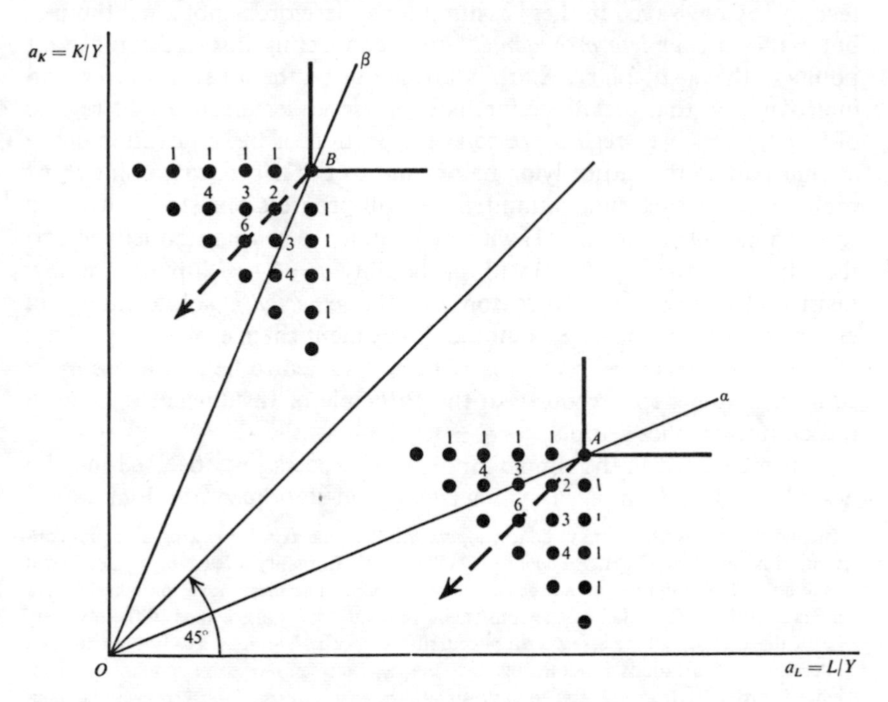

Figure 5(a) Unimpeded non-stationary random walks

changes that lower the unit labor requirements (a_L) *or* the unit capital requirements (a_K) in the production of a well-defined standard commodity. Formally, assume that (1) the changes in the coefficients, $-\Delta a_L$ and $-\Delta a_K$, must be effected by independent innovational 'trials'; (2) the outcome of each such trial is a uniform decrease in one and only one of these coefficients, say, a unit move toward one axis or the other in the

[1] Cf. William Feller, *An Introduction to Probability Theory and Its Applications*, Third Edition (New York: John Wiley and Sons, 1968), Ch. III, on basic concepts and theorems concerning random walks. Coin-tossing experiments, and the (Brownian) motion of gas molecules, are often described in the intuitively appealing terminology of random walks. The representation of technological change at the microcosmic level as an unimpeded random walk has been proposed recently by Nelson and Winter, 'Evolutionary Theory' (1973). I have benefited by reading the more detailed account of this feature of their model in R. R. Nelson, S. G. Winter and H. L. Schuette, 'Technical Change in an Evolutionary Model,' University of Michigan Institute of Public Policy Studies Discussion Paper, July 1973. See footnote 1 on p. 76, below, for further comments.

plane of Figure 5(a); (3) only one innovational trial is conducted at each *epoch.*[1] Thus, each useful modification is taken to be the outcome of a Bernoulli trial in which the input-saving *direction* of change is the random variable.

The appropriate representation of the discoverable micro-engineering technology envisaged in these assumptions therefore is not a continuum, but rather *a complete orthogonal lattice* connecting discrete equidistant points in the a_L–a_K plane. A firm starting out on the β-ray, however, and improving on that technique for its own use under an invariant regime of input prices, will regard the relevant portion of the innovation-space as that part of the lattice lying below and to the left of, say, point *B*. At each epoch (in each innovation trial) it will progress one step downward to the adjacent point ($x_i = 1$) with probability p, or one step leftward to the adjacent point ($x_i = -1$) with probability $q = 1-p$. Suppose we now insist that there is no good reason why the *expected* absolute change of either of the input coefficients should not be identical in every independent trial. As these changes have already been assumed to be possible only in equal absolute steps, recourse to the Principle of Insufficient Reason in this context implies setting $p = q = \frac{1}{2}$.

A firm engaged in the simple innovation process just specified may be described as performing a symmetrical, non-stationary random walk.[2]

[1] The word *epoch* is used here to denote a *point* on the time axis, borrowing a helpful convention adopted by Feller, *Probability Theory*, p. 73. In reality a learning sequence will have an extended duration, and the development of each actual innovation takes a finite interval of time. Nevertheless, the constituent innovational trials defined in the text may be idealized as being timeless, and thus occurring at epochs. This leaves us free to entertain alternative specifications about the real time elapsed between successive epochs (which are indexed ordinally, from the 'first' to the '*n*-th'). In the ensuing exposition, however, the time-rate of development of the system is not the central question under consideration. Nothing is said which denies the possibility that the time-rate of progress down some rays toward the origin of Figures 4 or 5 may differ from that down others, or that the *time-rate* may undergo systematic retardation – as would be expected in the case of movements along a learning curve.

[2] From an alternative formal standpoint, this model treats endogenous technological change as *a stationary* (*or homogenous*) *Markov chain.* Each point in the place of Figure 5(a) is a 'state' and the general Markov property is satisfied because, given the probabilities of transition between states at each epoch, the future development is determined completely by the present 'state' and is quite independent of the path by which the present 'state' was reached. The Markov chain is stationary (or homogenous), a special case of the general Markov feature, because the transition probabilities depend only upon the *epoch interval* separating the states in question; the transition probabilities remain the same at each successive epoch, rather than varying with the position of the epoch *vis-à-vis* some fixed point in the sequence. Feller, *Probability Theory*, Third Edition, Chapters XV, XVI, provides a thorough treatment of Markov chains. It will shortly be seen, however, that in order to generate quasi-neutral innovation paths for the individual firm, this model must be abandoned; in the model suggested in its place, in which past states do influence the future course of development, the general Markov property is absent.

Its movements trace out an 'innovation-path' formed of the orthogonal segments joining successive variants of the original production process (point B). In a sequence of n Bernoulli trials there are 2^n possible innovation-paths for the firm to take, each of which may be assigned probability 2^{-n}. The perambulation as a whole is *non-stationary* in the sense that these paths carry the firm inexorably away from the origin point at B, never returning it thither. The particular point or technical variant to which an innovation path of n segments leads the firm, that is to say, the outcome at the n-th epoch, is itself a random variable having a binomial probability distribution. Examination of Figure 5(a) will quickly reveal the emerging binomial pattern in the numbers at the points which are shown fanning out below B; one should be able to see that these numbers correspond to the total number of paths leading to each position, in the first through the fourth epochs.[1]

It is evident that at subsequent epochs these outcomes will remain distributed *symmetrically* around the vector which has a positive unitary slope and originates at the point of departure, B. Thus, although the variance of the distribution of outcomes increases at each successive epoch, the expected outcome always remains on that vector. Likewise, for a firm engaged in the endogenous modification of process A on the α-ray the *expected* variants all lie along another vector paralleling the 45° line but originating, in this case, at A.[2]

[1] At the fourth epoch there are 16 possible paths (2^4) leading to 5 ($= n+1$) distinct outcomes, which in general may be indexed by $k = 0, 1, \ldots, n$. The reader should satisfy himself that the frequencies of these outcomes obey the binomial rule:

$$f(k) = C^n_{kl}p^k q^{n-k}, \quad \text{where} \quad C^n_k = n!/k!(n-k)!, \quad \text{and} \quad p = q = \tfrac{1}{2}.$$

[2] For moderately large n the normal distribution gives a good approximation to the binomial, and we can therefore say that the firm's stepwise distance from its vector of unitary slope at the n-th epoch approximates the normal variate $N(0, \sqrt{n})$, with mean 0 and variance n.

I am grateful to Fred Nold for showing me that the formal derivation of this result is quite simple once the problem of a two-dimensional drifting random walk is restated as an equivalent problem involving a stationary random walk in one dimension.

Consider D_n, a measure of the difference at the n-th epoch between the *cumulative* change in a_K and the *cumulative* change in a_L, starting from a common point of origin:

$$D_n \equiv 1/h\left[\sum_i^n(-\Delta a_K)_i - \sum_i^n(-\Delta a_L)_i\right].$$

The constant h is simply a normalization factor, transforming the absolute change in either coefficient into a unit step. Note that D_n is a measure of 'distance': it measures the minimum number of permissible unit steps – either of $(-\Delta a_K/h)$ or $(-\Delta a_L/h)$ – required to return to a position on a vector which has a positive unitary slope and passes through the point at which the walk originated.

Now define $x \equiv 1/h(\Delta a_K - \Delta a_L)$, which permits us to write

$$D_n = \sum_i^n (x_i).$$

Quite obviously, the probabilistic model just described is a flop: it fails to generate even the primary characteristic of quasi-neutral localized technical progress for the individual firm. Except in the very special circumstances of the firm starting out with an original process on the 45° line, the central tendency in the resulting distribution of points or variant production techniques is *forced away from the original ray* and toward the closest axis. Equally obviously, this elementary defect can be remedied without abandoning the essential features of the model.

Suppose that instead of the expected *absolute* change in the coefficients a_K and a_L being identical, their expected *proportionate* change is held to be equal. This alternative specification could also have been defended by appealing to the Principle of Insufficient Reason at the outset, although one must grant that does not make it any the less arbitrary. In any event, the indicated revision at this stage of the exposition is easily affected by the simple trick of relabeling the axes of Figure 5(a) so that $\log(a_K)$ and

---

The definition of x will be satisfied if we interpret x_i as a linear combination of the incremental change, at epoch i, of the firm's Cartesian coordinates in the plane of Figure 5. Thus,

$$x = \begin{cases} 1 \text{ when } \{\Delta a_K/h, \Delta a_L/h\} \text{ is } \{0, -1\} \\ -1 \text{ when } \{\Delta a_K/h, \Delta a_L/h\} \text{ is } \{-1, 0\}. \end{cases}$$

The variate x is random and has the discrete probability distribution

$$f(x) = \begin{cases} p \text{ when } x = 1 \\ q \text{ when } x = -1, \qquad p = (1-q). \end{cases}$$

Figure 5(a) considers the case where we set $p = q = \frac{1}{2}$. At each epoch the expectation of x_i is then

$$E(x_i) = \sum x_i f(x_i) = 2p - 1 = 0,$$

and the variance is

$$\text{Var}(x_i) = E[x_i - E(x_i)]^2 = \sum [x_i - E(x_i)]^2 f(x_i) = (1)^2 p + (-1)^2 q = 1.$$

The *expected distance* (in steps from the vector of unitary slope) at the n-th epoch is obviously

$$\text{E(D)}'' = \text{E}\left[\sum_{u}^{!}(x^{!})\right] = \sum_{u}^{!} \text{E}(x^{!}) = 0^{\cdot}$$

Correspondingly, the variance of the distribution of D_n is

$$\text{Var}(D_n) = \left[\text{Var} \sum_{i}^{n} (x_i)\right] = \sum_{i}^{n} \text{Var}(x_i) = n.$$

because the assumed independence of the x_i's permits us to say that $\text{Cov}(x_i, x_j) = 0$, which makes the variance of the sum just the sum of the (unitary) variances.

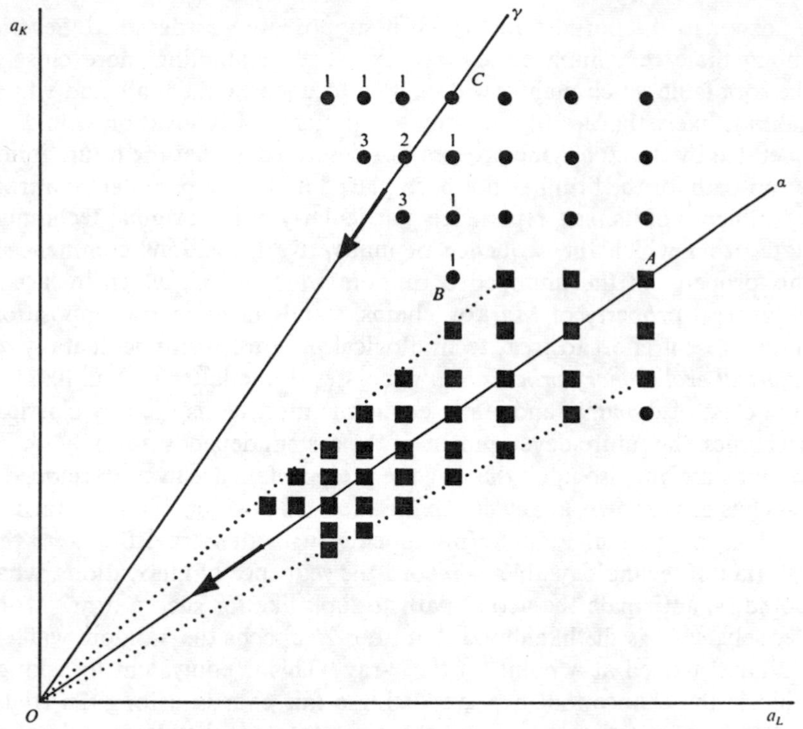

Figure 5(b) A random walk between elastic barriers

$\log(a_L)$ appear in place of the actual input coefficients.[1] Equivalently, in Figure 5(b) the orthogonal lattice – representing the underlying, discoverable technology – has been redrawn so that with $p = q = \frac{1}{2}$, as before, the expected change in position at each epoch now becomes a uniform *proportionate* movement toward the origin along whichever ray the firm happens to be on at that point in time. This same revision also serves to remove the troubling fact that the random walk depicted by Figure 5(a) could carry the firm to a position on one of the axes, or, still more preposterously, to the origin itself! In Figure 5(b), by contrast, the axes (and the origin) never can be reached.

Yet, with these more glaring defects removed, it may now be seen that there remain other respects in which the model (as revised) fails seriously

[1] The variate x in the preceding footnote may then be replaced by x^*,

$$x^* = 1/h[\Delta\log(a_K) - \Delta\log(a_L)],$$

and one may proceed analogously to determine the moments of the distribution $D_n^* = \sum_i^n x_i^*$. As a result of this modification $\mathrm{Var}(D_n)$ no longer increases directly with n – because now $\mathrm{Var}(D_n^*) = n$. But it does not appear that the logarithmic transformation serves to stabilize the variance of D_n, except possibly in the limit as $n \to \infty$.

to answer to the purpose for which it supposedly was designed. Several among these remaining difficulties are worth examining more closely. The root fault which plainly will be seen to underlie them all, and which makes it likely that locally non-neutral patterns of innovation would be generated by the stochastic process just described, is that the future innovation-path for the firm has not been placed under the persistent controling influence of its past experience, particularly of the original technique (state) from which the sequence of innovative transitions commenced. This property of the unimpeded random walk model, which in fact is the general property of Markov chains, stands in diametric opposition to the present effort to treat technological progress as fundamentally *an historical evolutionary process.*[1] In processes of the latter sort, unlike the large class of dynamic and static economic models inspired by classical mechanics, the future development of the system depends not only on its present state but also upon the way the present state itself was developed.[2]

As has been shown, at any epoch the *expected* position of a firm starting out from point C in Figure 5(b) coincides with the γ-ray. But were the historian of technology able to record the sequence of innovations, what should he anticipate the actual path to look like for such a firm? More precisely, what is the likelihood that after $2k$ epochs the 'random walker' will have arrived at a point on the γ-ray? This is equivalent to asking: What is the chance that a player using a fair coin in a long ($2n$ trials) coin-tossing experiment will have scored as many heads as tails after $2k < 2n$ trials? According to popular conception there is something called 'the law of averages' which insures that if the coin is fair a player will score more heads than tails about half of the time (for n trials), and more

[1] On the general Markov property, cf. footnote 2 on p. 72, above. I have thought it necessary here to qualify the adjective 'evolutionary' with 'historical,' because economists often identify as 'evolutionary' theories which make use of the Darwinian principle of *survival* of the (competitively) fittest – rather than those which incorporate the Mendelian principle of *heredity.* Both principles are represented in the present theory, a point already noted above in the text discussion of the 'Essentials of the First Step.' By contrast, Nelson and Winter, 'Evolutionary Theory' (1973) have recently described as 'evolutionary' a model of economic growth in which the use of economically efficient innovations emerges by 'natural selection' based on market survival, rather than from the rational maximizing efforts of firms. Yet, inasmuch as the Nelson–Winter treatment of the generation of technological innovations makes their model as a whole formally correspond to a Markov process, they necessarily fall short of embracing the genuinely historical conception of economic growth as an irreversible evolutionary development. In this respect their theory of secular change remains fundamentally neoclassical in spirit, even though their conception of micro-economic behavior departs from the neoclassical tradition.

[2] For some trenchant comments on the dubious inspiration which neoclassical equilibrium models of economic systems have drawn from Newtonian mechanics, cf. Nicholas Georgescu-Roegen, *The Entropy Law and the Economic Process* (Cambridge, Mass.: Harvard University Press, 1971), esp. pp. 1–22, 316–41. Conceptually, it should be observed, Markov processes are probabilistic analogues of the processes of classical mechanics.

tails than heads in the other half; further, the same 'law' ought to insure that during the game the lead will not infrequently pass back and forth between heads and tails. That being the case, the likelihood of observing a quasi-neutral perambulation, crisscrossing the γ-ray, should be quite high.

Rather startlingly, however, that is not the case at all. Unschooled intuitions in this context, more often than not prove very misleading. From the numbers adjoining the points in Figure 5(a) it may be seen that after the first two epochs the *frequency* of paths leading back to the central point of the distribution falls below $\frac{1}{2}$ and goes on declining at each successive epoch. The same holds true for the random walk depicted by Figure 5(b). More generally, an intriguing pair of theorems due to William Feller (the 'arc sine laws for last visits and sojourn times')[1] show that it is not at all unlikely that in a long coin-tossing experiment the cumulative number of heads (or tails) will hold the lead for practically the entire sequence of $(2n)$ trials. Indeed, if n is large it can be computed that with a 0·20 probability heads (or tails) will be in the lead for 97·6 per cent of the epochs!

This suggests a model based on purely random innovation could account for observations of an individual firm persisting in developing techniques that were located to one side – or to the other side – of the set of factor proportions with which it began. No theory of 'inducements,' no mechanism responsive to market signals *needs* to be found to explain the not infrequent appearance of extended single-firm innovation sequences through which some original process becomes more (or less) intensive in its use of capital *vis-à-vis* labor.[2] But by the same token, the unimpeded symmetrical random walk model under consideration cannot supply an *a priori* rationale for hypothesizing that the individual innovator's path is one that carries him downward, crossing and recrossing the original process-ray but staying always in its immediate neighborhood.

Essentially the same objection can be made against accepting a gambit borrowed from statistical mechanics. To do so one would in effect have to argue that even though the individual firm's course is not one of quasi-neutrality, in a population of such firms the 'representative innovator' will be moving on a strictly neutral path. It is true enough that if imitative

[1] *Probability Theory*, Third Edition, pp. 79, 82, and more generally, Ch. III, sections 3, 4, 5, *passim*.

[2] Feller, *Probability Theory*, Third Edition, p. 79 offers the following sobering illustration: 'Suppose that in a learning experiment lasting one year a child was consistently lagging except, perhaps, during the initial week. Another child was consistently ahead except, perhaps, during the last week. Would the two children be judged equal? Yet, let a group of 11 children be exposed to a similar learning experiment involving no intelligence but only chance. One among the 11 would appear a leader for all but one week, another as laggard for all but one week.'

innovation were disregarded and one imagined a competitive industry composed of many, independent, stochastic learning units of the kind we have been considering, at any epoch after starting from a common point of origin these random walkers would be found symmetrically distributed around the process-ray through that point. Half the population always will lie to the right, and half to the left, and this would hold true also in any stationary (or equilibrium) distribution toward which the system converged.[1] A firm whose *position* was 'representative' of this population would thus have to progress down the original process-ray.

The 'representative firm' in the foregoing argument is exactly the statistical construct that Alfred Marshall had in mind, and so its story might be of interest if technological innovations were generated by the *ensemble* of an industrial population, by the members of the industry acting *in concert* – that is to say, in the way that a population of producers generates the aggregate output of the industry. But such is quite clearly not so. Special circumstances aside, technological innovations are not simply 'added up' in the manner of the (standard) goods output of the firms belonging to an industry. The way one arrives at a useful characterization of the experience of the 'representative innovator' is, rather, to investigate the nature of the paths that are taken by, say, a *majority* of the members of the population.

It then follows that a relevant 'representative' (of the majority) is *not* the representative of the ensemble. For, the previously cited mathematical results due to Feller tell us that the majority of the symmetric random walkers will spend a disproportionately large part of their time off to one side (or the other side) of the original process-ray, instead of remaining on or repeatedly crossing it.

The foregoing considered how a group of stochastic learning units starting from a common point of origin would become distributed over the technology-space defined in Figure 5(b). It is equally appropriate to notice the awkward problem of two random innovators who start from widely different initial techniques, on the α-ray and the γ-ray for example, but in this model may wind up at an identical variant technique. The histories of firms originally operating at A and C in Figure 5(b) might be said to cease being relevant to their future course of innovation when

[1] Note that I have only conjectured that there exists a finite asymptotic variance for the outcomes of the process in Figure 5(b). Markov chains are described as being *ergodic* if they asymptotically reach an *invariant* probability distribution of states – an equilibrium distribution which is independent of the initial distribution. On 'ergodic chains,' cf. Feller, *Probability Theory*, Third Edition, pp. 393ff. Note that the symmetry property of the transient (and invariant) probability distributions in the sort of model presently under discussion will remain undisturbed if additions to the population (births of new innovating firms) occur at the center of the distribution, and disappearances (failures of firms) occur randomly.

purely by chance they come to occupy coincident positions, as at *B*. Thereafter the firms share a common innovation-space, as well as identical (stationary) probability distributions for transitions within it.

If a coincidence such as that at *B* is at all possible, a theorem due to Polya[1] establishes that the two innovators simultaneously performing independent symmetric random walks in two-dimensional space, with probability one will sooner or later (and therefore infinitely often) come to occupy a common position. The *certainty* of such chance 'collisions,' to use the metaphor of particle mechanics, is a feature of the present model that obviously runs counter to the intuitive sense of a localized neutral learning sequence. On the latter conception, the range of future technological developments ought i some meaningful fashion to be *restricted* by the firm's initial myopic selection of a process for the purpose of production.

To meet this objection one must really do more than insure that there is *some* positive probability of no 'collisions' taking place. That much could readily be arranged merely by describing each technique more realistically by a *vector* of many (and in any case more than three) input coefficients, so that the space in which the unimpeded random walk proceeds is envisaged as an *n*-dimensional surface, instead of a two-dimensional plane.[2] Yet, if the future technology generated by each learning entity is truly to unfold from, and hence be limited by, its past experience in production operations, we need a more radical revision of the random walk model – one that makes chance 'collisions' highly improbable (and not just probable) when two firms have started out from widely separated process-rays.

By this stage it should be apparent that all the difficulties we have been compelled to notice might have been circumvented by supposing, to begin with, that each of the rays from the origin of Figure 5(a) was quite closely surrounded by a pair of *elastic barriers*. Such impediments would restrict the locus of those random innovators whose learning careers began within them, and would lower the probability of 'collisions' by reducing the regions of common innovation-space for firms initially situated on different process-rays. The relevant domain of innovative action for the firm originally at point *A* might then be a 'discoverable fragment' of the orthogonal lattice – such as is (incompletely) indicated by the square, heavy dots leading toward the origin in Figure 5(b). The barrier metaphor thus provides a suitable reinterpretation of the dashed lines drawn around

[1] Cf. Feller, *Probability Theory*, Third Edition, pp. 360–2 for alternative formulations and a proof of Polya's theorem.

[2] From the theorem of Polya referred to in the text above, it also follows that whereas two independent actors performing a random walk in one or in two dimensions are certain to arrive concurrently at a given point infinitely often, for three-dimensional random walks there is a positive probability of their never colliding.

the β-ray in Figure 4, although in that case the small arrows suggest that non-orthogonal discrete movements on the plane are possible.

The nature and operation of the 'barriers' now envisaged deserve brief comment before we turn to consider deeper reasons for their presence. In the terminology of a one-dimensional random walk in a finite interval, if there is an *elastic* barrier at position (state) 2 then a particle may freely move from state 1 to state 0 with probability $(1-q) = p$, but with probability vq it is 'reflected' by the barrier and remains in state 1; with probability $(1-v)q$ it goes to state 2 where it is 'absorbed' and where, therefore, the walk terminates.[1] Unlike the unending perambulation that would occur between a pair of *reflecting* barriers, a random walk by a number of independent innovators confined within a given pair of elastic barriers has a positive probability of ending. And in general, the distribution of outcomes would not tend to pile up in the neighborhood of the barriers to the same extent that would be found were those barriers of the *absorbing* type, i.e., $v = 0$ in the preceding formulation.

For any individual participant in this 'elasticity restricted' random walk, the larger is the positive probability v pertaining to transitions from the states adjoining the barriers, the more sustained will be *expected* finite distance traveled along the α-ray toward the origin of the plane. Thus, in addition to the positioning of the hypothesized pair of elastic barriers, the probability v can be treated as *an important parameter of each ray.* Suppose all rays were surrounded by similarly positioned symmetric elastic barriers, but values of v closer to 1 typified the process-rays with the higher machine–man ratios. This situation would correspond to the previously discussed notion that nineteenth-century fundamental scientific and technical knowledge was such that the cumulative course of *mechanical* innovation proceeded more swiftly and surely than was the case with innovations involving (say) biological or chemical processes of production.

By endowing each firm in the system with (non-stationary) transition probabilities that depend partially upon its present state, but no less importantly upon the initial technical state (the production process-ray) wherein its learning sequence commenced, we have found our way back to a microcosmic representation of endogenous technological progress as

[1] Cf. the discussion of 'reflecting,' 'absorbing' and (more general) 'elastic' barriers in Feller, *Probability Theory*, Third Edition, pp. 42–3, 376–7. From the footnotes to pp. 73–5, above, it has already been seen that the two-dimensional random walk in Figure 5(b) may be transformed into a one-dimensional (stationary) perambulation – around an origin-position (or 'state' to use Markov chain terminology) corresponding to the original ray. Describing states by their distances (in minimum permissible unit steps) from the origin (state 0), Figure 5(b) envisages symmetric elastic barriers at states $+2$ and -2. As in the previously considered unimpeded random walk, when the particle is in state 0 it makes transitions to state 1 with probability p, and to state -1 with probability $q = 1-p$.

a locally neutral, stochastic process. Being truly historical, the latter is clearly non-Markovian.

What, then, creates the 'barriers'? What causes the underlying discoverable technology-space to be in effect fragmented into incompletely connected sub-lattices? At least a tentative answer may be found by going back to an idea already noticed in connexion with Chenery's rationale for the existence of a fundamental production function: the overall input proportions characterizing any one process (or linear production activity) represent a weighted average of the *different* input proportions that respectively characterize each member of a set of complementary sub-activities. The latter, underlying proportions in turn reflect the choices made among available specifications for the constituent elements of the total engineering design.

Now suppose it were possible to design a discretely more efficient way of performing the function that is specific to a single sub-component of an existing process, thereby reducing requirements (per unit of 'throughput') for the collection of inputs that happens to be peculiar to the sub-component. The implied reductions in factor use at the aggregate process level would be most pronounced (proportionately) for the factor(s) in which the 'improved' sub-activity characteristically is most intensive. If this engineering advance left the throughput capacities of the other elements of the design unaffected, however, the gains from the contemplated innovation may be severely circumscribed by 'technical interrelatedness' (complementarities) among the sub-activities.[1] Further efficiency gains in the initial factor-saving direction would thus require complementary advances in a different direction, leading to the restoration of something closer to the original balance among the proportions of various inputs used.

In the illustrative instance of a batch-brewing operation, the per quart capital-goods requirements for *containing* the liquor during fermentation might be reduced simply by employing a brewing vessel with a larger radius, since the costs of constructing a cylindrical container would tend to rise in rough proportion to its surface area (and hence with the square of the radius) while the volume contained will be increased as the *cube* of the radius.[2] Yet, the existing design of the cooling system, needed to

[1] For fuller discussion of the concept of 'technical interrelatedness,' and its larger implications – including those for the diffusion of technological innovations, cf. Paul A. David, 'The Landscape and the Machine: Technical Interrelatedness, Land Tenure and the Mechanization of the Corn Harvest in Victorian Britain,' Ch. 5 in D. C. McCloskey, ed., *Essays on a Mature Economy: Britain After 1840* (London: Methuen, 1971), which appears as Ch. 5 of the present volume.

[2] Cf. L. Bruni, 'Internal Economies of Scale with a Given Technique,' *Journal of Industrial Economics*, Vol. 12 (July 1964), pp. 175–90 for many industrial applications of the so-called 'two-thirds' rule.

regulate the rate of fermentation within the container, may make this larger volume of liquor less efficient to control, increasing the per quart cooling costs (for ice, or power to run a compressor) for larger batch sizes. In the limiting case, a point might quickly be reached where further scale-related savings (of the capital goods inputs characteristic of the storage sub-activity) would be worthless because, in effect, the marginal costs rise too rapidly in a complementary (cooling) sub-activity. Further progress in raising 'storage' efficiency only could be achieved by easing the constraint posed by the initial cooling design. To expand the capacity of the cooling sub-activity without increasing unit cooling costs would entail development of another new design, increasing the efficiency of the peculiar collection of inputs characteristic of cooling systems; were these two the only sub-activities in the total process, the second engineering innovation necessarily would tend to offset the aggregate level change in relative factor use implied by the first innovation, restoring the set of input proportions that initially characterized the process as a whole.

In a world where there are indivisibilities in the capacities of available engineering components – in terms of which the designs for processes are specified – the balance among components cannot be continuously adjusted; a sub-activity in which there is excess capacity under one arrangement may in the next design become the binding constraint upon the entire process. The more limited are the possibilities for continuous substitutions among sub-activities in which the foregoing conditions apply, the more likely it is that when the past *direction* of (local 'bias' in) factor-savings cannot be reversed, the persistence of one sub-activity as an irremovable constraint will terminate progress toward further raising the efficiency of input use for the process as a whole. In such a world, it is not so difficult to see that it is the technical 'facts of life,' the non-convexities present in micro-engineering designs themselves, that will cause a ray describing an initial process design to be surrounded by barriers to random innovation – the sort of barriers whose existence other considerations have already led us to imagine.

The foregoing conception may be of special relevance for understanding the evolution of production methods during the nineteenth century, since it can be fitted together most readily with a body of observations made by close students of the history of *mechanical* technologies, and of machine tool technology in particular.[1] The term 'compulsive sequences' recently has been employed by Nathan Rosenberg[2] to describe a microcosmic

[1] Cf., e.g., the studies of the gear-cutting machine, the grinding machine, the milling machine and the lathe, collected in Robert S. Woodbury, *Studies in the History of Machine Tools* (Cambridge, Mass.: The M.I.T. Press, 1972); Nathan Rosenberg, 'Technological Change in the Machine Tool Industry, 1840–1910,' *Journal of Economic History*, Vol. XXIII, 4 (December 1963).

[2] Cf. Rosenberg, 'Direction of Technological Change' (1969), pp. 3–11.

pattern of technological adaptations that frequently appears in the history of western economies since the industrial revolution of the eighteenth century. In these sequences, it would seem, *internal technical relations* among the elements of a production process – rather than external, market conditions – generate a succession of more or less obvious 'engineering challenges,' practical difficulties that serve to focus inventive effort in one direction or another at different moments in time. The need for resolution of each of the compelling problems typically is signaled by breakdowns, equipment malfunctions, or more generally speaking, by the materialization of physical bottlenecks of various highly specific kinds. As the bottlenecks occur *within* the production operation, they are not immediately translated into relative price movements in markets for the inputs. Instead, technological innovations are seen as emerging in response to the challenges they pose.

As an explanation of the *unremitting* character of innovative efforts, the foregoing schema is more plausible in circumstances where cultural and economic conditions have been conducive to the view that life itself is best conducted as an exercise in 'problem-solving,' rather than in problem-avoidance.[1] Since every challenge is thus presumed to elicit a technologically creative response, attention has been focused upon the task of understanding why a *sequence* of challenges comes to be generated. A rigorous explanation heretofore has not been forthcoming from historians of technology. But where an engineering challenge is conceived of as a 'collision' with one of our hypothesized elastic barriers – either terminating the innovative process or altering its direction – one perceives there is an intimate connexion between the generation of finite compulsive sequences and the generation of patterns of local quasi-neutrality in endogenous technological process.

Indeed, a hint of the presence of this connexion is furnished in the resemblance Rosenberg discerns between these innovational progressions and the vastly more macrocosmic sequences of 'unbalanced economic development' described by Albert Hirschman.[2] Recurring 'technical imbalances,' on this account, play the crucial role generating the succession

[1] Perhaps because the active agents in this drama are implicitly cast as 'innovating engineers,' Rosenberg's discussion does not entertain the logical possibility that people might become discouraged by the dawning realization that each difficulty overcome simply brings closer the onset of a new vexation. Yet, there seems little doubt that the possibility of their settling for a 'technologically quiet life' safely can be disregarded in the context of the present comparisons of American and British development during the nineteenth century. For an illuminating history of the cultural and professional milieu of American engineers, cf. Monte A. Calvert, *The Mechanical Engineer in America, 1830–1910* (Baltimore: The Johns Hopkins University Press, 1967).

[2] Albert O. Hirschman, *The Strategy of Economic Development* (New Haven: Yale University Press, 1958).

of compelling challenges to engineering ingenuity. And the conditions which must be envisaged as perpetuating the existence of these imbalances turn out to be the very ones that already have been held responsible for the presence of elastic barriers impeding random endogenous innovation. A vivid illustration of this congruence is provided by considering the sequence of machine tool design modifications that were made in order to take advantage of the great jump in the *potential* rate of metal removal resulting from the introduction of cutting tools formed from high-speed steel.[1]

> The final effect of this redesigning that was initiated by the use of high-speed steel in cutting tools was to transform machine tools into much heavier, faster, and more rigid instruments which, in turn, enlarged considerably the scope of their practical operations and facilitated their introduction into new uses. Much of the progress in machine tools resulted from the generation of imbalances between the machine itself and the cutting tool. Improvements in the cutting tool required machines of greater strength, rigidity, capacity to withstand stress, etc. Improvements in the design and operation of the machines, in turn, were useless without improvements in the properties of the cutting tool.

Beneath the surface resemblance, then, there is some structural similarity with the 'imbalances' between directly productive activities and social overhead capacity discussed by Hirschman. Both derive from some elements of indivisibility, or lumpiness in the available components from which the design or the total production process can be constructed. Consequently, any initial design will not be likely to succeed in 'balancing' all the components, and balances eventually established may be upset by the introduction of some component derived from an exogenous source – as in the case of high-speed steel's impact on the components of lathes, and milling machines which had been 'improved' within the constraints set by carbon steel cutting tools. Removal of some proximate constraint by modification of the component involved, then, occasions a discrete addition to capacity which cannot be fully utilized without suitable adjustment of other components. And so on, until balance is restored to the point that further exploitation of excess capacities is no longer attractive.

The foregoing model of 'unbalanced' (and quasi-neutral) technical progress now may be made more persuasive by realistically relaxing the assumption of perfect competition. This we shall do only to the extent of depicting an otherwise competitive enterprise as exercising temporary

[1] Rosenberg, 'Direction of Technological Change' (1969), p. 8. This was an alloy steel developed by Frederick W. Taylor and his associates at the close of the nineteenth century. In comparison with the carbon steel previously used, it enormously raised the red hardness of cutting tools, thereby making possible the removal of metal faster – by cutting at higher speeds and by taking heavier cuts.

monopoly control over the *information* regarding its own design innovations for sub-components of the basic production process currently employed by firms in its industry. When such an innovator comes to decide how much to expand capacity in the newly modified sub-activity, the market prices for the requisite investment goods will be those which reflect the pre-existing state of technological knowledge (accessible to the other firms in the industry). Those prices will not reflect the new uses to which the still-secret innovation in question permits those investment goods to be put. Thus, it generally would turn out that in making a concrete alteration in its own production operations, the information-monopolizing innovator is led to install a comparatively large addition to the capacity of that sub-activity. To be more precise, the capacity expansion can be described as 'unbalanced' also in the sense of exceeding the increment that would have been optimal (for minimizing short-run production costs) had knowledge of the new design been immediately disseminated, and had the prices of the requisite investment goods therefore been bid up instantly by the imitative investment actions of the innovator's product-market rivals.

Because an information-monopolizing innovator winds up having *comparatively* more capacity in the improved sub-activity (*vis-à-vis* the whole process) than other producers – and, due to indivisibilities, may actually have substantial 'excess' capacity there – the incremental gain to be reaped from innovations that would expand capacity *in the remaining, complementary sub-activities* turns out to be greater for him than it is for the rest of the industry's firms. In other words, there is a divergence between market prices and the relative shadow-prices which the innovating firm will perceive for the inputs that are intensively used by the yet-unimproved (bottleneck prone) sub-activities. Under the conditions we have envisaged, this relative price divergence creates a positive inducement to improve the efficiency of input use in the latter class of sub-activities, and thereby raises the probability that the firm's next innovation will reverse the overall input-saving 'bias' of the technical advances that immediately preceded it.[1] This is precisely what has been imagined to happen in the neighborhood of the elastic barriers surrounding the α-ray in Figure 5(b).

On the other side of the coin, were no member of the industry to exercise a monopoly over information, the competitive bidding for the

[1] The same situation may also bring to the firm's notice (private) information regarding spontaneous, or 'outside,' technological advances which happen particularly to raise input efficiency in the sub-activity that has currently emerged as the constraining element in production operations. Independent inventors of such devices will tend to seek out the firm, as a potential purchaser for whom their knowledge will have greater-than-average value.

means of installing a newly redesigned component would raise the costs of effecting such alterations in the production process, thereby checking the tendency toward a *comparative* expansion of the capacity of the improved sub-activity at an earlier point. Further, in (unrealistic) circumstances where the capacities of each of the sub-activities are infinitely *divisible*, complete cost-minimizing balance among them can be achieved without any excess capacity remaining.

Consequently, and perhaps not unexpectedly, when the basic technological and (information) market assumptions of the present model are *both* discarded, we are led straight back to the proposition upon which Fellner and Salter logically insisted: a rational (myopic) firm, using infinitely divisible factors of production whose prices it must take as parametrically determined, would move to a position of cost minimization at which it had no 'under-utilized' inputs, and where all its factors had to be regarded, at the margin, as both equally cheap and equally dear. But under those conditions, as has been implicitly demonstrated, the 'barriers' we have been contemplating hardly could arise. Each innovating entity would appear completely free to wander throughout the continuum of the discoverable technology space, and would have no economic inducement to do anything except wander thus, in a drifting, totally unrestricted random walk.

Some readers may find it deeply reassuring that these same conditions (infinite divisibility and truly perfect competition) were at one time supposed, by economists working in the neoclassical tradition, to preclude the possibility of price-induced biases in *global* technological advances. For they reappear in the foregoing non-neoclassical setting as wholly inimical to the existence of the restricted type of random walk that could generate the envisaged compulsive sequences of quasi-neutral, localized innovations. And, in the larger context of the present argument, it is a neutral microcosmic pattern of innovation that forms the crucial link between the historical configuration of relative factor-prices and the appearance of an *unintended* global 'bias,' or factor-saving drift in the ensuing course of a society's technological progress. Thus, in accepting the view of matters advanced here, we ultimately are obliged to reject *inter alia*, the *empirical* validity of the neoclassical argument by which Fellner and Salter dismissed the notion that a particular bias in the intended direction of inventive effort could be induced by the prevailing pattern of factor-scarcity relationships.

THE SECOND STEP: A CLOSING OF THE CIRCLE

We are now within sight of home. It remains only to indicate some plausible lines of connexion between America's comparative land abundance and the choice there of relatively capital-intensive techniques. For, a rationale then will have been provided linking all the 'factual' propositions of the Rothbarth–Habakkuk thesis.

The first argument to consider is essentially micro-economic in character. It turns upon an additional empirical proposition regarding the fundamental nature of machine technologies, and, more specifically, of the sort of mechanical technology which had come into reach by the early nineteenth century.

Let us begin, however, with a point on which there was agreement by the participants in the controversy over the neoclassical interpretation of the Rothbarth–Habakkuk thesis in its 'more machinery' version. It is not sensible to ignore the direct and indirect use of natural resources by industrial activities, nor the employment of reproducible capital in agricultural pursuits.[1]

The relevant question therefore becomes: what was the technical relationship among natural resources, labor, and capital inputs characteristic of industrial and of non-industrial production activities during the nineteenth century? There are two obvious styles of analysis within which the implications of the answers supplied to this empirical question might be explored. One, which will not be pursued here, would require some elaboration of the two-sector models of general equilibrium already considered in the discussion of Rothbarth's argument by Temin, Fogel, and Jones.[2] The other style of approach is of a *partial* equilibrium character.

It starts from the broad factual assertion that the comparative abundance of natural resources in America, and the nature of the available transport technology, implied that firms in most branches of production *there* faced low relative prices for direct and indirect natural resource inputs – i.e., land for the grain farmer, and wheat for the flour-miller. The implied comparisons here are those with the array raw materials–final product price ratios prevailing in Britain.[3]

To this now add an empirical assertion about the character of nineteenth-century mechanical technologies, namely, that the known principles by which it was possible to have machinery perform tasks otherwise executed by men, and more fully to mechanize such operations, as a rule required a greater input of 'land,' or the products of 'land' per unit of output. More

[1] Cf. Part I, above, footnotes 1–5, p. 27 for references.

[2] In particular one would need to consider the implications of different patterns of factor-intensity for the different sectors, under alternative assumptions about price and income elasticities of demand. Cf., e.g., Jones, 'A Three Factor Model' (1971).

[3] While I would hazard that most historians of the two economies will regard this as a reasonable general assertion, and could provide illustrations (as well as an occasional counter-example), it should be stressed that the necessary *systematic* comparison of the British and American price structures on different dates in the century has yet to be made. It would be a project well worth undertaking.

formally stated, the hypothesis is that the relevant fundamental production functions (FPF of Figure 3) for the various branches of industry and agriculture did not possess the property of being *separable* in the raw materials and natural resource inputs; instead, the relatively capital-intensive techniques we have been considering were also relatively resource-using.[1]

This generalization might be referred to as the 'Ames–Rosenberg hypothesis,' for those authors have made a persuasive case that comparative prodigality in the use of raw materials was the key to many of the early American triumphs of mechanization which so impressed visitors from abroad. Perhaps the most striking instances are those which Ames and Rosenberg have noted in wood-turning operations, such as the production of highly irregularly shaped musket stocks with the early (1818) version of the Blanchard lathe. But more generally, they maintain[2]

> there is evidence which suggests that the woodworking machines which were popular in America and neglected in England were not only labour-saving but also wasteful of wood. Their adoption in America and neglect in England may be attributable not only – or perhaps not even primarily – to differences in the capital-labour ratios in the two countries but rather to the cheapness of wood in the United States and its high price in England. If nineteenth-century machine processes were generally more wasteful of raw materials than the corresponding handicraft processes this fact would help to explain the longer persistence of handicraft methods in England than in the United States. We suggest that this may have been the case in woodworking machinery, where, by general agreement among all observers and commentators, the technological differences between England and the United States were the most remarkable.

Supporting evidence, drawn from quite different areas of industrial endeavor might be cited. Lars Sandberg[3] has shown that, in comparison with the input requirements of the mule spinning equipment which dominated the English cotton textile industry, American ring spinning machinery was more capital-intensive but – and this was so especially for yarn counts above 40 – also required a greater input of longer staple cotton (a more costly grade of raw material) per pound of yarn. Drawing

[1] This says that production processes of the micro-economic level did not possess (the Leontief condition of separability) the property of exclusive dependence of the relative margin productivities of capital and labour on the capital–labor ratio. Note, however, that this property was assumed for the *aggregate* U.S. Domestic Economy in the econometric study on which Figure 2, above, is based. Such conflict as this raises has more to do with the spirit of the earlier argument than with its formal correctness, inasmuch as the relationship between the aggregate and the specific industry functions is neither direct nor obvious.

[2] Ames and Rosenberg, 'Enfield Arsenal' (1968), pp. 831ff.

[3] Cf. Lars G. Sandberg, 'American Rings and English Mules: The Role of Economic Rationality,' *Quarterly Journal of Economics*, Vol. LXXXII, 4 (November 1968).

a parallel in the sphere of agriculture, we find that the efficient use of mechanical reapers required a level, stone-free farm terrain, arranged in large and regularly shaped enclosures – a specific natural resource input that at the midpoint of the nineteenth century was obtained much more cheaply (relative to the price of grain) in the United States than in the British Isles.[1]

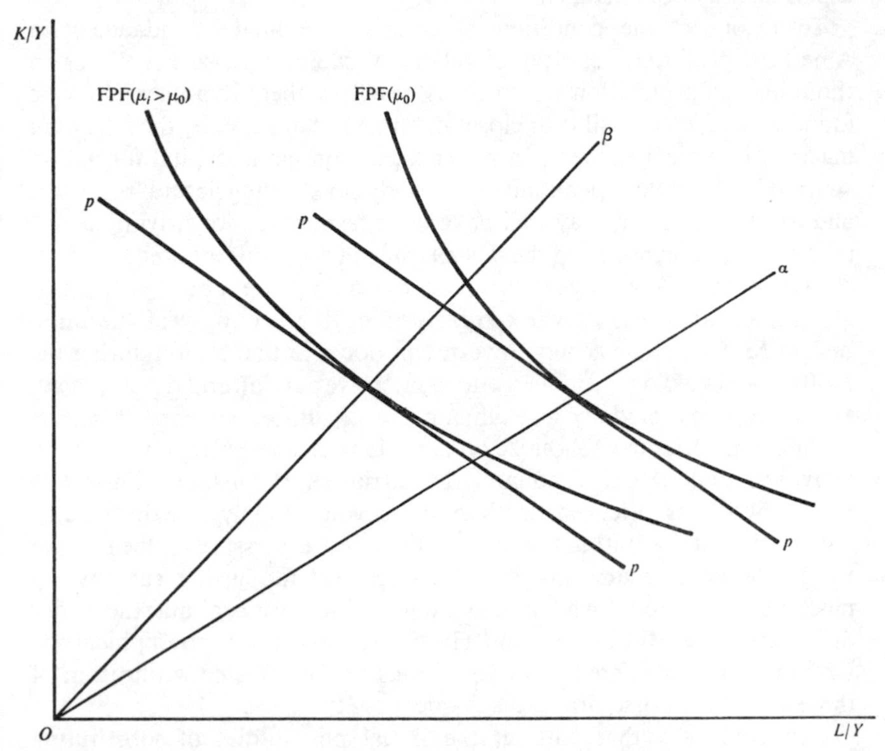

Figure 6 The Ames–Rosenberg hypothesis: Non-separability of resource inputs

From Figure 6 it may be seen that acceptance of the Ames–Rosenberg hypothesis would permit the formulation of an argument rationalizing the entire Rothbarth–Habakkuk thesis on the basis of a string of purely *technological* propositions. The fundamental unit isoquants describing the capital and labor requirements now must be distinguished according to the accompanying level of the 'materials' input coefficient (μ): along FPF(μ_i) all the techniques use *more* materials per unit of output than do techniques with the corresponding capital–labor proportions described by FPF(μ_0). Thus, even if the same labor–capital price ratio (*pp*) had faced producers

[1] Cf. my study, 'The Landscape and the Machine' (1971), which appears as Chapter 5 of this volume.

in Britain and America, the comparatively greater availability of natural resources would have suggested to some American producers the design, and to others the selection for use, of more capital-intensive methods. Technique β would be chosen in preference to methods like α. And from such choices immediately would flow the consequences for the global characteristics of the ensuing technological developments in each country, as has already been detailed.

To reconnect the condition of comparative land abundance with American producers' choices of relatively capital-intensive methods in those lines of production where the Ames–Rosenberg hypothesis may be found invalid, it is possible to close the circle of causation by the following macro-economic argument. In America, the on-going capital formation spurred by the greater possibilities of jointly substituting natural resources and capital for labor, may well have been responsible for driving up the relative price of labor from the *demand* side of the labor market.

Elsewhere,[1] I have suggested that a dynamic process built very much along such lines was at work transforming the economy of the antebellum Midwest. The general investment boom in that region during the 1850s – predicted on natural resource–intensive agricultural development, and rapidly executed by a profligate use of lumber stripped from the Michigan pineries to economize on labor in urban and rural construction activities – undoubtedly encountered recurring short-run labor constraints that proved less price-elastic than the savings supply schedule facing regional investors. Within this setting, the upward pressure on the relative wage rate would materialize as an inducement for further substituting machinery for labor. And in the event, it did bring about the wider diffusion of the McCormick and Hussey reapers in a topographical environment which offered ideal conditions for the efficient utilization of those and other horse-drawn mechanical contrivances.

One could say that without the initial possibilities of substituting combinations of natural resources and capital for labor, the initial condition of sparse settlement would have so depressed the rate of return on investment that the Midwestern development boom would never have acquired the momentum it did attain. As an incidental consequence, the induced substitution of machinery for labor in the agricultural and manufacturing activities of the region, undoubtedly would not have proceeded as far as it had gone on the eve of the Civil War.

[1] Cf. 'The Mechanization of Reaping in the Ante-Bellum Midwest,' Ch. 1, in *Industrialization in Two Systems: Essays in Honor of Alexander Gershenkron*, ed. H. Rosovsky (New York: John Wiley and Sons, 1966) which appears as Chapter 4 of this volume. The labor market aspects are examined in considerable detail in a still unpublished study: 'Industrialization and the Changing Labor Supply in a Region of Recent Settlement,' Stanford Research Center in Economic Growth, *Memoranda*, Nos. 25–6 (Stanford: August 1963).

We shall close on this characteristically American story. In its attention to the macro-economic side of the matter and to the dynamics of labor market adjustments, as well as in its concern about the affects of the region's initial comparative labor scarcity upon the rate of investment, it is perhaps not so removed in spirit from the tale Habakkuk originally sought to tell.

Generation

2 Learning by doing and tariff protection: a reconsideration of the case of the ante-bellum United States cotton textile industry[1]

Can learning by doing be held to have played a significant part in raising productive efficiency during the early growth of manufacturing industries in the United States? If there is indeed an adequate basis for regarding technical progress during the pre-Civil War period as an endogenous process depending crucially upon the accumulation of practical experience, what sorts of 'learning functions' best describe the forms in which that process manifested itself? And, in evaluating the impact of national commercial policies – be they historical or contemporary – what implications flow from the existence and the characteristics of learning effects in young industries, and possibly also in not so young industries?

My objective in this essay is not to settle the foregoing questions, but rather to incite others to give them closer consideration. For, although they have been posed here in a fashion immediately germane to the study of American industrialization in the nineteenth century, the core issue of the relationship between tariff policy and the technical characteristics of production in developing industries remains a subject of much broader current relevance. Anyone intrigued by the larger subject will, nevertheless, concede that the question of United States tariff policy in the ante-bellum era is one that stands in need of re-examination, in the light of the advances that have occurred in the pure theory of international trade, and in the study of American industrial history, since the time when the orthodox rendering of United States tariff history was first laid down by F. W. Taussig.

With a view to promoting, or provoking, such a reconsideration, this

[1] Before anything else, I wish to acknowledge my indebtedness to Robert B. Zevin, whose yet unpublished work applying a long-run learning function to data on an ante-bellum cotton textile firm inspired the econometric research reported in this study. Professor Zevin may be held responsible for having started this hare, though not for the course it has run. The subsequent direction taken by my thoughts on the subject of learning and tariff policy was inexorably (but pleasurably) influenced by many conversations with my colleagues Emile Despres and Ronald McKinnon. They ought not, however, to be blamed for any of my failures in faithfully absorbing their wisdom on the subject. The empirical investigation of learning effects was carried out in conjunction with the Stanford–S.S.R.C. Study of Economic Growth in the U.S., and benefited immeasurably from the insight of my colleague in that project, Moses Abramovitz. Finally, I wish to acknowledge the able computational assistance Harry Cleaver provided in connexion with this study.

article inquires first into the logic of resting the case for tariff protection of young industries upon the presence of learning effects, and draws notice to the implications that differing kinds of learning effects would carry for the design of government development policies. That motivation having been established for distinguishing among alternative forms of learning functions, the remaining sections of the study present some econometric evidence concerning learning by doing as a source of productivity growth in the New England cotton textile industry during 1834–60, a period in which that protected industry emerged as the major branch of United States manufacturing activity. On the basis of this evidence it is possible to venture at least a partial reassessment of the orthodox position with regard to the case of the early American textile tariffs.

TEXTILES, TARIFFS, AND TAUSSIG

The manufacture of cotton yarns and cloth, perhaps more than any other specific branch of industrial endeavor, has been afforded 'encouragement' in many countries through the device of protective tariffs. It is perhaps not too hard to see why, in the framing of national commercial policies since the early years of the nineteenth century, this particular activity was repeatedly slated for the treatment prescribed by pragmatic and theoretical dispensers of aid to 'infant industries.' Great Britain's achievement, having made the establishment of cotton textile mills an almost universally recognized symbol of economic progress, at the same time impeded easy imitation of her example; the very fact of Britain's industrial pre-eminence manifested itself in her ability to sell factory-made yarns and cloth cheaply in overseas markets, despite the necessity of importing the raw cotton from distant sources. In such an historical context, the growth of a domestic cotton textile industry might well appear in the eyes of would-be imitators to run a serious risk of being blocked by the mere accident of British industrial leadership, rather than by 'natural' causes.

Quite possibly it was with this particular dilemma in mind that a respectable 'Free Trader' like F. W. Taussig[1] exaggerated the case for protecting young industries, describing as its essential point the stipulation that the rise of the industry in question was being thwarted by conditions that were in some sense *temporary* and *artificial*, rather than 'natural and permanent.' Certainly the cotton textile industry was not far removed from his thoughts, for Taussig illustrated the admissible case for protective import duties by asking his readers to suppose that

[1] Cf. F. W. Taussig, *The Tariff History of the United States*, Eighth Edition (New York: Putnam, 1931), esp. p. 2.

(t)here is no permanent cause why cotton goods should not be obtained at as low cost by making them at home as by importing them;... But the cotton manufacture, let it be further supposed, is new: the machinery used is unknown and complicated, and requires skill and experience of a kind not attainable in other branches of production. The industry of the country runs by custom in other grooves, from which it is not easily diverted. If, at the same time, the communication of knowledge be slow, and enterprise be hesitating, we have a set of conditions under which the establishment of the cotton manufacture may be prevented, long after it might have been carried on with advantage.[1]

The intellectual pedigree of the argument was unassailable – it had been recognized by J. S. Mill – and the illustration was meant to be plausible. But this was still in the realm of theory. When Taussig embarked on an examination of the historical experience with the protection of young industries in the United States, he again turned first to the manufacture of cotton as it was 'the most important of them, on account both of its magnitude and of the peculiarly direct application of protection to it.'[2] Here, however, Taussig concluded that the tariff protection given the industry up to 1860 was largely without justification, because for most of the relevant period the manufacture of cotton textiles in the United States had ceased to satisfy the conditions of the infant industry argument.

One has to allow that the theoretical suppositions previously entertained might have been appropriate as a basis for encouraging the tentative, often commercially unsuccessful, steps taken toward establishing spinning mills on the English pattern in the years before 1808. Yet, Taussig stressed, a radical alteration of the position of the American textile industry was actually brought about by the intervention of diplomatic and military events. The cumulative effect of the Embargo Act, the Non-Intercourse Act, and then the formal commencement of hostilities with Great Britain, was to provide the domestic industry with extraordinary stimulation, insulating it completely from foreign competition, while compelling a concurrent withdrawal of loanable funds from the crippled maritime trades.

Having flourished in the sheltered, hot-house climate thus created during 1808–15, the new American factory textile industry as a whole was unprepared to withstand exposure to the withering blast of English cotton cloth exports consigned for quick sale in East Coast ports, frequently at auction, during the years 1816–19. The first national tariff support extended to domestic cotton textile production was consequently aimed at easing the readjustment that had been suddenly forced upon the high-cost firms lately propelled into existence. As Taussig saw the situation,

[1] *Ibid.*, pp. 2–3.

[2] Cf. *ibid.*, pp. 25–36, 135–42 for Taussig's discussion of the industry in the ante-bellum period. (The quotation appears on p. 25.)

such justification as could be found for the Tariff of 1816 would have to turn upon its effectiveness in smoothing an over-expanded industry's accommodation to the rigors of peacetime competition, rather than in the nurture of a puny infant. In the event, the specific duties levied on the coarse cloth under the Act did not begin to afford noticeable protection until after the sharp decline in the level of textile prices during 1819.

According to this account, even the period during which such transitional assistance was called for was short-lived and coincided with the decade of revolutionary technical and organization advances – from 1814 to 1824 – that saw the introduction of the power loom and the water-driven integrated spinning and weaving plant. 'Probably as early as 1824, and almost certainly by 1832, the industry had reached a firm position, in which it was able to meet foreign competition on equal terms.'[1] Clear indication of an improvement of the international competitive position of the industry was to be found in the fact that significant quantities of American cotton goods began to be sold in export markets, primarily in China and the Far East, during the 1830s. At least this was the case so far as coarse-goods were concerned.[2] It apparently required no reference to Lerner's[3] proposition regarding the symmetry of import and export taxes for Taussig to sum up his account of the textile tariff's impact in the years 1840–60 by remarking that an industry which regularly exported a substantial part of its output 'can hardly be stimulated to any considerable extent by protective duties.'[4]

For more than three-quarters of a century, the conclusions put forward by Taussig (1887–8) with regard to the ante-bellum protection of the cotton textile industry have held undisputed sway. It is a curious fact that,

[1] *Ibid.*, p. 136.

[2] On the eve of the Civil War the (f.o.b.) value of domestic exports accounted for something on the order of 8–10 per cent of the gross value of U.S. cotton goods production. Cf. U.S. Census Bureau, *Eighth Census of the United States* (1860), 'Manufactures,' pp. 733–42, for gross value of product, and Taussig, *Tariff History*, p. 142 for 1859–60 export values from the Reports of the Secretary of the Treasury on Commerce and Navigation. E. Baines (*A History of the Cotton Manufacture in Great Britain* [London, 1835], pp. 509–10) concluded, from a comparison of the costs of manufacturing cotton goods in the U.S. and England in 1832: 'It may be said that Americans are capable of rivaling the English in coarse and stout manufactures, ... especially, in an article called "domestics," which they consume largely and export to some extent; but that in all other kinds of goods, all of which require either fine spinning or hand-loom weaving, the English possess and must long continue to possess a great superiority; in other words, "the Americans cannot economically produce fine manufactures."'

[3] Cf. A. P. Lerner, 'The Symmetry Between Import and Export Taxes,' *Economica*, III (August 1936), pp. 308–13; also, R. I. McKinnon, 'Intermediate Products and Differential Tariffs: A Generalization of Lerner's Symmetry Theorem,' *Quarterly Journal of Economics*, LXXX (November 1966), pp. 584–615.

[4] Taussig, *Tariff History*, p. 142.

despite all the interest exhibited by the so-called 'new' economic historians in the initial phases of American industrialization, no overt move has been made to reappraise the nation's avowedly developmental early tariff policies. There has thus far been no effort at a systematic reconsideration of the entire body of nineteenth-century American commercial policies, an undertaking that would obviously benefit from the insights gained by the past two generations of international trade theorists. Although the economic history texts no longer treat the tariff question as the burning issue it once was, when they choose to say anything about textile tariffs they hew unswervingly to the orthodox Taussigian line: from 1824 onward the continued application of protective tariffs to the manufacture of cotton goods ceased to have any justification at all, let alone a justification on strict 'infant industry' grounds.[1]

The durability of these strictures seems all the more odd when one observes that, with each passing year, historians' views of the technological characteristics of the ante-bellum cotton textile industry have, paradoxically, assigned greater weight to the presence of those very same conditions upon which the infant industry argument was traditionally grounded. More and more space is being accorded to historical accounts emphasizing the role played in American industrialization by the realization of economies of scale and specialization, and the improvement of productive efficiency through 'learning' and 'training effects' in new branches of production. For quite some time now the cost-reducing consequences of growing firm size, functional specialization, and localization based on the introduction of the power loom and the organization of integrated mills on the Waltham system, have figured as predominant themes in treatments of the rise of the New England cotton textile industry in the ante-bellum era.[2]

More recently, Lance Davis and Louis Stettler[3] have provided evidence

[1] Cf., e.g., D. C. North, *Growth and Welfare in the American Past* (Englewood Cliffs, N.J.: Prentice-Hall, 1966), p. 78; G. C. Fite and J. E. Reese, *An Economic History of the United States*, Second Edition (Boston: Houghton-Mifflin, 1965), pp. 244–5. However, Taussig's views on the sources of the growth of the U.S. pig iron industry prior to 1860, including the role he assigned the tariff changes of the 1840s, have recently received some consideration. Cf. Peter Temin, *Iron and Steel in Nineteenth Century America* (Cambridge, Mass.: M.I.T. Press, 1964); Robert Fogel and Stanley L. Engerman, 'A Model for the Explanation of Industrial Expansion during the Nineteenth Century: With an Application to the American Iron Industry,' *Journal of Political Economy*, LXXVII, 3 (May/June 1969), pp. 306–28.

[2] Cf., e.g., the treatment of the subject by V. S. Clark, *History of Manufactures in the United States*, I (New York: McGraw-Hill, 1929), pp. 450–3; D. C. North, *The Economic Growth of the United States, 1790–1860* (Englewood Cliffs, N.J.: Prentice-Hall, 1961), Ch. IX, *passim*.

[3] Cf. Lance E. Davis and H. Louis Stettler, 'The New England Textile Industry, 1825–60: Trends and Fluctuations,' in *Output, Employment and Productivity in the United States After 1800*, National Bureau of Economic Research Studies in Income and Wealth, XXX (New York: 1966), esp. pp. 227–32. Davis and Stettler also attribute to economies of scale the

of on-going increases in productivity within the cotton textile industry, and have drawn attention to the importance of the succession of gradual technical and organizational improvements that characterized the industry's experience *during the second quarter* of the nineteenth century – in the wake of the more spectacular innovations just mentioned. Arguing against the notion that the American textile industry underwent a revolutionary transformation prior to 1824 but then lapsed into a protracted (and tariff-protected) period of comparative technical stagnation, Davis and Stettler note the continued development of new textile machinery, but also point out that the appearance of radically new machines did not comprise the whole story of technical advance. Throughout the period from 1825 to 1860, they contend, new refinements were worked out in the machine shop and then incorporated in the latest models of the old textiles machines – just the sort of technological progress that can be thought to have derived from the accumulation of experience with design modifications under actual conditions of mill operations.

Indeed, Robert Zevin, in a still unpublished paper,[1] has taken the logical next step by explicitly estimating a long-run 'learning function' for the production of plain cotton sheeting in integrated mills during the period 1823–60. Basing his econometric work on the records of a single Massachusetts firm – the Blackstone Manufacturing Company – which operated a number of spinning and weaving establishments, Zevin concludes that by specifying a production function in which the productivity of labor and capital increase with the growth of experience as measured by the cumulative past output of the firm, a satisfactory rationalization is provided for the relationship between time series observations of cloth production and inputs of labor and capital services.

Thus have the 'new' economic historians insidiously been arming those who would subvert the Taussigian orthodoxy. Perhaps this has not been their objective, but one may wonder whether it will soon be proclaimed that the evidence of continuing substantial economies of scale and learning effects constitutes an adequate, if belated, justification for the extended tariff subsidization of the United States cotton textile industry throughout the period from 1825 to 1860. In the face of this latent revisionist threat, perhaps some general remarks are in order here on the subject of learning by doing, and on the basis of the case for protection of infant industries

[1] Robert B. Zevin, 'The Use of a "Long Run" Learning Function: With Application to a Massachusetts Cotton Textile Firm, 1823–1860,' mimeographed for the University of Chicago Workshop in Economic History, 22 November 1968. See below for fuller discussion of Zevin's work.

---

differences they observe between average labor productivity in the large integrated firms in the cotton textile industry and average labor productivity in the U.S. industry as a whole during the period *c.* 1830–60. Cf. *ibid.*, p. 231. This point is discussed below.

as it might apply to the specific case of cotton textile manufactures in the United States before the Civil War.

LEARNING BY DOING AND THE INFANT INDUSTRY ARGUMENT

Two lines of development in the theoretical treatment of tariff questions merit at least brief comment, for both make it rather odd that the appraisal of American tariff history set down by Taussig should continue to command widespread acceptance. On the one hand, recent years have seen a rapid lengthening of the list of recognized and admissible exceptions to the case for Free Trade. Whereas at the opening of the present century 'Free Traders' would acknowledge the intellectual legitimacy of the infant industry argument and the optimum tariff case (under national monopoly conditions) for the imposition of duties, a much broader array of considerations are now called into play against the classic position that trade without tariffs would be superior to self-sufficiency as a national policy. If the conditions under which levying tariffs permits the attainment of an *optimum optimorum* remain quite restricted, taxes and subsidies on international trade flows become utterly respectable as the means of ensuring a constrained, or 'second best,' welfare maximum. Tariffs are generally seen as a potential instrument for partially, if not completely, rectifying 'distortions' in the pattern of free market resource allocation caused by the existence of imperfections in domestic product and factor markets, or by the technical characteristics of some production processes (that is, scale economies), or the presence of significant interdependencies in domestic consumption or production, as well as a means of countering the external trade distortions created by the imposition of tariffs and subsidies by other countries.[1] On this score a more sympathetic, pragmatic reconsideration of American tariff policies than that allowed us by Taussig would seem to be indicated.

At the same time, however, there has been an important change in perception of the argument required to justify tariff protection for infant industries.[2] In the passage quoted at the beginning of the preceding section, the grounds Taussig entertained for the legitimate erection of protection tariffs around young industries were, by present standards, far too generous – which is certainly not what one would expect from a man who ascribed positive moral virtues to a 'natural' (that is, free market

[1] Cf. Jagdish Bhagwati, 'The Pure Theory of International Trade: A Survey,' *Surveys in Economic Theory*, II, American Economic Association–Royal Economic Society (New York: St Martin's Press, 1967), esp. pp. 214ff.

[2] The extent of this shift is made fully evident by the recent appearance of a cogently argued piece by Robert E. Baldwin, 'The Case Against Infant-Industry Tariff Protection,' *Journal of Political Economy*, LXXVII, 3 (May/June 1969), pp. 295–305.

equilibrium) pattern of international specialization and trade. Nowadays, by contrast, advocates of such tariffs would be immediately asked to show that the reason domestic enterprise proved 'hesitating' (to use Taussig's phrase) did not simply reduce to the incapacity of entrepreneurs to perceive that once production operations had been established, and the terrors of pioneering a new industry had been braved, they would reap an ultimate reward in the form of a lowering of costs sufficient to permit them to compete advantageously with previously dominant foreign producers. If, in the normal course of events, the workings of market forces would generate a stream of high profit for entrepreneurs who establish a new branch of industry, there should be no call to encourage through blanket protective tariffs those few who have the wit to anticipate it – and no particular virtue in seeking further to reward those who have not.[1]

The crux of the social case for protecting infant industries from foreign competition – saddling consumers with a static welfare loss by raising the price of imports above the international supply price – cannot really be the lack of perfect knowledge on the part of entrepreneurs and investors; for, if the State somehow possessed information indicating the ultimate profitability of investing in the establishment of a new line of productive activity, would it not be more appropriate simply to make that information freely available? No, the problem comes down to the fact that even the most alert and foresighted entrepreneurs might well 'hesitate' to finance an initial interlude of commercial losses – investing the while in learning by doing, so to speak – if they could not expect to recover eventually in private profits the social benefits that would flow from their initiation of the new and unfamiliar industrial pursuit. Any expectations of such profit recovery would surely be unrealistic if production methods developed by costly trial and error could readily be copied, at negligible expense, by subsequent entrants to the industry, or applied by firms in entirely different industries. In short, were there no elements of 'externalities' in the consequences of learning from experience, and thus no cause for the divergence of the social from the private rate of domestic transformation, what real justification could be offered for using tariffs to subsidize the accumulation of experience in a particular line of productive activity?[2]

[1] This assumes that foresighted entrepreneurs have access to perfect capital markets, and can therefore finance the 'learning period' during which their costs might remain above those of well-established foreign competitors. Capital market imperfections would give rise, understandably, to a range of inefficiencies in resource allocation; some of those would involve learning and training effects, i.e., the inadequate allocation of resources to activities in which such effects were present, and might be partially rectified by tariff intervention. One aspect of this broad set of problems is given further notice below, but otherwise the discussion of the implications of learning by doing for tariff policy abstracts from the second-best situations created by factor market distortions.

[2] On the crucial importance of the existence of externalities to the 'modern' infant industry

Thus, a proper restatement of the infant industry argument for tariffs deriving from the existence of irreversible scale effects must go beyond the mere citation of prospective benefits which, as Taussig saw, might be expected to flow from the acquisition of practical experience,

> where the industry is not only new, but forms a departure from the usual track of production; where, perhaps, machinery of an entirely strange character, or processes hitherto unknown, are necessary; where the skill and experience required are such as could not be attained in the occupations already in vogue.[1]

The crucial point on which one must be satisfied is that the contemplated learning and/or training effects will constitute externalities; their impact must not be confined to the reduction of production costs, attended by commensurate increases in profits, for those particular enterprises in which the vital experience was first acquired.[2] But inasmuch as the application of *effective* protection for any one branch of industry entails discriminating against the remainder, the entire case, in addition, has to be argued on the relative and not the absolute merits: the external benefits derived by subsidizing a specific line of production through taxing imports must at least outweigh the benefits foregone as a consequence of curtailing learning and training opportunities elsewhere in the economy.

As if the foregoing hurdles were not already hard enough for a candidate for protection to clear, it should now be noticed that in not all instances

[1] Taussig, *Tariff History*, pp. 3–4.

[2] In the case of training of workers in a transferable skill, the rational firm will not anticipate being able to go on paying their employees less than their marginal value products in order to recapture a return on the training costs. Instead it will fall to the workers to finance their acquisition of expertise in the new line of production. Quite apart from the problems that would arise if workers had to rely upon imperfect capital markets to finance this investment, it seems appropriate to ask whether it would actually be rational for the workers to invest in such readily transferable – and hence generally obtainable – training. While the first cadre of workers to do so might enjoy quasi-rents on their skill until the growth of the industry, and its trained labor force, multiplied the opportunities for training, is it not to be expected that the falling cost of training will create a situation rapidly approaching that of free entry into the occupation, and that the market value of the original investment would be drastically reduced in this process? If this is the case, the workers who are asked to finance the generally transferable portion of their training are in much the same position as that of pioneer firms who worry about subsequent entrants depriving them of a return upon their initial investment in learning. The question remains, however, as to whether imposing a tariff – rather than directly subsidizing training in transferable skills – is a desirable way of coping with this problem.

---

argument, cf. J. E. Meade, *Trade and Welfare* (Oxford: Oxford University Press, 1955), p. 256: H. G. Johnson, 'Optimal Trade Intervention in the Presence of Domestic Distortions,' in R. Baldwin, *et al.*, *Trade, Growth and the Balance of Payments* (Amsterdam: North-Holland Publishing Co., 1965); Bhagwati, 'Pure Theory of International Trade,' p. 218.

in which learning effects are present and generate comparatively great externalities, is the imposition of tariff protection for the activity automatically indicated policy. Logically, one ought first to establish whether or not subsidization of any level of domestic production is an appropriate means to the promotion of learning and training in the activity under consideration. The issue just raised is an empirical one and, in essence, concerns the determination of the proper specification of the argument of the putative learning function: supposing that the rate of decrease of marginal production cost is a positive function of the rate of accumulation of experience, what then is the relevant measure of 'experience'? And how effectively will the acquisition of the right sort of experience be fostered by the subsidization of domestic production through tariff protection? From the viewpoint of retrospective policy evaluation, it would not be sufficient to establish the existence of incompletely internalized, quantitively appreciable learning effects in the cotton textile industry during the ante-bellum era; by examination of the nature of the learning process itself, one must, in effect, determine the relevance of the tariff as a policy instrument for the promotion of effective experience.

Looked at from this vantage point, the theoretical and empirical literature devoted to learning by doing appears remarkably insensitive to the differing policy implications of the alternative specifications which have been suggested for 'learning curves.' Thus far the main concern has been to establish the existence of a link between experience and the level of productivity, without giving too much thought to the particular index of experience that happens to prove empirically most satisfactory for that purpose.

In the production of airframes – the activity in which the existence of a stable experience–cost relationship was first regularly observed[1] – unit variable costs were found to decline with the growth of production experience measured in terms of cumulated output in the particular production run. Werner Hirsch[2] referred to the same index of experience in studying the behavior of unit costs of producing new machine tools, finding that unit costs fell by approximately 20 per cent with every doubling of cumulated output of the given type of machine. Parallel evidence supporting this specification of the experience index is provided by Leonard Rapping's[3] investigation of learning effects in the construction of Liberty

[1] Cf. Harold Asher, *Cost Quantity Relationships in the Airframe Industry* (Santa Monica: The R.A.N.D. Corporation, R-291, July 1956): also, A. Alchian, 'Reliability of Progress Curves in Airframe Production,' *Econometrica*, XXXI (October 1963), pp. 679–93.

[2] Werner A. Hirsch, 'Manufacturing Progress Functions,' *Review of Economics and Statistics*, XXXIV (May 1925), pp. 143–55. Cf., also, by the same author, 'Firm Progress Ratios,' *Econometrica*, XXIV (April 1956), pp. 136–43.

[3] Leonard Rapping, 'Learning and World War II Production Functions,' *Review of Economics and Statistics*, XLVII (February 1965), pp. 81–6.

ships in the United States during World War II: the growth of cumulated Liberty output of the yard, rather than the mere passage of time, accounted for the recorded production gains which could not be ascribed to rising inputs of labor and capital. The implication flowing from this specification of the argument of the learning function is quite straightforward: inasmuch as a tariff would provide a subsidy for domestic production, it would thereby directly subsidize the accumulation of experience relevant to efficiency growth.

On the other hand, the form in which the learning by doing hypothesis was elaborated by Kenneth Arrow,[1] and which has thus had an important impact upon subsequent theoretical treatment of learning phenomena, proposed that the installation of capital equipment provides the vehicle for learning in the economy: cumulative gross investment was therefore made the measure of experience. This alternative view may well have been suggested by the fact that Arrow envisioned technical progress as coming in the form of improvements in the design of 'machinery,' so that the supposed reductions in the variable input (labor) requirements per unit of output produced on successive vintages of machinery might be thought of as being dependent on the accumulation of production experience in 'machine-building,' rather than in 'machine-using.' In other words, if the capital goods sector is the seat of endogenous technical progress, it is appropriate to notice that the cumulative investment experience of the economy coincides with the cumulative production experience of the capital goods sector. A recent study by Eytan Sheshinski[2] concludes that, in accounting for factor productivity changes observed at the SIC 2-digit industry level, cumulated gross investment serves as a more satisfactory index of experience than does cumulated output. Considerable doubt remains regarding the meaningfulness of Sheshinski's cross-section tests for the presence of learning effects in data pertaining to regions between which information and factors of production are known to move quite readily, and there is ground for skepticism about the accuracy of the measure employed by this study as proxies for the alternative experience variables considered. It is clear, nevertheless, that the inference to be derived from Sheshinski's empirical conclusion regarding the form of the learning function is at odds with his general assertion that learning effects 'fundamentally' provide a basis for the protection of infant industries.[3]

[1] Cf. Kenneth J. Arrow, 'The Economic Implications of Learning by Doing,' *Review of Economic Studies*, XXIX (June 1962), pp. 155–73.

[2] Cf. Eytan Sheshinski, 'Tests of the Learning by Doing Hypothesis,' *Review of Economics and Statistics*, XLIX (November 1967), pp. 568–78.

[3] Cf. *ibid.*, p. 568 n. 1, on the implications for international trade of the existence of 'irreversible' economies of scale, i.e., those which 'appear with every act of investment and do not disappear subsequently.'

The econometric findings reported by Sheshinski are based on cross-section observations

If efficiency gains did flow from an increase of cumulated gross investment in, say, the cotton textile industry, would it not have been more to the point to subsidize such investment directly than to tax imported cotton textile products? Indeed, if the crucial thing was the opportunity afforded firms in the domestic textile machinery industry to acquire greater experience (supplying gross investment goods), the fruits of which could be embodied in technically superior equipment designs, it becomes appropriate to notice that, in the absence of other measures, putting a tariff on textile imports is equivalent to placing a tax on the domestic production of (textile) machinery. The fact that in the particular historical context under consideration the early textile firms built their machinery in their own shops complicates the story without altering the essential point: protection of cloth production could encourage machinery production through *income* effects within the industry, but the relative price effects of the tariff would tend to lead initially integrated firms to shift resources away from their machine shop activities toward specialization in the production of cloth.[1]

Is there, then, no basis for infant industry tariffs when learning effects flow from experience in the carrying out of gross investment rather than production? Such a conclusion is, perhaps, too extreme. If capital markets in the economy under consideration were characterized by serious imperfections which impeded foresighted entrepreneurs' access to loanable funds, and if gross domestic investment promoted efficiency growth through learning effects, then the imposition of a tariff might be defended on the ground that it shifted income toward firms in the protected industry and thereby permitted more *internally financed* investment to be under-

[1] Of course, the story of the emergence of specialized machinery producers, like the Saco–Lowell shops and the Whitin Machine Works, from early cotton textile mill machine shops is usually told in terms of the spin-off of activities subject to economies of scale. But perhaps the role of the textile tariff, in inducing new entrants to the industry (and existing smaller mills) to specialize in cloth production and depend upon a few 'outside' shops for their equipment, warrants more attention than it has received. Cf. G. S. Gibb, *The Saco–Lowell Shops, Textile Machinery Building in New England, 1813–1849* (Cambridge, Mass.: Harvard University Press, 1950); T. R. Navin, *The Whitin Machine Works Since 1831* (Cambridge, Mass.: Harvard University Press, 1950), esp. Ch. I, II; D. C. North, *The Economic Growth of the United States*, pp. 10, 161–2.

---

for manufacturing industries in U.S. states in 1957, and mixed cross-section time series data for selected manufacturing industries in several countries during 1950–60. The former part of the study takes gross book values of the 1957 capital stock of the industry (in each state) as a measure of cumulated gross investment, without any justification for equating the two quite distinct concepts. Unfortunately, nowhere in the article or the appendix on 'Sources of the Data' does Sheshinski describe the derivation of the cumulated output and cumulated gross investment variables used in the cross-section time series study; there is some doubt whether the cumulations begin prior to 1950.

taken. The rationalization just offered is analogous to one that is sometimes suggested in the case in which irremovable capital market imperfections thwart attainment of optimum plant size in a branch of industry subject to decreasing costs: tariffs are one means of subsidizing the internally financed expansion of firms in the industry.[1] Of course, this line of defense for tariffs has to be thrown up along the 'second best' frontier and, assuming always that the capital market imperfections could not be eliminated or directly manipulated by the State to the advantage of the decreasing cost industry, one would still be called upon to show that the tariff was not inferior to some other fiscal device which subsidized internally financed capital formation in the industry without first having to distort the structure of domestic prices.[2]

Yet to suggest that the foregoing style of argument could be persuasively invoked in defense of offering tariff protection to cotton textile producers – inevitably, at the expense of other activities – would require the suppression of some significant bits of American economic history.[3] In the first place, the large integrated cotton textile mills of the Massachusetts type – where the problems of indivisibilities of requisite investment outlays might have been more imposing – occupy a unique place among pre-

[1] The importance of finding means of easing the constraints imposed by capital market imperfections in situations where indivisibilities are present in production processes that offer greatly reduced marginal costs has received recent emphasis in discussions of economic development policies by R. I. McKinnon and E. S. Shaw. Cf., e.g., R. I. McKinnon, 'On Misunderstanding the Capital Constraint in LDC's: The Consequences for Trade Policy', in *Trade, Balance of Payments and Growth*, eds. J. N. Bhagwati *et al.*, 1971.

[2] This latter stricture applies equally to the proposal that tariff support be provided as a means of permitting firms in the industry to pay higher wages and thereby make it possible for workers internally to finance their training in industrial skills that are transferable, and consequently not financed by rational firms. It is clear that the purpose would be served equally well by other fiscal measures that effected the same net redistribution of income in favor of workers engaged in textile production; welfare theorists would naturally prefer to do this by lump sum grants to the workers involved. Moreover, it is debatable whether raising the domestic price of the industry's product by means of tariffs would have the desired result of helping to overcome the problem of financing labor-force training when capital markets are imperfect. Quite possibly the higher product prices would simply induce firms to expand their employment of untrained (lower marginal productivity) laborers. Cf. Baldwin, 'Tariff Protection,' p. 301, on the last style of objection to tariffs.

[3] The following discussion is largely based upon the research of Lance E. Davis, 'Stock Ownership in the Early New England Textile Industry,' *The Business History Review*, XXXII, 2 (Summer 1958), pp. 204–22; 'The New England Textile Mills and the Capital Markets: A Study of Industrial Borrowing 1840–1860,' *Journal of Economic History*, XX (March 1960), pp. 1–30. Cf. also Federal Reserve Bank of Boston, *A History of Investment Banking in New England*, Annual Report for 1960 (Boston, 1961), Ch. I. For further material regarding the balance sheet positions and investment financing practices of the cotton textile companies whose production operations are the focus of attention in the present study, cf. Paul F. McGouldrick, *New England Textiles in the Nineteenth Century, Profits and Investment* (Cambridge, Mass., 1968), pp. 14–18, 134–8.

Civil War industrial ventures in their successful heavy reliance upon initial equity financing from local (primarily mercantile) groups of investors. Almost all the original capitalization of the integrated mills was provided by stock sales, with individual holdings ranging from $2,500 to $200,000. Moreover, while no one would seriously contest the assertion that to speak of *the* United States capital market in the ante-bellum era is to speak of a fiction, the cotton textile industry was far from being peculiarly disadvantaged by the fact that markets for loanable funds remained highly personal, as well as geographically and institutionally segmented. It is certainly relevant to remark on the relatively advanced development of bank and non-bank financial intermediation in New England – the region in which the cotton textile industry was concentrated – and to observe that the industry as a whole became a major user of credit, with the Massachusetts-type mills emerging as especially voracious borrowers. Instead of having to rely upon internal finance, the latter found that even in the local capital market persisting imperfections worked to their advantage. As a result of the large industrial lending operations of intermediaries like the Massachusetts Hospital Life Insurance Company and the Provident Institution for Savings in the Town of Boston, which had been established as philanthropically motivated institutions intended to encourage habits of thrift among the poor, during the 1840s and 1850s the integrated cotton textile firms benefited from the availability of long-term loans at the anomalously low rates of interest stipulated by the still extant usury laws.[1]

To some degree the apparently favored capital market position of New England's large cotton textile mills may have reflected the margin of security which the existence of the tariffs afforded to investors in the industry. But, if it is not possible to establish that in the absence of protection from imports the terms on which the domestic producers of cotton

[1] This accepts L. E. Davis's conclusion that respect for the Massachusetts usury law's 6 per cent ceiling, on the part of the quasi-public, philanthropic intermediaries, had the effect of segmenting the market for loanable funds and keeping the long-term rate persistently below the short-term interest rate during the 1840s and 1850s. Barbara Vatter ('Industrial Borrowing by the New England Textile Mills, 1840–1860,' *Journal of Economic History*, XXI [June 1961], pp. 216–21), on the other hand, has argued that interlocking directorates between financial and textile firms were primarily responsible for the low rates charged the latter by the former. But, as Davis ('Mrs. Vatter on Industrial Borrowing: A Reply,' *Journal of Economic History*, XXI [June 1961], pp. 222–6) has observed, there is no clear evidence of systematic customer discrimination favoring the textile industry in the rates charged by either commercial banks, savings institutions, or insurance companies such as Massachusetts Hospital. It should be noted, nevertheless, that with the long-term rate legally held below the free market equilibrium there would be ample opportunity, in the rationing of funds by the non-bank intermediaries, for the web of interlocking directorates to protect the interests of the textile mill owners at the expense of other supplicants for loans.

cloth could have obtained finance would have remained as advantageous, neither can one point to peculiar capital market conditions which could justify the singling out of this branch of industry for compensatory tariff support.

We are led back, therefore, to the more conventional line of justification for infant industry tariffs as the means of promoting the acquisition of technical knowledge and skills of a sort especially beneficial to society, and obtainable primarily through direct experience in production activities. One final, somewhat paradoxical point remains to be noted before we ask how the case for the ante-bellum cotton textile tariffs would fare on this criterion. Where knowledge derived from production experience can be readily applied by new entrants to the industry, as well as by the less experienced among existing firms, the more persuasive the apparent *ex ante* basis for tariff subsidies, and the less likely it is that the *ex post* case for protection will turn out to be correspondingly strong. This is because it may really be necessary to subsidize only the acquisition of production experience by a relatively few 'pilot plants' in the industry – so long as one insures that the firms in question cannot extract a rent which would impede the dissemination of that information throughout the industry and, for that matter, the entire economy.

Where the learning process involves experimental determination of optimum plant layouts, or work organization, or operating speeds and machinery maintenance schedules, or similar best-practice information that can be introduced rather straightforwardly elsewhere – perhaps by managers, superintendents, and foremen initially trained in the 'pilot' establishments – the effect upon the industry-wide level of productivity deriving from the growth of cumulated output in those few leading establishments is likely to be very potent. It will clearly not redound fully to the advantage of the owners of those pioneer firms, and so will provide grounds for subsidization of their production experience. The situation contemplated is not without historical relevance here: most of the early textile machinery-builders in southern New England had received their training under Samuel Slater and David Wilkinson.[1] Slater's protegés, however, seized the opportunity to profit from their experience and left his employ in large numbers. They either became itinerant, free-lance machinists who moved about from town to town assisting with the starting up of new mills, or else set up shop on their own to manufacture specialized machinery items for which they discovered an adequate market existed within the region. Thus, the former machine-shop superintendent of Slater's Providence Steam Cotton Mills was one of the original owners of the Providence Machine Company, and Slater's disciples numbered

[1] Cf. Navin, *Whitin Machine Works*, pp. 10, 206, 269. Wilkinson was Slater's American-born brother-in-law, and an exceptionally clever blacksmith.

among the founders of other important textile machinery firms such as the Fales and Jenks Machine Company, the Bridesburg Manufacturing Company, and the Cohoes Iron Foundry and Machine Company.[1]

It was previously stressed that the basis for infant industry protection evaporates if the knowledge supposedly gained through production experience cannot be transferred at all but, instead, has to be generated afresh by repetition of that experience. For, under such circumstances, whatever benefits the experience might convey could be completely internalized and would not give rise to a gap between the private and the social rates of domestic transformation. Now, however, it is seen that at the opposite pole, where the transfer of knowledge gained in the leading firms occurs swiftly and at a little cost, the indiscriminate subsidization by tariffs of any and all domestic production in the industry would turn out to be a gross extravagance. Where such conditions obtained, it might be anticipated that without protection the accumulation of a margin of redundant experience within the industry would manifest itself in a progressive weakening of the effect of increments in cumulated industry output upon the industry-wide level of productivity. This *ex post* reduction of the strength of observable aggregate learning effects is likely to occur, *a fortiori*, when the imposition of a tariff encourages extension of domestic production well beyond the levels required to sustain the training of an effective technical-managerial cadre.

In its empirical implications, the last point reduces to another question involving the specification of the learning function. Supposing it is experience in production activity (rather than in gross investment) which counts, then beyond what point should additions to the stock of experience be regarded as largely repetitive, and therefore essentially redundant, in the learning process? It is quite possible that a more satisfactory, stable relationship between learning and efficiency growth may be found by choosing a measure of production experience which is less comprehensive than that provided by cumulating the output of all firms under consideration. The policy implications of such a finding would be no less immediate and, in the present context, would probably be most sympathetically received.

If closer attention to the available evidence revealed that a more restricted measure of experience were applicable in the case of the early

[1] It is likely that in the (northern) section of the industry where integrated spinning and weaving mills were more predominant, the Boston Manufacturing Company's original mills at Waltham fulfilled a parallel training function. Moreover, it seems plausible that while the illustration supplied in the text concerns the training of machinery-builders, the pioneer firms also played a significant role in the education of cadres of future superintendent and mill managers. Close historical studies of the latter type of training effects would be of considerable interest, but still remain to be done in the case of the early cotton textile industry.

cotton textile industry, it would afford an easy reconciliation of seemingly conflicting positions regarding the protection of that young industry in the United States. On the one hand, empirical support would be provided for maintaining that externalities in the form of learning and training effects derived from regular commercial mill operations did play a very significant part in the reduction of costs and the rapid expansion of cotton cloth production in the ante-bellum era. But on the other hand, the same evidence would make it appropriate to reassert the orthodox Taussigian view that the cotton textile tariff, as a blanket subsidy to domestic producers, was largely unjustified.

Led on by the possibility of effecting such a resolution simply on the basis of the production relations observed in the industry, we shall be preoccupied in the remainder of this study with the econometric problems of establishing the existence and form of a suitable long-run learning function for the manufacture of coarse cotton cloth in integrated mills during the years following 1824. Almost inevitably, the ulterior purpose will seem at times to have become utterly lost from view amidst the welter of subsidiary historical and methodological issues that occupy the ensuing discussion. Before plunging onward into this cliometric thicket, therefore, it would be well to fortify ourselves with both a brief glance backward at the considerations that motivate this particular line of investigation, and a look ahead toward the main conclusion for which I shall argue.[1]

The point of departure for any evaluation of America's nineteenth-century tariff policies must be an acknowledgement that the infant industry argument when properly stated is an argument for tariff protection on 'second best' grounds: the optimal solution is presumed to lie beyond the reach of the State. While the imposition of an unjustified tariff is to be condemned, the necessity of a resort to the use of infant industry tariffs as means of combating 'market failure' is, at very best, to be lamented. Patently, the award of direct subsidies raised by lump sum levies – or even certain forms of income tax – would always be judged to be better than the imposition of excise taxes. For the latter occasion a static loss of welfare by distorting the price structure facing domestic consumers. Thus, if there are externalities in the form of learning effects justifying public subsidization of private productive activities, recourse to tariffs must be regarded as an inferior means of achieving that end – so long as some other instrument for affecting the subsidy payments was available.

Availability in this connexion is not simply a matter of the legal

[1] The following paragraphs in this section incorporate material from P. A. David, 'The Use and Abuse of Prior Information in Econometric History: A Rejoinder to Professor Williamson on the Antebellum Cotton Textile Industry,' *Journal of Economic History*, Vol. XXXII, 3 (September 1972), pp. 708–10.

powers entrusted to the State to arrange transfer payments; some attention has to be given also to the costs of effecting such transfers. In the absence of outright legal prohibitions of differential taxation of domestic industries, a case might still be made that direct subsidization was too awkward to arrange because of the sheer number of enterprises involved in domestic textile production – and the even larger number of establishments, dispersed as they were over a region within which internal communications were not all that good. Equally, inasmuch as substantial subsidy payments might be required, the tariff could recommend itself – and excise taxes have done so historically – as an effective means by which a fledgling government, lacking elaborate administrative and enforcement machinery, might raise the required revenues.

To counter such an argument (without entering upon elaborate considerations involving balancing the tariff-occasioned welfare losses against the costs of arranging equivalent producer subsidies by some other fiscal device) it is suggested that attention be directed to the technical question of how many productive, learning entities it was really necessary to subsidize. If the essential learning process could be carried on in a small number of pilot plants, and the knowledge gained therein diffused readily to private domestic producers, the practical arrangement of direct subsidies to 'learning by doing' would be rather easier to contemplate. Certainly this is the case by contrast with situations in which 'learning' takes the form of, say, general labor force training.

Previous writers have been remiss in failing to recognize that the specific nature of the learning activity – as represented in the form of the 'learning function' – would be germane to decisions about the sufficiency of the justification offered for extending subsidies via tariffs. The econometric inquiry reported here, by contrast, takes pains to show that although learning by doing was a significant source of productivity growth in the industry, most of the subsidies afforded the cotton textile producers under the tariffs were subsidies to redundant learning – redundant in the sense that the experience acquired by the myriad firms that entered the industry, and the multiplication of mills, was largely superfluous for the improvement of the technology of power-loom cloth production after 1825. In sum, what emerges from the present study is an attack upon the often unstated premise of the second-best argument for tariff subsidization of legitimate infant industries, viz., that publicly supported pilot plants could not achieve substantially the same technological advances.

At an early point in its history the United States government saw its way clear to establishing National Armories, under whose auspices much knowledge was acquired about the technology of mass producing firearms. Is it inconceivable that a small number of subsidized integrated cotton mills might have been similarly established – 'National Manufactories,' perhaps, for the ostensible purpose of furnishing sailcloth to the Navy? The evidence will be shown to be consistent with the view that had such a group of pilot plants been directly subsidized, they could have constituted an equally effective (and socially less costly) substitute for the policy of taxing the consumption of imported cotton goods in order to afford any and all domestic producers the chance to learn for themselves by 'doing.'

FRAMING TESTS OF THE APPLICABILITY OF THE LEARNING HYPOTHESIS

A general statement of the nature of the production relationships prevailing in activities characterized by learning effects proves useful as a first step in formulating quantitative tests of the relevance of the learning by doing hypothesis to the case of cotton textile production in the ante-bellum United States. Then, when we have taken note of the confines within which the limitations of available data oblige us to proceed, a more concrete form can be given to those hypotheses which it seems important and feasible to submit to statistical testing.

A general production model incorporating long-run learning effects

Although initial applications of the concept of learning to production cost projections considered (short-run) cases where physical production facilities remained fixed – at least for the length of the production run with a given 'model' – the idea can readily be extended, and has been applied to long-run contexts in which learning entities (firms) are free to vary input proportions in the course of altering their physical plants.[1] For present purposes the discussion may be restricted to situations in which the efficiency of inputs used in the production of a standard item is affected by 'learning,' as well as other, exogenous processes, in a Hicks-neutral fashion. That is to say, the effects of the accumulation of experience, like those of exogenous technical progress, are presumed to leave the relationships among the marginal productivities of the several inputs quite undisturbed. We represent this by considering that an output (X) produced using flows of capital services (K), labor services (L), and purchased materials (M), according to

$$X = A(G, \tau)\, F(K, L, M), \qquad (1)$$

where the function $F(\cdot)$ is twice-differentiable and exhibits the usual properties of a production function.

[1] Rapping ('Learning'), Sheshinski ('Tests of Learning By Doing'), and Zevin (unpublished work), all have formulated tests of the hypothesis in long-run terms. The general statement of the production model employed here, with some modifications, follows that given by Sheshinski, 'Tests of Learning by Doing,' pp. 569–70.

In this general model of production relations, the level of neutral input efficiency is governed by the function $A(\cdot)$, and thus depends upon G, an index of the experience acquired by the production entity, as well as upon calendar time, which is indexed here by τ. Let the elasticities of efficiency with respect to 'experience' and with respect to time be defined, respectively, as

$$\lambda \equiv (\partial A/\partial G)\,(G/A),$$

and

$$\beta \equiv (\partial A/\partial \tau)\,(1/A).$$

Then, differentiating (1) with respect to time, it is readily established that the rate of overall productivity growth (measured as the difference between the output growth rate, X, and the weighted sum of input growth rates, f) is, at each instant in time, a linear function of the rate of growth of 'experience,' g:

$$x - f = \beta + \lambda(g), \tag{2}$$

where lower case letters are used to denote the variable's rate of change through time, thus,

$$x \equiv (\partial X/\partial \tau)\,(1/X);$$
$$g \equiv (\partial G/\partial \tau)\,(1/G);$$
$$k \equiv (\partial K/\partial \tau)\,(1/K), \text{ etc.,}$$

and the measure of aggregated input growth rates is

$$f \equiv [(K/F)\,(\partial F/\partial K)]k + $$
$$+ [(L/F)\,(\partial F/\partial L)l + $$
$$+ [(M/F)\,(\partial F/\partial M)]m.$$

Now consider a specification of the efficiency function $A(\cdot)$ which will leave the elasticity parameters appearing in equation (2) unaltered by the passage of time:

$$A(\tau) = A_0\{\exp(\beta\tau)\}\,\{G(\tau)\}^{\lambda}. \tag{3}$$

This specification embraces two limiting cases. First, when $\beta = 0$ we have the situation in which the productivity of the inputs engaged in the activity follows the usual form of learning function that has been employed by all the empirical studies of learning phenomena cited in the preceding section,

$$A(\tau) = A_0 G^{\lambda}. \tag{4}$$

At the opposite extreme the learning coefficient, λ, is zero and the productivity of the inputs rises, presumably for reasons exogenous to the particular production operations under consideration, at the steady rate indicated by β.

The foregoing maintains that when a proper index of experience, G, is considered, there are neither increasing nor decreasing returns to 'learning'; were it possible to make additions to the relevant stock of experience at the steady rate $g = \lambda$, productive efficiency would also rise steadily. But that means it would not be possible to distinguish statistically between the effects of learning by doing and exponential technological change deriving from exogenous sources; if the index of G itself takes an exponential path, equation (3) collapses to

$$A(\tau) = A'_o \{\exp(\beta'\tau)\}$$

where

$$A'_o = A_o G_o, \quad \text{and} \quad \beta' = (\beta + \lambda\gamma).$$

Thus, if the customary unbounded form of learning function is adopted for purposes of econometric analysis, any hope of being able to discriminate between the two hypothesized sources of efficiency growth must ride on the divergence from an exponentially rising trend of some inherently plausible index of experience.[1]

Alternative measures of experience

Two such experience-indexes will be considered here. The first is already thoroughly familiar: it takes cumulated output to be the relevant measure. On this basis, where output is growing at the steady rate x, the growth rate of G so defined (G_1) would fall continuously through time toward a lower (asymptotic) limit at x itself. Stating the matter formally, given the definition of experience

$$G_1(\tau) = Q(\tau) = \int_o^\tau X(t)\mathrm{d}t, \tag{5}$$

when $X(t) = X_o\{\exp(xt)\}$, upon evaluating the definite integral in equation (5) and forming the expression for the rate of growth of G_1, we obtain

$$g_1(\tau) = X(\tau)/Q(\tau) = x[\{\exp(x\tau)\}/\{\exp(x\tau)-1\}]. \tag{6}$$

[1] Of course, one might alternatively specify that there were decreasing returns to experience, so that the level of efficiency in any activity rose – perhaps along some logistic or Gompertz function – toward an upper asymptote. With experience rising at a constant rate under such circumstances, the rate of efficiency growth would fall continuously, reaching zero in the limit. As is shown below, it is quite possible to generate the latter efficiency-path with an 'unbounded' form of learning function, such as that in equation (4), by a suitable choice of the experience index, G. We shall therefore avoid the added econometric complexities that would accompany experimentation with alternative algebraic forms for the learning function.

Note, then, that the ratio in the brackets on the right-hand side of this expression is initially greater than unity, but that $g \rightarrow x$ as $\tau \rightarrow \infty$. *In the limit*, therefore, even under conditions of steady growth of output (additions to the stock of experience), the learning effect on productivity becomes $\lambda g = \lambda x$, and is indistinguishable statistically from either the effect of exponential exogenous technical progress or the effect of economies of scale.

Before that limit is reached, however, one may hope to sort out these effects. Indeed, as has been previously pointed out, considerable success has attended econometric estimation of the learning coefficient when the definition of experience described by equation (5) has been employed in conjunction with the functional specification given by equation (4). Perhaps the most notable aspect of past work in this field is the conciliance among the values of learning coefficients obtained in different studies: estimates of λ typically are found to lie in the range between 0·2 and 0·5, clustering around 0·33. The implication is that costs are reduced by something in the near neighborhood of 20 per cent with each doubling of cumulated output.[1]

A second, less conventional index of experience can now be considered: *T*, the accumulated time during which the decision-making unit, or units, engaged in the new branch of production. A line of argument for entertaining *T* as quite possibly a more relevant measure of experience has been put forward by Robert B. Zevin:

Consider a profit maximizing firm which produces more than one unit of its product simultaneously on different production lines or different machines or in different plants. At any given moment all of the simultaneous production processes will be carried on with the combination of techniques which the firm believes to be optimal. The amount of time required for the firm to conclude that an alteration in the mix of techniques is desirable could only be shortened by the extent to which simultaneous

[1] If the index of productivity is $\Pi(\tau)/\Pi(0)$, the 'learning curve' can be interpreted as implying that

$$\Pi(0)/\Pi(\tau) = [Q(0)/Q(\tau)]^{\lambda} \equiv C(\tau)/C(0).$$

Where $C(\tau)/C(0)$ is the index of costs at constant input prices. Thus, doubling cumulated output, we have: $(0{\cdot}5)^{0{\cdot}33} \sim 0{\cdot}80$. In some studies with production facilities held fixed the productivity index has related only to labor inputs, whereas, in others where physical facilities were also changed, the productivity measures were of the total rather than the partial type. Cf. the estimates from Hirsch, 'Firm Progress Ratios.' Rapping, 'Learning,' reports significant learning coefficients of 0·29 using cumulated output of Liberty ships (including current production), and coefficients of 0·23 to 0·34 using cumulated Liberty ship output including only half the current year's production. The original observation of a 'learning curve' in the case of airframes production – relating only to the behavior of labor costs – served to define a learning curve as one in which the coefficient λ was below unity. Cf. Asher, *Cost Quantity Relationships*; also, A. Alchian, 'The Reliability of Progress Curves.'

production enabled the firm to distinguish random components of results from those depending on choice of techniques. This effect may be insignificant compared with the length of time required to devise a superior set of techniques derived from production experience. Moreover, the really determining factor may be the length of time required to introduce various improved techniques which involve differently trained personnel, different spatial arrangement of equipment or housing of equipment, or possibly, different equipment. Under such circumstances increase in [cumulated output] caused by the proliferation of simultaneous production would have no learning impact on the firm.[1]

Another formulation of essentially the same points can be offered, which serves to bring out in more rigorous fashion the relationship between the 'Zevin-index' and the conventional cumulated output index of production experience.[2] Starting from the notion that all current additions to production experience beyond some technically determined amount (call it X^*) are really redundant in the learning process, we seek to cumulate just the relevant portions of the actual production flow and identify that sum with the stock of 'effective' experience in existence at time τ:

$$G_2(\tau) = Q^*(\tau) = \int_0^\tau X^*(t)\mathrm{d}t. \tag{7}$$

But now suppose it is not possible to add really new information to the store of experience, except by continuing to operate the basic production unit, be it a plant or an individual machine, at its designed capacity rate, X_0^*, commencing from the date of initial installation at $t = 0$. The flow of non-redundant additions to the stock of experience is therefore

$$X^*(t) = X_0^*, \quad \text{for all} \quad t = 0, \ldots, \infty, \tag{8}$$

whence it becomes apparent – upon evaluation of the integral in equation (7) – that Q^* must be strictly proportional to the amount of time elapsed since the initiation of the learning process:

$$G_2(\tau) = Q^*(\tau) = X_0^*\{\tau - 0\} \equiv X_0^*\{T(\tau)\}. \tag{9}$$

[1] Cf. Zevin (unpublished work), pp. 21–2.

[2] After this section had been drafted, my attention was drawn to a very recent paper by William Fellner, 'Specific Interpretations of Learning by Doing,' *Journal of Economic Theory*, I (August 1969), pp. 119–40. In this article Fellner (p. 124) distinguishes two 'versions' of the learning hypothesis:

'Why assume that experience in the relevant sense is acquired by *doing more* than one has so far done, regardless of the length of time it takes to do more (Version A); and why not assume that experience is acquired by *doing it longer*, regardless of the steepness of the rise of cumulated output (Version B)?'

Fellner's Version B is, of course, the cumulated time formulation suggested – with an argument for its plausibility – by R. B. Zevin, to whom I have here awarded (unclaimed) priority.

From this it directly follows that

$$g_2(\tau) = X^*(\tau)/Q^*(\tau) = 1/T(\tau), \tag{10}$$

which coincides with the expression obtained for the rate of growth of the G_1–index of experience defined by equation (5) – under the special conditions in which the output growth rate is $x = 0$. Thus, when elapsed time is the measure of experience, learning effects cease *in the limit* to sustain the growth of productive efficiency; this qualitative difference has nothing to do with the form of the learning curve, along which the elasticity of efficiency with respect to experience is assumed to remain undiminished.

In the context of our earlier discussion of the case for infant-industry tariffs, the distinction between the two experience-indexes just considered takes on special significance. If a firm's productivity depends upon the length of time during which it has been engaged in some new line of activity, because the mere duplication of identical production units will not result in enhanced opportunities for acquiring new information, subsidizing an expansion of the scale of the firm's production by sheltering it from 'foreign' competition would clearly not be indicated as government policy. Of course the growth of a multiplant organization could be socially beneficial, even if every plant were identical, in that it might well give rise to economies of scale. But unless those scale economies were external to the firm, as well as to the individual plants, one would be hard put to justify fostering such expansion through taxation of imports.

Thus, if the existence of learning effects in general is revealed by the movement of the rate of productivity growth in accordance with equation (2), the following relation describes a specific form of learning which offers no comparable justification for the extension of tariff protection to the activity in question:

$$e(\tau) = [x(\tau) - f(\tau)] = \beta + \lambda/T(\tau), \tag{11}$$

subject to the restriction $0 < \lambda < 1$. In this expression, given first-order homogeneity of $F(\cdot)$, the measured rate of productivity growth can be read as the rate of Hicks-neutral technical progress, $e(\tau)$, arising from both exogenous and endogenous processes.

Concrete specification of cotton textile production functions

To implement the general line of empirical investigation suggested by the previous sections requires giving a concrete form to the hypothesized relationship between output and inputs. In the following, use is made of the particularly convenient specification offered by the Cobb–Douglas production function. Momentarily suppressing the questions already raised about the exact nature of the efficiency term $A(G, \tau)$, we may consider the working hypothesis: the yardage of cotton cloth (Y) produced at time τ in integrated spinning-weaving establishments depended upon the current rates of inputs of capital services gauged in spindle-hours (SH), of labor services measured in standard quality man-hours (MH), and of raw cotton (C), according to the function:

$$Y_\tau = \{A(G, \tau)\} \{(SH)_\tau\}^{\alpha_S^*} \{(MH)_\tau\}^{\alpha_L^*} \{(C)_\tau\}^{\alpha_C^*}. \tag{12}$$

Running yards, rather than the weight of cloth produced, has been found to serve as quite satisfactory measure of aggregate cotton cloth output by Davis and Stettler, Zevin, and other students of the ante-bellum New England textile industry.[1] Although the yardage figures here represented by (Y) are not in principle homogeneous with respect to either widths of cloth or the number of counts, it has been found that for Massachusetts-type mills producing mostly low (under 16) count goods the ratio of aggregate heterogeneous yardage (Y) to estimated standard yardage (Y^*) remained quite stable over the period 1830–60.[2] Since the output (and input) data employed in the present study pertain to firms in precisely that category, it would be legitimate to regard the recorded yardage measure (appearing as the dependent variable in equation (12)) as an index of Y^* – the output of cotton sheeting of a standard width and weight to which the specified production function pertains.

Raw cotton was not the only intermediate input purchased by integrated mills, but it was far and away the most important among such items in cost terms; and outlays for purchased materials, in turn, represented a very substantial proportion of the total value of shipments from United States cotton textile establishments. In 1860, for example, the ratio of the industry's aggregate value of purchased materials to aggregate value of product was 0·51.[3] To ignore so significant a class of inputs in

[1] Cf. Davis and Stettler, 'The New England Textile Industry,' pp. 218–19; Zevin (unpublished work), pp. 10–11. Paul F. McGouldrick also has employed running yardage measures (cf. McGouldrick's 'Comment' in *Output, Employment and Productivity in the U.S. After 1800* (N.B.E.R. Income and Wealth Conference, Vol. XXX), p. 240; Robert G. Layer (*Earnings of Cotton Mill Operatives, 1825–1914* [Cambridge, Mass.: Committee on Research in Economic History, 1955]) maintained that pounds of cloth provided a superior output measure. Poundage figures nominally take account of count differences but ignore width variations in cloth.

[2] At least this is the case for the sample of mills for which both yardage and poundage figures were obtained by Davis and Stettler ('The New England Textile Industry,' p. 219, n. 9). The latter source (*ibid.*, Table A-2, p. 238) provides annual output estimates in standard yards of 14s × 14s, 48 × 48, 36-inch wide brown sheeting, and corresponding heterogenous yardages for a group of nine firms engaged primarily in manufacturing low count goods: Hamilton, Boston, Suffolk, Tremont, Lawrence, Nashua, Naumkeag, Jackson, and Pepperill. The (Y) output series employed here (see Tables 1 and 2) pertains to the first six firms of this group.

[3] Cf. underlying figures in *U.S. Eighth Census*, 'Manufactures,' pp. 733–42. Detailed costs available in textile company manuscripts from the era under review, disclose that cotton

describing the production of cotton goods would be unjustifiable, but this does not imply that it is essential for us to secure figures of the weight of cotton consumed (C) corresponding with the data on the volume of cloth manufactured. During the 1825–60 period the proportionality between the weight of raw cotton consumed and the weight of cloth produced is observed, both at the aggregate level and at the level of individual firms, to have undergone no significant alterations.[1] Now the weight of cloth (Y') will simply be a scalar transformation of a standard (width and count) measure of yardage produced (Y^*), an acceptable proxy for which, as we have already noted, is provided by running yardage figures.[2]

The latter equivalences can be formally represented as

$$Y'_\tau = kY_\tau^* = kq(Y)_\tau \tag{12a}$$

and in conjunction with the stability generally observed in the weight ratio ($c \geqslant 1$) defined by

$$C_\tau = cY'_\tau, \tag{12b}$$

allow us to express the input variable as a scalar transformation of the available output measure:

$$C_\tau = mY_\tau, \quad \text{where} \quad m \equiv ckq. \tag{12c}$$

In the present circumstances, the legitimacy of making just this substitution proves especially helpful, affording us an easy way to bridge the serious empirical gap left by the absence of any observations for the weights of raw cotton consumed in producing the available output figures, Y.[3]

As the labor input variable (MH) is defined in *homogenous* man-hours,

[1] Davis and Stettler ('The New England Textile Industry,' p. 218, n. 8) refer to 'the rather constant proportion between cotton input and cloth output (regardless of the quality of output).' Zevin (unpublished work, p. 16, n. 10) reports that in the case of the Blackstone Manufacturing Co., the ratio of the weight of cloth (Y') to (C) the weight of raw cotton, increased only very slightly over the period 1823 to 1860, from 0·87 to 0·92, i.e.,

$$C(\tau) = C(0)\ Y'(\tau) \exp\{-0{\cdot}0015\ \tau\}.$$

[2] Cf. McGouldrick, *New England Textiles*, p. 145, for use of a fixed yards-per-pound conversion factor ($4{\cdot}36 = 1/k$) which would be roughly appropriate for the average count level of the cloth produced by the New England mills covered in the present inquiry.

[3] Note that it is actually not necessary that (m) be strictly constant. So long as the variations in m are neither serially correlated not correlated with any of the other arguments of the production function for Y^*, one can replace C by ckY^* and then substitute qY for Y^*, obtaining a least-squares regression model in which the variation around $E(M\tau) = m$ will manifest itself, acceptably, as errors in the *dependent* variable: ($\ln Y\tau$). Use is made of this below.

---

costs were over 90 per cent of the cost of all purchased materials. Cf. McGouldrick, *New England Textiles*, p. 144. It should be noted that in the nineteenth century power was treated, by and large, as a capital expense, rather than as a purchased item, in the New England textile company accounts.

its movements through time may have departed appreciably from the path indicated by the average number of (heterogeneous) worker-days (L) employed annually. Other investigators of productivity change in the ante-bellum cotton textile industry, no doubt constrained by the same paucity of statistical information that confronts the present study, have sought to make do with the latter, simpler index of labor inputs.[1] Yet, there are at least two factual grounds for our refusing to settle for so crude a substitution, especially as we are concerned with the longer-term trend growth of the productivity of labor (and other) resources devoted to cotton textile manufacturing.[2]

In the first place, the average duration of the textile factory operative's daily toil appears to have undergone a steady reduction during the decades under review.[3] Whereas a 13-hour workday is suggested by the handful of reports available for firms in the cotton and woolen textile industries in 1830, detailed reports for typical Lowell cotton mills in 1844 point to an average workday lasting 12 hours and 15 minutes; the decline continued until, by 1860, 11 hours was established as a common workday for textile mills. According to these figures, particularly after allowance is made for some degree of cyclical incomparability among the three dates to which the observations refer, the trend rate of reduction in the length of workday (h_L) appears to have been pretty much the same throughout

[1] Zevin (unpublished work), works with output per worker measures of labor productivity, and explicitly assumes the ratio of women and children employed to the total number of textile operatives in the Blackstone Manufacturing Co. labor force remained constant at its average 1823–60 level of 0·88. Davis and Stettler ('The New England Textile Industry,' p. 221) mention changes in hours of work but conduct their discussion of labor productivity on the basis of movements of output per worker.

[2] In the short run the problem is complicated by the fact that whereas initially women and children were predominant in the labor force of the spinning and weaving departments, adult males were employed within the mills on the construction and installation of machinery, as well as in activities more immediately connected with current cloth production – such as carding, dressing, or machinery maintenance. Variations in the extent of the capital formation activity taking place within the mills, which would not of course be registered in the movements of (cloth) output at the time, would thus be likely to cause abrupt alterations in the ratio of adult males to women and children employed. For this reason such *short-run* sex composition shifts are probably best ignored in a study focusing upon cloth production (rather than the joint activities of cloth production and machine construction). That, essentially, is the approach adopted by Zevin (unpublished work, pp. 9–10) who constructs an index of labor inputs based on the number of women and children alone, assuming that the relationship between those workers and males engaged in current cloth production remained constant. Yet this solution to the short-run problem unfortunately leaves out of account the effects of longer-run alterations in the make-up of the cotton textile work force, and it is with the latter that we shall here be concerned.

[3] Cf. Stanley Lebergott, *Manpower in Economic Growth* (New York: McGraw-Hill, 1964), p. 48, for the evidence cited in the text. Confirmatory data is available in Layer, *Earnings of Cotton Mill Operatives*, p. 43.

these decades, and probably ran in the near neighborhood of the 0·5 per cent per annum average annual rate of decrease recorded over the 1830–60 interval as a whole.[1] Thus, a reasonable first approximation to the trend of man-hours worked would be provided by: $\{L(\tau)\}\{h_L(\tau)\} = \{L(\tau)\}\{h_L(0)\}[\exp(\psi_L\tau)]$, where $\psi_{L1830-60} = -0{\cdot}0055$.

But this still makes no distinctions among different grades of workers. For one thing, it recognizes no difference between a man-hour and a 'woman-hour' of labor, which is potentially a serious oversight in an industry where relatively low-paid females initially formed a very large part of the labor force. The point is that the sex-composition of the work force did not remain unchanged over the long-run: between 1832 and 1900 female employment in Massachusetts' cotton textile mills fell from 158 per cent to 91 per cent of male employment.[2] This secular transformation accompanied a narrowing of the differential between male and female wage rates (actually average earnings) which seems to have proceeded at roughly the same pace in the cotton textile industry as in other branches of manufacturing, particularly the woolen textile and boot and shoe industries, where women were extensively employed during the antebellum era. It is therefore tempting to treat the secular compositional shift observed within Massachusetts' cotton textile labor force as reflecting the process of input substitution in response to an exogenous relative expansion of the supply of the 'higher quality' (male) workers.

The question is whether average labor quality improvements arising from such developments taking place outside the cotton textile industry were quantitatively important enough to require recognition as a source of secular increases recorded in production per unit of undifferentiated labor. An affirmative answer would imply that if we are restricted to time series observations of labor input measured not in standard quality units but in heterogenous worker-hours (LH), an allowance should be made at least for the augmentation of the latter at some constant, non-negligible secular rate ($\beta_L \geqslant 0$), by representing the standard man-hour input variable as

$$MH_\tau = \{L_\tau h_L(0)e^{\psi_L\tau}\}e^{\beta_L\tau}. \tag{12d}$$

[1] The chronology of cyclical fluctuations in the New England textile industry prepared by Davis and Stettler ('The New England Textile Industry,' Table 6, p. 224) locates both 1830 and 1860 as falling between troughs and peaks, whereas 1844 is a peak year. If the length of the workday tended to vary positively with the movement of output over the cycle, the mid-point to peak change (1830–44) would understate the secular rate of fall in hours, while the cyclical peak to mid-point change (1844–60) would tend to overstate it. Note, then, that from the data cited in the text, we may compute the following rates of change in the duration of the workday; $\psi_{1830-44} = -0{\cdot}0042$; $\psi_{1844-60} = -0{\cdot}0067$; $\psi_{1830-60} = -0{\cdot}0055$.

[2] Cf. Stanley Lebergott, 'Wage Trends, 1800–1900,' in *Trends in the American Economy in the Nineteenth Century*, N.B.E.R. Income and Wealth Conference, XXIV (Princeton, N.J.: Princeton University Press, 1960), pp. 459–60. On the quite distinct problem of short-run variations in sex-composition, see footnote 2 on p. 121, above.

In support of entertaining some such trend allowance for exogenous improvements in labor quality, we may consider just the impact of the alteration of the sex-composition of the work force during the latter two-thirds of the nineteenth century. Stanley Lebergott[1] has drawn attention to the apparent fact that the long-run elasticity of substitution between female and male cotton textile workers was apparently close to unity: despite the rapid fall in the female–male worker ratio already noticed over the 1832–1900 interval, the accompanying movement of the male–female earnings differential was sufficient to hold the women's share of the Massachusetts cotton mills' wage bill unchanged at approximately 0·4. The constancy of the latter share suggests that we might boldly assume that relative wage rates reflected the relative marginal productivities of the sexes, and thus regard its complement (0·6) as a parameter more-or-less accurately indicative of the elasticity of aggregated standard quality cotton textile labor inputs with respect to the employment of higher quality labor services provided by males. Then, on the implicit assumption that the function describing standard quality labor service was first-order homogeneous in the two classes of labor (males and females), Lebergott's figures can be used to calculate the rate of increase of standard quality labor per (undifferentiated) worker that was entailed by the substitution of male for female workers in Massachusetts' cotton textile mills.[2] Over the long

[1] Cf. Lebergott, 'Wage Trends, 1800–1900,' p. 460.

[2] The calculation is made as follows:

Let μ represent the ratio of male workers in 1900 to male workers in 1832, and let ϕ be the corresponding ratio for female workers. As females represented 0·613 of the Massachusetts cotton textile mills' workers in 1832 (according to the figure already cited in the text for the ratio of females to males at that date), the following equation gives the relation between μ and ϕ required to maintain the size of the total work force, in numbers of workers, unchanged between 1832 and 1900:

$$\mu = (1/0{\cdot}387)(1-0{\cdot}613\,\phi).$$

But, from the change in the female–male worker ratio, we also know that over the 1832–1900 interval,

$$\mu = (1{\cdot}58/0{\cdot}91)\phi = 1{\cdot}74\,\phi.$$

Solving these two equations yields: $\mu^* = 1{\cdot}358$ and $\phi^* = 0{\cdot}781$.

Now, let Z be an index of the 'quality' of the average worker in the industry. The proportional growth of standard labor inputs per worker employed is thus:

$$Z_{1900}/Z_{1832} = S_m(\mu^*-1)+(1-S_m)(\phi^*-1)+1.$$

Since the elasticity $(\partial Z/\partial L_m)(L_m/Z)$ is estimated as the male share in the cotton textile wage-bill, $S_m = 0{\cdot}6$, we find

$$Z_{1900}/Z_{1832} = 2{\cdot}127,$$

which corresponds to an average rate of growth of worker quality of $\beta_L = 0{\cdot}0112$ per annum over the period 1832–1900.

interval from 1832 to 1900 the average annual rate of labor quality improvement on this account alone works out at slightly more than 1 per cent per year, a far from negligible magnitude in the present context. Had the same rate prevailed in the ante-bellum decades, it would have more than counterbalanced the direct effects of the shortening of the workday upon the input of labor in the manufacture of cotton goods. The available data is too fragmentary to permit conclusive statements on this score. Yet an analysis of this evidence, presented as an addendum to this essay, does indicate a 1 per cent per annum appreciation in average labor quality over the interval between 1830–2 and 1850, and an average annual rate that was not much below that for the 1830–60 period as a whole.

There is a long tradition which treats spindlage statistics, specifically figures giving the number of spindles in place on various dates, as indexes of fixed capacity in the cotton textile industry.[1] Surprising though it may be, from a technical standpoint this practice seems well justified; recent students of the operations of integrated 'Massachusetts-type' mills in the ante-bellum period have remarked upon the existence of a 'high degree of rigid complementarity' between the number of spindles in use and the physical amounts of capital equipment being employed at other stages of the production process.[2] But for present purposes interest lies with the flows of services provided by the mills' fixed capital equipment, rather than with the sum of resources locked up in the machinery stock. We are therefore fortunate in being able to base our capital input index upon statistics pertaining to the average number of spindles *in daily operation* throughout the year (S), rather than simply the average spindlage in place.[3]

[1] Taussig thought that spindles were 'the best single indication of the extent and growth of such an industry as cotton manufacture' (*Some Aspects of the Tariff Question* [Cambridge, Mass.: Harvard University Press, 1915], p. 265), but the relationship between spindles and output of cloth in the U.S. industry was far from fixed even over the long run. Cf. Davis and Stettler, 'The New England Textile Industry,' pp. 227ff., for discussion of this point, and also the data assembled in Table 8, below.

[2] Cf., e.g., Zevin (unpublished work), p. 10. Switching between count levels would unbalance the machine stock (and the labor force employed) within a given integrated mill, because the number of machines required in the opening and carding department per spindle, and the number of looms required per spindle, were both decreasing functions of the yarn count. The number of looms per spindle also decreased with the number of twists per inch. (Cf. McGouldrick, *New England Textiles*, p. 29.) These considerations served to narrowly restrict the range of counts of the cloth produced by a single integrated mill. But by the same token, for cloth of a given count and twist, or for a standard mix of different-grade fabrics produced by a group of integrated mills, the number of spindles serves as an adequate index of the entire complement of machinery employed.

[3] Cotton textile firms recorded the number of spindles operated in each mill on a daily basis, and such information has been used by Zevin (unpublished work, p. 10) to secure an index of annual capital services flows for the Blackstone Manufacturing Co. Davis and Stettler do not explicitly state that their estimates of annual cloth output per spindle for a sample of six firms (which is utilized in the present study) pertain to spindles *in operation*, but the

While this is a major step, it does not completely suffice to transform a stock measure into a service flow index. While the *normal* number of days of mill operations in the year (d) appears to have remained fixed in the near neighborhood of the 311 days conventionally cited in contemporary sources, the length of the factory day was contracting during the period 1830–60.[1] Were we to suppose that this contraction, like the shortening of the mill operative's workday, had occurred at a more or less steady trend rate (ψ_s) over the course of these decades, a reasonable approximation to the desired spindle-hours input variable would be provided by

$$(SH)_\tau = S_\tau\{d \,.\, h_s(0)\}e^{\psi_s \tau}. \tag{12e}$$

It is important to pause at this point to notice that the specification of capital inputs in terms of spindle-hours (and complementary machine-time) implicitly places the subject of secularly increasing equipment operating speeds in the domain of 'efficiency improvements.' This seems entirely reasonable, although there may be those who would equate higher speeds of operation with more intensive utilization of fixed equipment and therefore maintain that, on the contrary, such changes should not be permitted to reflect themselves as increases in the efficiency of the inputs.[2] The point that has to be made is that in the ante-bellum cotton textile industry, at least, lower operating speeds did not simply represent less than complete utilization in the usual 'excess capacity' sense. Prior to 1828 the operating speed of the New England textile mills was narrowly circumscribed by the mechanical difficulties encountered with the English

[1] On the number of normal operating days (implied by the six-day work week), cf. Lebergott, 'Wage Trends, 1800–1900,' p. 479. Davis and Stettler ('The New England Textile Industry,' p. 227) comment on the reduction of the workday in this connexion, and report that when their output per spindle estimates (*ibid.*, Table 8) are adjusted to a common workday – an adjustment whose details are not described – the 13 per cent decline in output per spindle recorded in the decade following 1840 is no longer observable.

[2] I rather suspect this latter position would appeal to D. W. Jorgenson and Z. Griliches, who, in their study ('The Explanation of Productivity Change,' *Review of Economic Studies*, XXXIV [July 1967], pp. 249–83) of the post-World War II U.S. economy, undertake to adjust capital input measures for *secular* alterations in equipment utilization as reflected in the changing relationship between electric motor capacity and electricity consumption for electric motors (in manufacturing industries).

---

spindlage figures that can be derived from those partial productivity data and the corresponding aggregate output series (see Table 8 below for sources) exhibit a pattern of short-period fluctuations which clearly reveal that they cannot refer to the number of spindles in place. For example, the estimates of S (for the six-firm group described in Table 8, below) show a drop of 7 per cent between 1837 and 1839, and of 16 per cent between 1853 and 1854 – both movements being very abrupt and coinciding with business recessions.

gear drive. In that year, however, Paul Moody's application of the principle of the belt drive to textile production raised greatly the potentially attainable average operating speed.[1] Now rather than find ourselves in the peculiar position of denying that the subsequent diffusion and further refinement of belt drive systems constituted 'technical progress' in the New England textile industry – because it merely permitted higher average rates of equipment utilization (measured by the rate of spindle revolutions) without raising the productivity of a spindle revolution – we have chosen to gauge the annual flows of capital services not in spindle revolutions, but in terms of operating spindle-hours (SH).

Obviously this in no way inhibits us from thinking of exogenous technical progress impinging upon textile producers in the form of machinery design changes, which specifically enhanced the effectiveness of the input of the average spindle-hour. If such improvements augmented the 'quality' of the average set of spindles (and complementary equipment) at a steady rate through time, say β_s, whereas endogenous learning effects raise the efficiency with which all inputs were being used, the full production given by equation (12) can be written out as:

$$Y_\tau = A_0^* G_\tau^{\lambda *} \{(SH)_\tau e^{\beta_s \tau}\}^{\alpha_s^*} \{(MH)\tau\}^{\alpha_L^*} \{(C)_\tau\}^{\alpha_c^*}, \tag{13}$$

where the experience-index, G, is given by either equation (5) or (9).[2]

Had we available to us a set of corresponding observations on cloth yardage and each of the input variables in equation (13), the foregoing few pages might well have been dispensed with. But since the best that can be done at the moment is to assemble comparable time series for Y, S, L, and the cumulated output and cumulated time measures of experience, Q and T, it becomes extremely useful to notice that in conjunction with the input specifications set forth by equations (12a) through (12e), equation (13) implies:

$$Y_\tau = A_0 G_\tau^{\lambda} e^{\xi \tau} (S_\tau)^{\alpha_s} (L_\tau)^{\alpha_L}, \tag{14}$$

where the net exponential time trend extracted from (13) is given by

$$\xi^* = (\beta_s + \psi_s)\alpha_s^* + (\beta_L + \psi_L)\alpha_L^*, \tag{14a}$$

[1] Cf. Gibb, *The Saco–Lowell Shops*, pp. 76–80; Davis and Stettler ('The New England Textile Industry,' p. 230) point out that changes in average operating speeds within the industry continued to disturb the observed spindle–output ratio in the years following 1828, which form the period under examination here.

[2] With the Cobb–Douglas form all input-augmenting technical changes can be expressed as equivalent Hicks-neutral efficiency improvements. Thus, in equation (13), the compound coefficient ($\alpha_s^* \beta_s$) would take the place of the rate of exogenous 'disembodied' technical change, β, in equation (3).

and the 'reduced form' parameters of (14) are,

$$\xi = \xi^*/(1-\alpha_c^*),$$
$$\lambda = \lambda^*/(1-\alpha_c^*),$$
$$\alpha_s = \alpha_s^*/(1-\alpha_c^*),$$
$$\alpha_L = \alpha_L^*/(1-\alpha_c^*), \qquad (14b)$$

and the constant,

$$A_o = A^*(m)^{\alpha_c^*}(d)^{\alpha_s}\{h_s(0)\}^{\alpha_s}\{h_L(0)\}^{\alpha_L}.$$

Now it can readily be seen that by taking natural logarithms of both sides of equation (14), we arrive at a linear regression model from which least-squares estimates of the 'reduced-form' production function's parameters may be immediately secured:

$$\ln Y_\tau = \ln A_o + \alpha_K\{\ln S_\gamma\} + \alpha_L\{\ln L\} + \lambda\{\ln G_\tau\} + \xi\tau + \eta$$

where η is a randomly distributed error term. As a practical matter, in view of the serial correlation problems that are likely to arise from the presence of strong time-trends in all the variables, it proves advantageous to work with a modification of this equation in which the logarithm of yardage produced per man-day appears as the dependent variable. Dropping the disturbance term, we may estimate

$$\ln\{Y/L\}_\tau = a_0 + a_1\{\ln S_\tau\} - a_2\{\ln L_\tau\} + a_3\{\ln G_\tau\} + a_4\tau, \qquad (15)$$

considering as alternative specifications,

$$\ln G_{1\tau} \equiv \ln Q_\tau \quad \text{or} \quad \ln G_{2\tau} \equiv \ln T_\tau,$$

and bearing in mind that the estimated regression coefficients are related to an underlying production function via the parameters of (14):

$$a_0 = \ln \hat{A}_0$$
$$a_1 = \hat{\alpha}_s$$
$$a_2 = (1-\hat{\alpha}_L)$$
$$a_3 = \hat{\lambda}$$
$$a_4 = \hat{\xi}. \qquad (15a)$$

This last regression equation (15) is precisely the form that Robert Zevin undertook to estimate using time-series observations compiled from the records of the Blackstone Manufacturing Co. for the years 1823–60. Zevin did not seek, however, to connect it with the sort of underlying production flow relationship set forth here in equation (13). Of course, it is quite evident that the structural parameters of (13) are not immediately retrievable from the reduced-form coefficients estimated in equation (15). On the other hand, by imposing the restriction that the length of the factory-day and of the operatives' workday changed at the same rate

$$\psi = \psi_s = \psi_L, \tag{16}$$

and mobilizing fragmentary data of the kind already considered to independently fix the magnitudes of β_L, ψ, and α_c^*, one may contrive to solve (14a), (14b), and (15a) for the remaining structural parameters of the underlying production function – α_s^*, α_L^*, λ^*, and β_s.

Yet it is not essential to go to those lengths. Stopping short of an effort to identify the true production function uniquely up to a scalar constant, our principal questions concerning the existence of learning effects, the specific form of the learning function, and the presence or absence of reversible economies of scale in the manufacture of cotton coarse-goods before the Civil War, all may be adequately answered on the basis of the ersatz production parameters simply secured by estimating equation (15).

THE ECONOMETRIC EVIDENCE

Our passage from the sublimely general statement of the learning by doing hypothesis contained in equation (1), to the ridiculously concrete 'reduced-form' version of the production function presented by equation (14), necessarily has been guided at some critical points by the accessibility of historical data that would permit econometric application of the model. From the recent work of Davis and Stettler[1] on the ante-bellum cotton textile industry, it is possible to extract an internally consistent set of annual estimates of the cloth yardage produced (Y), and the man-days of labor (L) and average operating spindlage (S) employed by a small, quite homogenous group of New England cotton textile companies, whose members were all actively engaged in the manufacture of coarse-goods throughout the period 1834–60. From the same source, moreover, one can obtain the statistics needed in fashioning reasonably accurate measures of the collective production experience of these firms, in terms of both their cumulative cloth output (Q) and their cumulative years of operating experience (T), on every date within the 27-year period.

But while the latter information is available for each constituent firm of the sample, Davis and Stettler's work provides only aggregate estimates

[1] Cf. Davis and Stettler, 'The New England Textile Industry,' especially pp. 231, 234–6. The labor input and spindle input estimates, (L) and (S), respectively, have to be derived from the partial productivity series (Y/L) and (Y/S) presented by Davis and Stettler, and an aggregation of company production estimates into the corresponding yardage output (Y) of the sample group. For convenience, the series obtained from this source are presented in Table 8 below, along with the corresponding experience-indexes (Q) and (T).

of the labor and spindle inputs used by the entire group. Consequently, the group and not the firm must be taken as the unit of observation for purposes of estimating equation (15). Some preliminary comments are thus in order concerning the nature of the companies represented in the sample, the resulting characteristics of the particular segment of the ante-bellum cotton textile industry to which the econometric findings presented here pertain, and the sort of interpretations it will be appropriate to place upon this evidence.

Characteristics of the six-firm sample group

The six textile companies identified in Table 1 were typically multi-plant enterprises, and together they operated some seventeen mills by the early 1840s.[1] All of them concentrated on the manufacture of coarse-goods (under 16 counts), that being the portion of the cloth–quality spectrum in which, it should be recalled, the American cotton textile industry first established an international competitive position. The mills operated by these firms were of the more modern 'Massachusetts type' which, in contrast with the small 'Rhode Island-type' spinning mill, integrated spinning and power-loom weaving within a single, large establishment. In some instances they incorporated ancilliary facilities such as bleacheries and print-works. By all standards of comparison, short of actual measures of productive efficiency which we do not possess, it appears that the statistics assembled for this sample afford us a look at the experience of a leading group of enterprises in the leading segment of the United States cotton textile industry.

Quite apart from the fact that the Massachusetts-type mill added power looms to its outfit of spinning machinery, consideration of the spindlage figures alone suggests the comparatively large size of the average mill owned by the sample firms. In Table 2, the second panel reveals that in 1831 and 1839 the number of spindles per mill in the sample group was more than three times greater than either the corresponding average for cotton mills in Massachusetts, or the average for the nation as a whole. Indeed, it would seem – from the data provided in the same place for 1849 and 1859 – that the average *mill* in the sample had over twice the number

[1] Cf. Davis and Stettler, 'The New England Textile Industry,' Table A-2, p. 238, for the number of mills in their sample of firms producing low-count cloth. Note that Jackson (whose recorded annual yardage output did not increase notably in the decade after 1833) had accounted for one mill when it entered in 1832 (*ibid.*, Table A-1, pp. 234ff), whereas by 1843 Naumkeag and Pepperill did not yet have mills in production. The remaining six firms, and the implied number of mills in 1843 (i.e., 17), constitute the sample group under discussion here. Brief sketches of the organizational histories, and the characteristics of the operations of the individual companies are available in McGouldrick, *New England Textiles*, Appendix A, pp. 219–21.

TABLE 1 Members of the sample group of 'Massachusetts-type' cotton textile firms engaged in production during 1834–60

Firm name	First recorded production		Recorded years of production experience in 1860, $T(1860)$	Annual output rate, in millions of yards		Cumulated output, in millions of yards, at year's end:	
	Date	Annual output in thousands of yards		1839	1860	1834	1860
Boston	1815	76	46	2·862	4·573	28·810	107·189
Hamilton	1828	2,810	33	5·528	12·917	22·404	234·280
Suffolk	1832	2,396	29	4·857	8·455	10·529	161·421
Tremont	1832	615	29	6·735	11·169	10·354	207·837
Lawrence	1833	1,104	28	11·018	19·183	7·277	339·064
Nashua	1826	321	35	9·033	15·265	21·851	282·606

Source: Underlying data for the respective companies in Davis and Stettler, 'The New England Textile Industry,' Table A-1, pp. 234–6.

TABLE 2 Production characteristics of the group of mills in the six-firm sample, compared with the cotton textile industry in the state of Massachusetts, New England, and the United States

Output (in millions of yards)				
	1828/30	1838/40	1848/50	1858/60
Sample group	6·539	38·708	50·355	65·763
Massachusetts	53·199	141·749	248·824	384·742
New England	102·267	366·631	591·238	824·157
Spindles per mill				
	1831	1839	1849	1859
Sample group	6,073	6,794	8,370	9,912
Massachusetts	1,359	2,145	(6,048)[a]	(7,711)[a]
United States	1,567	1,842	(3,284)[a]	(4,799)[a]
Output per man-year (in thousands of yards)				
	1831	1837	1849	1859
Sample group	10·618	12·823	16·215	19·878
Massachusetts	5·938	6·395	10·399	10·800
United States	3·707	n.a.	7·796	9·410
Spindles per man-year				
	1831	1837	1849	1859
Sample group	44·1	49·4	57·2	68·2
Massachusetts	25·5	28·6	44·9	43·6
United States	20·0	n.a.	35·9	43·0

[a] The Federal Censuses reported spindlage for firms only.

Sources:

Output, from Davis and Stettler, 'The New England Textile Industry,' Table A-1, pp. 234–6; Table 4, p. 221.

Spindles per mill, from *ibid.*, p. 231, n. 22.

Output per man-year (output per man per year), from *ibid.*, Table 9, p. 231.

Spindles per man-year (output per man-year divided by annual output per spindle), derived from *ibid.*, Table 9, p. 231.

of spindles operated by the average United States *firm* engaged in cotton textile production, including multi-mill companies among the latter. Still another reflection of the comparative dimensions of these plants is to be read in the uppermost panel of Table 2, where it may be seen that during the early 1840s the group's seventeen mills must have been responsible for at least a tenth of the total annual cotton goods yardage being produced throughout New England.

Considered as a group, the six companies in the sample undoubtedly occupied a position of leadership with respect to labor productivity:

according to the estimates assembled in Table 2 (third panel), throughout the 1830–60 period the group's output of cloth per man-year remained twice the average reported by the census of the industry in Massachusetts, and even further above the nationwide average. There is no basis for supposing that a difference of that magnitude was attributable to the sample firms' comparative emphasis upon lower quality cloth, nor can their workday be presumed to have been longer than the industry average. But on the other side of the ledger, it should be noted that in basing their estimates of man-*year* productivity for the sample upon an unrealistically small number of working days per year, Davis and Stettler purposely sought to bias this comparison against the conclusion that average labor productivity was relatively high in the sample mills.[1]

Although numerous writers have been tempted to ascribe the Massachusetts-type mills' labor productivity advantage to economies of scale, in view of their comparatively larger size, such a conjecture cannot yet be supported with the available data. On its face, moreover, the suggestion overlooks the greater capital-intensity of the production process being carried on by the firms represented in the sample. The figures showing spindles per man-year in the lower panel of Table 2 provide some indication of the sample firms' relative position in that dimension: it is striking to observe, for example, that in 1859, when the average yardage produced per man-year in Massachusetts reached the level attained by the sample group some twenty-eight years earlier, the number of spindles per worker in Massachusetts also just matched that found in the sample mills in 1831. But the picture conveyed by these data really cannot do justice to the possibility that capital-intensity differences, rather than pure scale economies, accounted for the higher labor productivity characteristic of the mills in the sample group. The spindle-worker measure most seriously understates the magnitude of the cross-section difference between the capital–labor ratios characteristic of the integrated mills comprising the sample and those of the average Massachusetts or United States textile factory.[2] This inadequacy, however, occasions no

[1] Cf. Davis and Stettler, 'The New England Textile Industry,' pp. 230–1. As pointed out in the following footnote, however, the bias on this score cannot have been large.

[2] The implication is that were the data from the bottom two panels of Table 2 to be made the basis of a scatter-diagram showing the relationship between labor productivity (on the ordinate) and capital intensity, the observations relating to the sample would have to be displaced (*vis-à-vis* the rest of the points in the scatter) rather far to the right but only slightly upward. The small upward displacement would take into account the approximate 15 per cent understatement of the relative labor productivity standing of the sample firms, resulting from Davis and Stettler's assumption of a work-year of 265 days (rather than the conventional 311 days) in converting the output per man-day figures for the sample into output per man-year – the latter being the basis on which the census labor productivity data for Massachusetts and the U.S. are available.

difficulty in the context of the time-series study conducted here: the rather strict complementarity that appears to have been maintained *within integrated mills* between the number of spindles and the amount of other machinery employed, means that for this type of factory the spindle–worker ratio provides virtually as good an index of the movement of capital-intensity through time as it would for the class of mills specializing in yarn production.

While it is not possible to pursue the question of the importance of scale economies further on a cross-section basis, the radical difference between the final products of the small Rhode Island-type mills and those of the Massachusetts-type mills operated by the sample of leading firms does raise doubts as to the existence of a meaningful answer. On the other hand, the homogeneity of the type of mills to which the sample data refer clearly renders it appropriate to try to ascertain whether or not economies of scale were to be derived by producing power-loom cloth in multi-mill enterprises. Determining the degree of homogeneity of the production function (14) estimated from the observations for the sample group will, in other words, be equivalent to establishing the presence or absence of scale economies external to the integrated mill and internal to the group of firms comprising the sample. Fortunately, we shall also be in a position to juxtapose such findings for the collection of sample firms with those reported by Robert Zevin's study of the Blackstone Manufacturing Co. – a thoroughly comparable Massachusetts enterprise, operating integrated spinning and weaving mills – during the years 1823–60.[1] As the general characteristics of the Blackstone mills appear to have closely resembled those of mills in the Davis–Stettler six-firm sample, and as in both cases the available observations aggregate the outputs and inputs of more than a single mill, it is to be expected that, in the absence of all scale effects external to individual mills, the input coefficients of the production function would turn out to be much the same when estimated from either of the two sets of data.[2] The latter comparison should therefore afford a strong additional test of the hypothesis that significant reversible scale economies existed beyond the level of the individual cotton textile factory.

The characteristics of the sample group carry two further implications that are germane to the interpretation of the econometric findings about the sources of efficiency growth. The sample's homogeneity in the type

[1] Cf. Zevin (unpublished work), pp. 8–9. See Table 3, below for direct comparisons with Zevin's regression results.

[2] The learning coefficients, unlike the exponents of the labor and capital and raw materials inputs in Cobb–Douglas functions, may be subject to an aggregation bias even under conditions of constant returns. Furthermore, there is no necessity for the firm and group learning coefficients to be identical even when the rest of the production function is found to be the same, and account is taken of any aggregation bias. These matters receive fuller attention in the text below.

of mill represented, and the persisting position of its members among the industry's leaders in regard to size, capital-intensity, and unit labor costs, suggests that during the 1834–60 period (when all the firms in the group were operating at least one mill) changes in the relationship between average-practice and best-practice *within* the sample were not so likely to have been a significant source of advances in the average level of efficiency recorded by the entire group. And if, as therefore seems most probable, efficiency gains largely reflected changes in best-practice methods that spread rapidly among this collection of mills, it is also reasonable to think that a substantial number of those improvements derived from the experience being acquired by these leading firms, rather than from 'learning' that was taking place elsewhere, in a less progressive segment of the industry. Putting this another way, the application of the notion of learning by doing to account for the measured input efficiency growth of an aggregation of production units admittedly creates worrisome ambiguities about exactly what was being 'learned,' how, and by whom. But most of the questions thus occasioned can be more plausibly answered within the framework of the basic learning hypothesis when, as in the present instance, the group of producers happens to be drawn from the portion of the industry that was operating closest to the existing technological frontier.

The preceding discussion has focused primarily upon the characteristics of the mills represented by the sample data. But it should be evident that in a long-run context, where physical facilities as well as production personnel may be changed, it is the firm rather than the mill that must be regarded as the relevant 'learning entity.' This implies that at a given point in time the individual firm's experience should be gauged – as Q_j, for the j-th firm – by cumulating its previous output regardless of the number of mills among which that total happened to have been distributed; or, alternatively, as T_j, by cumulating the j-th firm's years of operating experience, without regard to the number of separate establishments it happened to have operated during that time. The experience-index Q for the sample group may thus be obtained by aggregating the cumulative production records of the constituent mills, whereas the alternative index T for the group simply records the cumulative firm-years of operating experience.

Summing up the years of recorded production experience for each of the six companies (see Table 1), by the close of 1860 the sample group *in toto* had acquired 200 firm-years of recorded experience in the manufacture of cotton cloth.[1] Although the group's senior member, the Boston

[1] It should be understood that the entries in the third column of Table 1 relate not to the strict age of the companies (commencing with the year of their respective organizations), but instead give the number of years of recorded production experience. Inspection of the

Manufacturing Co., harked back to the industry's halcyon days of wartime expansion and had pioneered in the development of the integrated mill at Waltham, the other firms had come into operation only in the late 1820s and the early 1830s. In fact, four of them – Hamilton, Suffolk, Tremont, and Lawrence – can be regarded as the children of the brief episode following passage of the Tariff of Abominations, when tariff rates for cotton goods stood at their pre-Civil War maxima. In order to avoid clouding the interpretation of the group experience-indexes by permitting their movements to reflect the addition of new learning entities to the group, as well as the further accumulation of production experience by previously established firms, the ensuing analysis of the sample data is restricted to the observations pertaining to the years after 1833.

Rather than imposing an encumbrance, this temporal delimitation means that the present econometric findings will refer specifically to the portion of the industry's ante-bellum history which, according to traditional views, offered no justification at all for the retention of infant-industry tariffs. As far as the experience of the typical member of the sample is concerned, the group data for the 1834–60 interval are appropriate in testing the relevance of the learning by doing hypothesis for those firms that began to acquire production experience in a tariff-protected market, well after the revolutionary technological advances of 1814–24 had established the manufacture of power-loom cloth in integrated mills as a new branch of American industrial enterprise. Thus reassured, we may at last turn to the econometric evidence.

The presence of learning effects

The results obtained using the six-firm sample data to estimate several variants of equation (15) by least-squares regression analysis are displayed in Table 3, where these findings can be compared with the regression equations estimated from the records of the Blackstone mills by Robert Zevin. That the production of cotton goods by these firms was subject to significant learning effects is a general conclusion that emerges most forcibly from this body of evidence.

To begin with, one may ask whether acceptable explanations of output per man-day cannot be secured when the learning coefficient, λ^*, is held

---

output levels in the first years for which figures could be obtained by Davis and Stettler – as shown in Table 1 – makes it evident (upon comparison with initial production figures of other firms and subsequent output levels for the firms in question) that while there may be errors in the base year (1834) estimates of Q and T, due to missing initial production records and fractional years of experience, such errors cannot be very serious. For further details regarding approximate dates of incorporation and the initiation of production, cf. McGouldrick, *New England Textiles*, p. 4 (Table 1) and Appendix A.

TABLE 3 Comparison of regression estimates for the group of six New England firms (1834–60) with Zevin's regression estimates for the Blackstone Manufacturing Co. (1823–60)

Regression equation with ln Y/L the independent variable	Subject of observations	Coefficient of determination adjusted for degrees of freedom: $\overline{R^2}$	Durbin–Watson statistic: D–W	Estimated coefficients, with t-statistics[b] of the independent variables:					
				Constant	ln S	ln L	τ	ln Q	ln T
I	Blackstone[a]	0·93	n.a.	+0·6644 n.a.	+0·6737 (4.28)**	−0·9042 (−4·91)**	+0·0230 (2·64)*	—	—
	Six-firm sample	0·92	1·07	+6·2540 (7·73)**	+0·1078 (0·69)	−0·4634 (−4·93)**	+0·0226 (4·73)**	—	—
II	Blackstone[a]	0·95	n.a.	−1·5806 n.a.	+0·5455 (4·03)**	−0·6270 (−3·68)**	−0·0018 (−0·19)	+0·2303 (3·73)**	—
	Six-firm sample	0·95	1·57	+4·6250 (6·22)**	+0·2719 (2·07)*	−0·4908 (−6·73)**	−0·0006 (−0·08)	+0·2085 (4·06)**	—
III	—	—	—	—	—	—	—	—	—
	Six-firm sample	0·95	1·52	+4·0670 (4·67)**	+0·2607 (1·93)	−0·4705 (−6·28)**	−0·0028 (−0·36)	—	+0·3845 (3·77)**
IV	Blackstone[a]	0·95	n.a.	−1·4105 n.a.	+0·5334 (4·56)**	−0·6280 (−3·75)**	—	+0·2224 (4·97)**	—
	Six-firm sample	0·95	1·57	+4·6750 (10·88)**	+0·2637 (3·09)**	−0·4899 (−6·95)**	—	+0·2049 (7·52)**	—
V	Blackstone[a]	0·96	n.a.	−0·8417 n.a.	+0·3749 (2·68)*	−0·4557 (−2·49)*	—	—	+0·7354 (5·21)**
	Six-firm sample	0·95	1·52	+4·3410 (10·15)**	+0·2259 (2·41)*	−0·4681 (−6·40)**	—	—	+0·3530 (7·16)**

Notes and Sources:

[a] See Zevin (unpublished work), Table I, for regression result pertaining to the Blackstone Manufacturing Co., in the period 1823–60.

[b] t-statistics are shown (in parentheses below regression coefficients) for two-tail tests of the null hypothesis that the respective coefficients are zero: * indicates the hypothesis is rejected at the 95 per cent level; ** indicates rejection at the 99 per cent level.

equal to zero in the underlying production function (13), and the experience-indexes are therefore omitted completely from the analysis, as in regression equation *R*-I. The results are not satisfactory, except to proponents of the learning by doing hypothesis. True, a respectably high proportion of the variance is explained by equation *R*-I; but for the sample group the estimated spindle-coefficient of the reduced-form production function ($a_1 = \hat{\alpha}_s = 0{\cdot}1078$) is not significantly different from zero. Further, confirming our sense of unease, the degree of autocorrelation in the residuals, signaled by the low Durbin–Watson statistic, strongly suggests the presence of a specification error in the estimating equation. The picture presented by *R*-I is no more acceptable in the case of the Blackstone mills: there it is the reduced-form coefficient for the labor input variable *L*, instead of that for *S*, which turns out not significantly different from zero.[1]

Now, in comparison with the results just examined, the regression equations which include one or another of the experience-indexes among the independent variables are rather more presentable. Inasmuch as we are concerned for the moment simply to establish the existence of learning effects as a significant determinant of productivity growth in the sample group of firms, and not to decide between alternative forms for the learning function, it suffices to observe that in every instance the estimated learning coefficient of the reduced-form production function (14) is found to be less than unity and significantly greater than zero.[2] This, of course, directly

[1] As can be seen from the *t*-statistics (given in parentheses beneath the respective regression coefficients) in Table 3, the coefficient a_2 in *R*-I: Blackstone *is* significantly different from zero at the per cent error level. But what is relevant is whether the labor input coefficient in the production function, $\alpha_L = (1-a_2)$, is statistically significant. Thus, we must ask whether or not it is possible to reject the null hypotheses $H_0: a_2 = 1$. The *t*-statistics computed for this test in the case of the Blackstone regression equations are:

	Blackstone equations			
	R-I	*R*-II	*R*-IV	*R*-V
t-statistic, $H_0: a_2 = 1$:	0·52	2·19*	2·22*	2·97**

Single and double asterisks indicate rejection at the 5 per cent and 1 per cent levels respectively. From Table 3 it is readily seen that in all the regression equations estimated from the sample group data, the same null hypotheses can be rejected easily at the 1 per cent level. The appropriate *t*-statistics for the test of ($H_0: a_2 = 1$) may be readily computed from information in Table 3, since

$$t_{1-a_2} = [(1-a_2)/(a_2)](t_{a_2}),$$

where t_{a_2} is the statistic appearing in parentheses below the regression coefficient of (ln *L*) in Table 3. All significance levels cited here refer to two-tail tests.

[2] Specifically, the coefficients of (ln *Q*) in equations *R*-II and *R*-IV of Table 3, and the coefficients of (ln *T*) in equations *R*-III and *R*-V are all found to be significantly different from zero at the 99 per cent confidence level. This is true not only of the equations to the six-firm sample data, but holds equally for Zevin's estimates based on the Blackstone

contradicts the presupposition from which the form of equation *R*-I was derived, and the evidence of the existence of learning effects thereby provided for the sample group of leading textile firms can thus be added to the findings reported by Zevin's study of the mills owned by the Blackstone Manufacturing Co.

The introduction of (ln Q) or (ln T) not only adds a significant variable to the analysis, it also alters other aspects of the regression results for the better, making it even more difficult to resist accepting the existence of learning by doing as a source of productivity growth in the cotton textile industry during the period under review. As might be anticipated, the addition of these variables in *R*-II and *R*-III, respectively, raises the adjusted proportion of the variance explained above the 92–3 per cent accounted for by *R*-I. The higher $\overline{R^2}$-values, however, are maintained when the number of independent variables is brought back down to four by the suppression of the exponential trend-term, as in equations *R*-IV and *R*-V. This latter finding is not entirely surprising when it is noticed that in *R*-II and *R*-III the presence of the alternative experience variables (ln Q and ln T, respectively) greatly reduces the estimate of the net exponential trend, forcing ξ in those equations to assume values slightly below zero – although not significantly so. Indeed, it is the latter statistical insignificance of the calendar time index, τ, which justifies imposing the constraint $\xi = 0$ and re-estimating equations *R*-II as *R*-IV and *R*-III as *R*-V.

One may ask whether the apparent displacement of any significant exponential trend-term by the experience variables is a generally plausible result. Actually, in view of the interpretation which (14) and (15) indicate must be placed upon the reduced-form parameter ξ, even a significant negative value would not constitute an *a priori* implausibility. It would merely imply that the impact of exogenous improvements in the quality of the average textile operative and/or in the productivity of the average item of 'machinery' was more than offset by the effect of the shortening of working hours upon the volume of cloth produced per man-day. Certainly it need not suggest there was technological retrogression – any more than the disappearance of a net time-trend should be read as implying the absence of exogenous sources of productivity growth. Quite the

---

data, at least for those cases (*R*-II, *R*-IV and *R*-V) in which corresponding regressions are reported. See Table 3. As the preceding footnote points out, the results of *t*-tests of the null hypothesis that the coefficient λ is less than unity may be read indirectly from the *t*-statistics reported in the table, but it should be remembered that in this case the appropriate level of significance will be that indicated for a single-tail test. On this basis the null hypothesis $H_0: \lambda \geqslant 1$ can be easily rejected for every coefficient of (ln Q) or (ln T) that appears in Table 3.

TABLE 4 Alternative implications of the ξ-estimates derived by fitting 'learning' models to data for the six-firm group

Regression model	Value of time coefficient, ξ	Restriction: $\psi_s = \psi_L = \hat{\psi}_{1830-60} = -0{\cdot}0055$		
		Assuming $\beta_L = 0$: implies $\beta_s =$	Assuming $\beta_L = \beta_s$: implies $\beta =$	Assuming $\beta_s = 0$: implies $\beta_L =$
II	−0·0006	0·0136	0·0048	0·0073
IV	0	0·0161	0·0055	0·0083
III	−0·0028	0·0059	0·0020	0·0029
V	0	0·0184	0·0055	0·0078

Source: See equation (14b) for general expression for ξ; text for estimate of average annual rate of change in length of the workday, $\hat{\psi}$; Table 3 entries for estimates of ξ, α_s, and α_L from regressions pertaining to the six-firm sample.

contrary, so long as the steady reduction of the duration of the working day is accepted as a matter of fact, stipulating that $\xi = 0$ is tantamount to asserting that counterbalancing developments of a correspondingly exogenous character were taking place: the estimates of α_s and α_L obtained by fitting the constrained versions of the model, that is, *R*-IV and *R*-V, obviously can be used to determine implicit range for the putative underlying rates of exogenous labor-input and capital-input augmentation. In Table 4 these rates of primary input augmentation are calculated on the supposition, suggested by evidence previously considered, that the length

TABLE 5 Tests for scale effects in the aggregate production function for the six-firm group

Regression model	$(a_1 - a_2) = \dfrac{(\hat{\alpha}_s^* + \hat{\alpha}_L^* + \hat{\alpha}_c^* - 1)}{(-\hat{\alpha}_c^*)}$	$\hat{\sigma}_{a_1-a_2}$	*T*-statistic $H_0: (a_1 - a_2) = 0$	Degrees of freedom
II	−0·2189	0·1274	−1·72[ab]	22
IV	−0·2262	0·1288	−1·75[ab]	23
III	−0·2098	0·1325	−1·58[b]	22
V	−0·2422	0·1375	−1·76[ab]	23
V[c]	−0·2544	0·1485	−1·71[b]	22

[a] Since, with $df = 22$, $t_{0\cdot05} = 1{\cdot}717$ for a single-tail test, we can reject H_0: $(\alpha_s^* + \alpha_L^* + \alpha_c^* - 1) \geqslant 0$ at the 95 per cent level in the cases thus marked.
[b] H_0: $(\alpha_s^* + \alpha_L^* + \alpha_c^* = 1)$ cannot be rejected (on a two-tail test) at the 95 per cent level.
Source: $(a_1 - a_7)$ from Table 3; $\hat{\sigma}^2_{a_1-a_2} = \hat{\sigma}^2_{a_1} + \hat{\sigma}^2_{a_2} - 2\,\mathrm{cov}(\ln S, \ln L)$; $t = (a_1 - a_2)/(\hat{\sigma}_{a_1-a_2})$.
V[c] denotes the re-estimation of regression model *R*-V after correcting for autocorrelation in the residuals, the results of which are given in the text.

of the workday for men and machines alike was decreasing, at the average annual rate of 0·55 per cent, more or less steadily throughout the period from 1830 to 1860. As may be seen, the coefficients derived from the regression equations in which significant learning effects appear are in every case entirely consistent with non-negative rates of change in the exogenously determined 'quality' of the cotton textile industry's labor force (β_L) and of its capital equipment stock (β_s), although it is not conceivable that such changes were particularly rapid.[1]

From a technical econometric viewpoint, the fact that the inclusion of the experience-indexes serves to raise substantially the Durbin–Watson statistics for the four equations estimated from the group data is, perhaps of greater importance.[2] It is true that some ambiguity remains as to whether or not the residuals of equations *R*-II, III, IV, and V are wholly free of significant autocorrelation, but such doubts as this may occasion can readily be allayed. A method frequently employed in coping with serious problems of residual autocorrelation[3] can be utilized in this instance to demonstrate that the tests of the regression coefficients' statistical significance, reported for the sample group in Table 3, are not materially affected by biases arising from that quarter. First, estimation of the autoregressive structure by fitting the stochastic model

$$u_\tau = \rho(u_{\tau-1}) + \sigma_\tau$$

to the residuals ($\hat{u}_\tau$) of regression equation *R*-V, provides an estimate of the coefficient $\hat{\rho} = 0{\cdot}2032$ – which itself is not significantly greater than zero. The latter is, nevertheless, the best estimate of rho, and it therefore can be used in transforming all the variables of *R*-V by optimal differencing, thereby obtaining

$$\begin{aligned}
\Delta y_\tau &= \ln(Y/L)_\tau - \hat{\rho}\ln(Y/L)_{\tau-1},\\
\Delta s_\tau &= \ln(S)_\tau - \hat{\rho}\ln(S)_{\tau-1},\\
\Delta l_\tau &= \ln(L) - \hat{\rho}\ln(L)_{\tau-1},\\
\Delta t_\tau &= \ln(T)_\tau - \hat{\rho}\ln(T)_{\tau-1}.
\end{aligned}$$

[1] It is unfortunately not possible to make meaningful statements about the statistical significance of the implied values of $\hat{\beta}_s$ and $\hat{\beta}_L$ presented by Table 4, since the distributions of the associated errors of estimate are quite unknown.

[2] Zevin's paper (unpublished work), Table I, does not report results of any tests for the presence of significant autocorrelation in the residuals of the equations he estimated from the Blackstone data. The discussion of this matter here is consequently confined to the regressions for the six-firm sample in Table 3.

[3] Cf. J. Johnston, *Econometric Methods* (New York: McGraw-Hill, 1963), pp. 194–5; E. Malinvaud, *Statistical Methods of Econometrics* (Chicago: Rand McNally, 1966), pp. 420ff.

On the basis of these transformed variables the model can then be re-estimated, thus:

(R-V^c)

$$\Delta y_\tau = \underset{(10\cdot17)^{**}}{3\cdot660} + \underset{(2\cdot50)^{*}}{0\cdot2552}\,\Delta s_\tau - \underset{(-6\cdot70)^{**}}{0\cdot5096}\,\Delta l_\tau + \underset{(5\cdot78)^{**}}{0\cdot3266}\,\Delta t_\tau$$

$$\overline{R^2} = 0\cdot9196; \qquad D\text{–}W = 1\cdot788.$$

In this equation the Durbin–Watson statistic is now high enough to warrant dismissing the presence of positive autocorrelation with 95 per cent confidence.[1] And yet it may be seen that the absolute and relative magnitudes of the regression coefficients obtained in R-V^c stand little changed from those reported by Table 3 for the *R*-V: Group regression. The reduced-form learning coefficient ($\lambda = 0\cdot3266$) once again is statistically significant – this time at virtually any level of confidence one might care to specify – and the corresponding production function coefficients $\hat{\alpha}_s$ and $\hat{\alpha}_L$, like those derived from *R*-V, are *both* significantly different from zero.

The last-mentioned point deserves fuller attention, for it touches on what is perhaps the most critical respect in which taking learning by doing into account leads to an improvement of the regression results. Whereas the coefficient $a_1 = \hat{\alpha}_s$ was found not to be statistically significant in equation *R*-I for the sample group, the estimates of the same coefficient provided by fitting *R*-II, *R*-III, *R*-IV, and *R*-V to the group data can in each case be said with at least 95 per cent confidence to be significantly greater than zero.[2] Notice too that although the reduced-form parameter estimate of $\hat{\alpha}_L$ ($= 1 - a_2$) turned out to be not significantly different from zero when *R*-I was fitted to the Blackstone Co. data, using the same body of observations it is possible to reject the null hypothesis ($H_0 : \alpha_L = 0$) with 95 per cent confidence on the basis of Zevin's equations *R*-II or *R*-IV, and with 99 per cent confidence on the basis of his equation *R*-V.[3] Since all the estimates of α_L obtained from the sample group regressions, like all the estimates of α_s derived from Zevin's regressions for the Blackstone mills, are strongly significant, we may conclude that in order to

[1] With 22 degrees of freedom, and three continuous independent variables, the 5 per cent significance points of *D–W* are 1·05–1·66. Note that the *D–W* corresponding to *R*-V in Table 3 falls as far short of the upper 5 per cent point (1·66) as does the *D–W* for any of four equations with significant learning effects.

[2] Note that the test appropriate to the statement made in the text is a single-tail *t*-test. Although the null hypothesis $a_1 = 0$ cannot be rejected on a two-tail test at the 95 per cent confidence level in the case of *R*-III, Table 3 indicates that this more exacting test is passed by the a_1-coefficients of the other variants of the learning model fitted to the sample group data.

[3] See footnote 1 on p. 137 above, for the results of tests of this hypothesis.

retain the basic notion that both labor and capital were significant factors in the manufacture of power-loom cloth by leading New England firms before the Civil War, it is necessary to have in mind a long-run production function in which the efficiency of those inputs was being irreversibly raised by the accumulation of production experience.

The absence of conventional scale economies

Having been led to discard *R*-I in favor of the set of regression equations which allow for the existence of irreversible scale effects, or learning by doing, it is of interest to inquire briefly what the latter equations reveal concerning *reversible* scale effects, which is to say, concerning conventional economies or diseconomies of scale encountered beyond the level of the operation of individual integrated cotton mills. In the context of the historical question of the justification for the infant-industry tariffs levied by the United States on imported cotton-goods, the relevant issue is whether or not there were increasing returns to scale that were external to individual textile *firms*. Primary attention therefore focuses upon the null hypothesis that the aggregate production function for the sample group of firms was homogenous of degree one or higher.

The strict constant returns hypothesis H_0: $[\alpha_s^* + \alpha_L^* + \alpha_c^* = 1]$ must refer initially to the restriction on the sum of the capital, labor, and raw materials exponents in the underlying Cobb–Douglas production function (13). But as this initial hypothesis – call it $H_0(1)$ – may readily be shown to be equivalent to the proposition, $H_0(2)$, that the algebraic sum of the 1n (*S*)- and 1n (*L*)-coefficients in regression model (15) is identically zero, the question may be conveniently investigated from the findings summarized in Table 3 without any need actually to identify the parameters of the underlying production function.[1]

No support for tariff arguments grounded on the existence of significant scale effects can be found in the outcome of the statistical tests reported by Table 5. There it is seen that the estimated sums of exponents fall in the range between 0·76 and 0·88, well short of the unitary value indicated in the null hypothesis $H_0(1)$. Maintaining a 95 per cent confidence level,

[1] To prove that $H_0(1) = H_0(2)$: $[a_1 - a_7 = 0]$, we may begin by noting from (15a) that

$$a_1 - a_2 = \alpha_s - (1 - \alpha_L) = \alpha_s + \alpha_L - 1.$$

However, from (14b) we also know that

$$\alpha_s + \alpha_L - 1 = \frac{\alpha_s^*}{1 - \alpha_c^*} + \frac{\alpha_L^*}{1 - \alpha_c^*} - 1 = \frac{\alpha_s^* + \alpha_L^* + \alpha_c^* - 1}{(1 - \alpha_c^*)}.$$

If we accept the premise that α_c^* is positive but less than unity, rejection of $H_0(2)$ therefore must imply rejection of $H_0(1)$.

one cannot reject the hypothesis that the group production function was characterized by constant returns on the basis of equations *R*-II and *R*-III; at the 97·5 per cent confidence level with the same equations it becomes impossible to reject the hypothesis that the group function was subject to *constant or decreasing* returns to scale. The latter is equally true of the results obtained with regression equations *R*-IV and *R*-V, although in those cases constant returns to scale can be rejected, with 95 per cent confidence, in favor of decreasing returns. Re-estimating the last equation as $R\text{-V}^c$, in order to purge the residuals of any significant measure of autocorrelation, does nothing to upset these findings regarding the absence of economies of scale: in Table 5 it is seen that the sum of the exponents thus obtained $(1-[\alpha_1-\alpha_2] = 0{\cdot}7456)$ falls still further below unity, but as the standard error of this estimate is somewhat increased, it remains impossible to reject the constant returns hypothesis at the 95 per cent confidence level.

While there are some hints that the growth of the aggregate scale on which this collection of New England firms operated was subject to increasing costs, possibly arising from the progressive exhaustion of readily available supplies of higher quality labor in the region, the hints are not very strong. Certainly no parallel indications of increasing costs – much less any signs of decreasing costs – are in evidence at the level of the single firm; the sum of $\hat{\alpha}_s$ and $\hat{\alpha}_L$ as estimated by Zevin from the Blackstone data lies much closer to unity, making it quite impossible to reject the null hypothesis that there were constant returns to firm scale in this branch of the ante-bellum textile industry.[1]

Acceptance of the proposition that constant costs prevailed, both at and beyond the level of operations of these multi-mill firms, carries a twofold implication for the controversy over the social desirability of the tariff protection received by the industry. Of course, it directly undercuts any attempt to rest an infant-industry justification for the tariff on the ground that there were significant positive externalities to be obtained in the form of reversible industry-scale effects. Rather less obviously, perhaps, it also renders the results provided by regression equation *R*-V (or $R\text{-V}^c$) more attractive than those offered by equations *R*-II or *R*-IV in Table 3. In the former the estimates of $\hat{\alpha}_s$ and $\hat{\alpha}_L$ found for the group of leading firms are in much closer agreement with the values of the corresponding parameters estimated by Zevin for the Blackstone Co., a similar, isolated firm – just as they should be under conditions of

[1] From Table 3 it may be seen that the sum of the reduced-form parameter estimates $\hat{\alpha}_s$ and $\hat{\alpha}_L$, yielded by the Blackstone regressions is 0·9185 for *R*-II, 0·9045 for *R*-IV, and 0·9192 for *R*-V. Zevin (unpublished work, p. 12) reports only the result of a test of the constant returns hypothesis for the model estimated by *R*-IV, but it is clear that the other two regressions tell the same story.

strictly constant costs.[1] The findings of this section therefore indirectly suggest that the long-run learning model based upon cumulated production time as the measure of experience is more appropriate to the history of this segment of the industry than is the version of the learning model based upon cumulated output – a suggestion which undermines the case for tariff subsidization of domestic cloth production as a means of generating beneficial externalities through learning by doing. Other considerations, however, must be allowed a part in determining the important choice between alternative forms of the learning function.

A priori *considerations and the choice between learning functions*

Consulting the summary measures of goodness-of-fit presented for equations *R*-II through *R*-V in Table 3 proves, unfortunately, to be of scant help in deciding whether cumulated output, Q, or cumulated years of firm operation, T, is the more appropriate of the two available measures of production experience in the leading segment of the United States cotton textile industry.[2] As far as the $\bar{R}^2$ s of the sample group regressions are concerned, there is nothing in the choice between *R*-II (or *R*-IV) and *R*-III (or *R*-V), since in each case only 5 per cent of the variance is left unexplained. The same criterion applied to the Blackstone regressions would award a slight, but hardly a noteworthy advantage to equation *R*-V, since it accounts for 96 per cent of the variance, whereas equation *R*-IV explains 95 per cent.

In such situations it is sometimes possible to decide between contending variables – here $\ln(Q)$ and $\ln(T)$ – by including both within a single

[1] As pointed out earlier, this is a terribly stringent requirement which presupposes that Blackstone was indistinguishable from the representative member of the six-firm sample. Yet, it should be noted that Zevin's estimates for the Blackstone production function refer to the period 1823–60, which is not quite the same interval as that for which the sample group estimates are available. Consequently one ought not make too much of the fact that while the $a_1(=\hat{\alpha}_s)$ coefficients for Blackstone and the sample group are closest in *R*-V – 0·3749 vs. 0·2259 – they are still not statistically identical like the corresponding estimates of $\hat{\alpha}_L(=1-a_2)$ in that equation. The significance test referred to here is a t-test of the significance of the difference between the regression coefficients from the two *R*-V equations, using the pooled estimate of the standard error computed from:

$$\sigma^2_{a_1,\,\text{pooled}} = \left\{\left(\frac{\hat{\sigma}^2_{a_1}}{N}\right)_{\text{Blackstone}} + \left(\frac{\hat{\sigma}^2_{a_1}}{N}\right)_{\text{group}}\right\}$$

where $N_{\text{Blackstone}} = 33$ and $N_{\text{group}} = 22$ represent the respective number of degrees of freedom associated with the estimated errors of the regression coefficients.

[2] As far as Durbin–Watson statistics go, the differences between *R*-IV and *R*-V are negligible, and, in any case the slightly higher degree of autocorrelation in the residuals of *R*-III and *R*-V is not worth worrying about – as has been shown by the comparison of *R*-V^c with *R*-V in Table 3.

equation and allowing one to oust the other. But on this occasion the tactic is of little avail. Zevin reports the following regression for the Blackstone mills:[1]

$$\ln(Y/L)_\tau = \underset{}{-0{\cdot}8836} + \underset{(2{\cdot}17)^*}{0{\cdot}3846 \ln(S)_\tau} - \underset{(2{\cdot}11)^*}{0{\cdot}4666 \ln(L)_\tau}$$

$$+ \underset{(1{\cdot}12)}{0{\cdot}6808 \ln(T)_\tau} + \underset{(0{\cdot}09)}{0{\cdot}0174 \ln(Q)_\tau},$$

from which it is seen that as the movements of the experience-indexes are so similar, they divide the effect between them and neither is found to be statistically significant. Unfortunately, the problem of multicolinearity is still more serious in the sample group data. For, as a consequence of the aggregation of more than one firm's outputs, the amplitude of the short-term variations of Q around its trend is reduced, further increasing its resemblance to the smoothly rising alternative index of experience, T.[2] The one illuminating point to be gleaned from the foregoing experiment with the Blackstone data is that even though all the regression coefficients are reduced in relation to their respective standard errors, the estimated values themselves closely conform to those obtained when $\ln(T)$ alone is considered (see *R*-V), and not to the estimates provided by the alternative specification in *R*-IV. But while this is suggestive, it is far from conclusive.

We are therefore placed in the position of having to choose between alternative specifications for the experience-index largely on the basis of *a priori* notions concerning the appropriateness of the magnitudes of the reduced-form production parameters implied by the two sets of regression equations, *R*-II, IV and *R*-III, V (or V^c).[3] Here there are two sets of considerations that seem especially relevant. The first has to do with the labor and capital input coefficients, while the second concerns the magnitudes of the learning coefficients. Both leave one inclined to accept cumulated firm-years of production time as the better measure of experience in the manufacture of power-loom cloth, along with all the policy implications that would logically follow from that empirical finding.

[1] Zevin (unpublished work), Table I. For the equation cited here $\bar{R}^2 = 0{\cdot}95$, slightly lower than that found when the ln Q-term is deleted. Cf. Table 3, *R*-V: Blackstone.

[2] The simple correlation coefficient between ln (Q) and ln (T) for the sample group (1834–60) is $R = 0{\cdot}999$.

[3] The magnitude of ξ, and the difference between the constrained and unconstrained estimates will not be of concern here. We have already noted that with respect to the inferred values of the underlying structural parameters $\hat{\beta}_s$ and $\hat{\beta}_L$, Table 4 (above) reveals there is little to choose between *R*-IV and *R*-V. Moreover, none of the relationships between $\hat{\beta}_s$ and $\hat{\beta}_L$ suggested by the regression equations can be dismissed as implausible. It is therefore necessary to look elsewhere for guidance.

At the conclusion of his path-breaking study, having drawn attention to the *R*-V regression results derived using $\ln(T)$ as the experience-variable, Zevin expressed a decided preference for the more conventional form of learning function (involving *Q*) on the ground that *R*-IV suggested more plausible *relative* magnitudes for the elasticities of output with respect to labor and capital. And, indeed, reckoned in relation to the combined primary factor elasticity, the size of the labor input elasticity estimate yielded by Zevin's equation *R*-IV is

$$\frac{\hat{\alpha}_L^*}{\hat{\alpha}_L^* + \alpha_s^*} = \frac{\hat{\alpha}_L}{\hat{\alpha}_L + \hat{\alpha}_s} = \frac{1 - 0{\cdot}6280}{0{\cdot}3720 + 0{\cdot}5334} = 0{\cdot}411,$$

whereas the average share of the wage-bill in gross value added (which Zevin was able to compute from the original Blackstone Co. records for 1823–60) turns out to have been 0·414.[1] So exact is the match, it is not surprising that it should have been immediately seized upon as confirming *R*-IV. Yet on second consideration, this very coincidence between the estimated relative elasticity of output with respect to labor and labor's share in value added, serves to cast serious doubt upon the specification underlying regression equations *R*-II and IV; for such correspondence is to be expected only where the markets for labor and capital are equally competitive.[2]

In Zevin's judgement, however, the Blackstone Manufacturing Co. was actually a monopsonist in the local (Worcester County, Massachusetts) market for industrial labor – a point that has been made more generally about the position in which the large, Massachusetts-type mills of this period found themselves.[3] Had Blackstone exercised its monopsony power – and we are given no cause for supposing it would not – the upshot would have been a tendency for the prevailing average wage rate to be set below the marginal value productivity of labor to the firm. This can be seen immediately from Figure 7.

But in such circumstances the wage-bill's proportion of total production costs clearly must have fallen short of the elasticity of output with respect

[1] Cf. Zevin (unpublished work), Table II, p. 13.

[2] Under conditions of perfect competition in all (product and factor) markets, labor and capital would receive an equilibrium rate of remuneration equal to their respective marginal value productivities, and the elasticities of output with respect to these inputs would exactly equal the shares of the total product each received. But the equality of *relative* shares (s_i) with the relative elasticity coefficients, $(\alpha_L)/(\alpha_L + \alpha_s) = (s_L)/(s_L + s_s)$, requires only that the real rates of factor remuneration be in *uniform proportion* to their respective marginal physical productivities. The latter condition is compatible with *equal* degrees of imperfection in all factor markets, as well as with equally perfect competition. Note, therefore, that the burden of the following argument is that the labor market was *comparatively* imperfect.

[3] Cf. Zevin (unpublished work), p. 15; Lebergott, 'Wage Trends, 1800–1900,' pp. 450–2, 454.

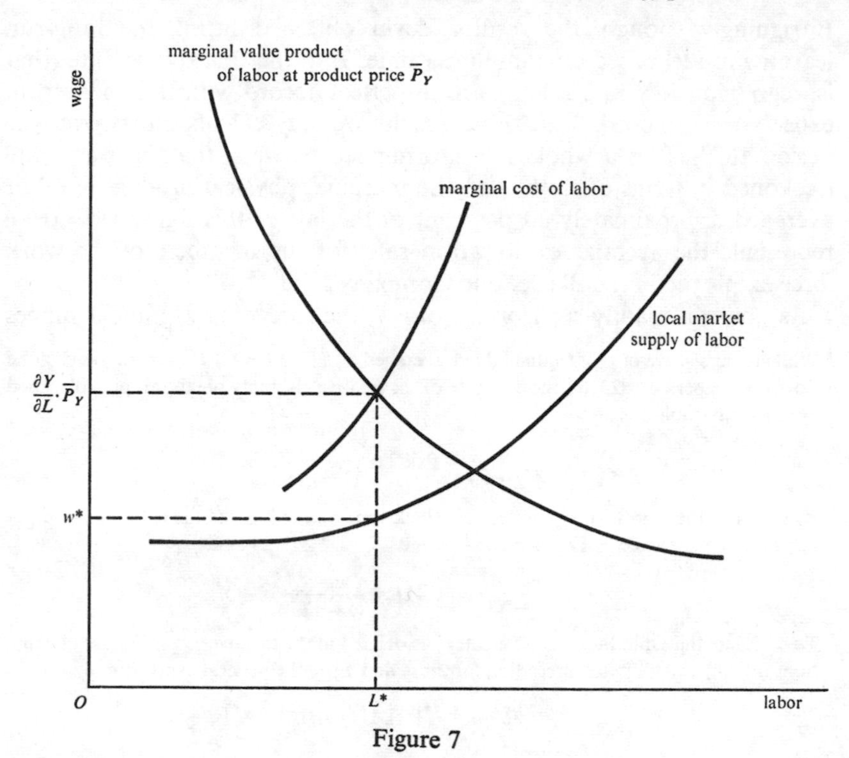

Figure 7

to the input of labor services, as determined by the underlying production function. And so long as Blackstone remained a price-taker in the market for raw cotton and other purchased intermediate inputs, one might also anticipate that labor's share in the firm's gross value added would have been less than the ratio $\alpha_L^*/(\alpha_L^*+\alpha_s^*)$, rather than identically equal to it.[1]

[1] If homogenous labor, L', is paid less than its marginal value product (supposing the firm to be a price-taker in the *product* market, though a monopsonist in the labor market), we have $v > 0$ in

$$w = [(\partial Y/\partial L')-v]P, \tag{i}$$

where w is the money wage rate, and P is the price of cloth. The share of labor in total costs is therefore

$$s_L = (wL'/PY) = [(\partial Y/\partial L')-v](L'/Y). \tag{ii}$$

But as we know from (13) that

$$\alpha_L^* = (\partial Y/\partial L')(L'/Y), \tag{iii}$$

it follows immediately that $s_L > \alpha_L^*$. Now suppose that the firm treats the price of its purchased material (cotton), P_c, as a parameter, so that profit-maximization subject to (13) leads to $P_c/P = (\partial Y/\partial C)$, and therefore to:

$$1-s_c = 1-(P_cC/PY) = 1-\alpha_c^*. \tag{iv}$$

Intriguingly enough, the results Zevin obtained fitting the long-run learning model based on cumulated time, *T*, to the Blackstone data (that is, regression *R*-V in Table 3), are in perfect accord with this theoretical expectation.[1] Indeed, from those results we may calculate that over the period 1823–60 as a whole the gap implied between the real wage rate (reckoned in terms of cloth) and the marginal physical product of labor averaged approximately 30 per cent of the latter: this figure ($V = 0{\cdot}30$) represents the average 'exploitation rate' for the members of the work force employed in the Blackstone Company's mills.[2]

As one can readily see from Figure 7, the foregoing argument hinges

[1] While labor's share of gross value added averaged ($s_L/(1-s_c) =$) 0·414 for the Blackstone Co. over the period 1823–60, the estimate of the relative elasticity of labor inputs obtained from *R*-V in Table 3 is

$$\frac{\hat{\alpha}_L}{(\hat{\alpha}_L+\hat{\alpha}_s)} = 0{\cdot}592 > \frac{s_L}{(1-s_c)}.$$

By contrast, the regression estimated for Blackstone on the basis of the cumulated output measure of experience, *R*-IV, was seen to yield

$$\frac{\hat{\alpha}_L}{\hat{\alpha}_L+\hat{\alpha}_s} = 0{\cdot}411 < \frac{s_L}{(1-s_c)}.$$

[2] To calculate the exploitation coefficient, *V*, as defined in the text, we can make use of equations (ii), (iii) and (v) in the preceding footnote and write the general expression

$$\frac{s_L}{(1-s_c)} = \alpha_L^*/(\alpha_L^*+\alpha_s^*)-[v(L'/Y)]/[\alpha_L^*+\alpha_s^*].$$

This can be rearranged as follows, noting the relationship between the ratio of structural input coefficients and the ratio of the reduced-form parameters:

$$\frac{s_L}{(1-s_c)} = \frac{\alpha_L}{(\alpha_L+\alpha_s)}\{1-v/(\partial Y/\partial L')\}, \qquad \text{(vii)}$$

which is readily solved for $V = v/(Y/L')$, thus,

$$V = 1-\left[\frac{(s_L)/(1-s_c)}{(\alpha_L)/(\alpha_L+\alpha_s)}\right]. \qquad \text{(viii)}$$

Substituting the estimate of the relative labor elasticity coefficient derived from regression *R*-V: Blackstone, and the average 1823–60 labor share in gross value added, we find:

$$V = 1-(0{\cdot}414/0{\cdot}592) = 0{\cdot}301.$$

If the underlying production function is first-degree homogenous, which the results discussed in the previous section indicate was likely to have been true for Blackstone, the share of value added in total cost of production is

$$1-s_c = (\alpha_L^*+\alpha_s^*). \qquad \text{(v)}$$

In conjunction with the result obtained from (ii) and (iii), (v) permits us to assert the proposition cited in the text:

$$\frac{s_L}{(1-s_c)} > \frac{\alpha_L^*}{(\alpha_L^*+\alpha_s^*)} = \frac{\alpha_L}{(\alpha_L+\alpha_s)}. \qquad \text{(vi)}$$

on the presumption that the labor supply schedule facing the firm was not perfectly elastic. Consequently, if the continuing expansion of a large employer like the Blackstone mills progressively exhausted what had been for a while a rather elastic reserve of labor, and compelled the firm to seek additional hands farther and farther afield, bidding against more attractive alternative employments, the initially small gap between the marginal physical product of labor and the real wage (v) would have grown more pronounced – especially as the more inelastic portion of the labor supply schedule was encountered. Looking at the geographical and temporal patterns in the induced diffusion of comparatively capital-intensive innovations through the New England textile industry before the Civil War, Zevin has ventured to surmise that

> the critical point [at which the local labor supply curve rapidly became inelastic] is reached first in southern New England around Providence in the late 1820's and moves north with the industry over time, reaching southern New Hampshire and Maine by the end of the 1830's.... In Worcester County, Massachusetts, where Blackstone was located, these considerations [of the timing of the introduction of comparatively labor saving innovations] point to the year 1830 as the approximate location of this critical point.[1]

It is therefore significant to observe, especially as further corroboration is thus provided for the interpretation proposed here, that the temporal movements in the 'exploitation rate' derived for the Blackstone labor force using regression equation *R*-V are broadly congruent with Zevin's conjecture. Whereas for the three decades following 1830 the estimated average exploitation rate was approximately 0·32, its average value for the years 1823–30 was only a bit more than half as large.[2]

Now from none of this need it follow that New England's cotton textile workers were being more severely 'exploited' in the popular sense of that term. Quite conceivably, the possibility of effectively setting local market wage rates below the private marginal value product of labor rendered it worthwhile for large firms to attempt to cope with their growing labor recruitment problems by raising the productivity of the available workers through investment in training, as well as by adopting techniques which substituted more conventional forms of capital for labor. The provision of training in skills which could not be perfectly transferred to other employers – either because the skills involved were technically specific

[1] Zevin (unpublished work), p. 15.

[2] According to the Blackstone records assembled by Zevin (unpublished work, Table II, p. 13) wages as a proportion of gross value added averaged 0·489 in the years 1823–30, and 0·404 during 1830–60 – with little secular change taking place within the latter interval. Thus, using equation (viii) of the previous footnote, we calculate V for 1823–30 as $(1-0{\cdot}489/0{\cdot}592) = 0{\cdot}174$, and $V = (1-0{\cdot}404/0{\cdot}592) = 0{\cdot}318$ for 1830–60.

to the firm's operations or, perhaps more plausible, as a result of the combination of a local labor market monopsony position (created by high costs of entry for potential competitive employers) and the imperfect mobility of labor – would be recompensed through formally 'exploiting' the trained workers, paying them less than the value of their marginal product to the firm.

Inasmuch as we have found strong evidence that some form of learning was taking place in these leading textile firms, it hardly seems such a bold step to entertain the idea that on-the-job training in incompletely transferable skills was also part of the picture during this period of the industry's development, and to suggest that the apparent gap separating the marginal physical productivity of labor from the real wage might appropriately be viewed as a rental charge on the firm's investment in such training. Viewed from this vantage point, the 30 per cent average exploitation rate estimated for the period 1823–60 does not seem improbably large; it is only what would be required if an outlay on 'training' equal to 90 per cent of the worker's post-training annual marginal product had to be (quickly) amortized over a four-year period and had to yield an opportunity cost rate of return to the firm of 10 per cent per annum.[1]

The implications flowing from the existence of comparatively pronounced labor market imperfections have carried us rather far toward establishing the credibility of the reduced-form parameters (α_L and α_s) which Zevin estimated by fitting equation *R*-V to the Blackstone Co. data. By the same token, the alternative specification underlying regression equations *R*-II and IV, although originally preferred by Zevin, now falls under suspicion as quite probably understating the relative elasticity of Blackstone's cloth output with respect to the firm's input of labor. In the light of this reversal, it should be observed that, as far as concerns the

[1] By employing the standard annuity formula for an asset of finite life, we may write:

$$v = Ir(1-e^{-rA})^{-1},$$

where (I) is the real value of the investment in training one worker, (v) is the real value of the uniform annual repayment extracted by exploiting the worker at the rate $V = v/(\partial Y/\partial L')$, r is the annual opportunity cost rate of return required by the investing firm, and A is the period over which (I) must be amortized. The higher the likelihood of losing a trained worker, for any cause, the shorter the amortization period on which the training firm will insist.

Now the above expression may be divided through by ($\partial Y/\partial L'$), and solved for (I) in terms of r, A, V – which last has already been estimated at $V = 0{\cdot}301$ for the period 1823–60 – and ($\partial Y/\partial L'$) itself. Assuming the values $r = 0{\cdot}10$ and $A = 4$, one finds

$$I/(\partial Y/\partial L') = (0{\cdot}301)/(0{\cdot}333) = 0{\cdot}904,$$

the relationship referred to in the text. It should be emphasized that this calculation is meant as nothing more than a check on the numerical plausibility of the estimated rate of exploitation implied by regression *R*-V: Blackstone.

absolute and relative magnitude of the labor elasticity coefficient α_L, there is a rather close correspondence between equation *R*-V for the Blackstone Co. and *all* the regressions allowing for learning effects which have been estimated here for the group of six leading firms.[1] The estimates of $\{\alpha_L^*/(\alpha_L^*+\alpha_s^*)\}$ derived from the latter set of equations – *R*-II through *R*-V^c – all cluster closely around 0·66, pretty much in agreement with the corresponding ratio (0·59) yielded by equation *R*-V: Blackstone, but thoroughly at odds with the relative labor elasticity estimates obtained from the regressions in which Zevin employed cumulated output, *Q*, as the measure of Blackstone's experience.

Therefore, if we seek both a conciliance of the single firm and group production function coefficients – along the lines to be anticipated in the presence of constant returns to scale – *and* a set of estimated input coefficients that remain plausible when viewed in relation to prior information about the circumstances of the individual firm's operations, our acceptance of cumulated time (*T*) as the historically relevant measure of production experience is clearly indicated by the evidence thus far examined.

Further support for this conclusion is to be derived by considering the estimated learning coefficients themselves. The original motive behind our concern with the choice between alternative indexes of production experience – *Q* versus *T* – suggests we should first inquire whether the learning effects of a given percentage increase in cumulated output were notably weaker than those deriving from an equal percentage increase in the temporal duration of the firm's or the group's production experience. There can be no question that as far as the Blackstone Manufacturing Co. was concerned this was in fact the case: the learning coefficient referring to the experience-index, *Q*, is ($\hat{\lambda} = 0{\cdot}2224$), only one-third the size of the coefficient referring to the index *T*.[2] Although the differences between alternative learning coefficients estimated for the sample group are rather less pronounced, these too imply significantly larger values for λ^* when *T* replaces *Q* as the argument of the learning function.[3]

[1] Cf. footnote 1, p. 143, for comparison of the group and single firm estimates of $\hat{\alpha}_L$ in equations *R*-V.

[2] Notice that as the suppressed exponent α_c is the same in *R*-IV and *R*-V for the Blackstone Co., the *ratio* of the reduced-form parameter estimates will reflect the relative magnitudes of the underlying learning coefficient in the production function (13); that is,

$$\frac{\hat{\lambda}_{\text{IV}}}{\hat{\lambda}_{\text{V}}} = \frac{\hat{\lambda}_Q^*}{\hat{\lambda}_T^*},$$

where the subscript on the estimated overall learning coefficient $\hat{\lambda}_i^*$ refers to the argument of the learning function.

[3] The significance tests referred to are *t*-tests of the form

$$t = \frac{(\hat{\lambda}_T - \hat{\lambda}_Q)}{(\hat{\sigma}^2_{\lambda_T} + \sigma^2_{\lambda_Q})^{\frac{1}{2}}}(N)^{\frac{1}{2}},$$

Now it should be clear that the rising volume of power-loom cloth being manufactured in integrated cotton mills, by both the individual firm and the group of six leading textile companies, kept cumulated output mounting more quickly than the sum total years of experience in this particular line of production.[1] From the findings just examined, however, it would appear that some considerable portion of that expanding domestic output merely afforded the sort of repetitive experience which was for all intents and purposes redundant in its effects on costs.

Yet another point to be noted in support of the less inclusive experience-index (T) concerns the absolute sizes of the alternative learning coefficients. While all the learning effects disclosed by the regression equations in Table 3 are statistically significant, the coefficients referring to the experience-index Q lie at, if not below, this parameter.[2] Looking first at the reduced-form parameter estimates for λ from equations R-II and R-IV, there is close agreement among the results for Blackstone and the sample group: the four regression coefficients of $\ln(Q)$ fall in the narrow limits between 0·20 and 0·23.

Were it the case that only the primary inputs used in cloth production had been subject to Hicks-neutral efficiency improvements due to 'learning,' these estimates ($\hat{\lambda}$) would indicate the size of the learning

[1] From Table 8 it can be calculated, for example, that the annual growth rate of Q exceeded that of T (for the six-firm group) by as much as 10 percentage points in the latter half of the 1830s (1836–9), by 3·2 percentage points on average during the most of the next decade (1843–5), and by 2·2 percentage points on average – with little annual variation – throughout the remainder of the period (1850–60). It might be pointed out in this connexion that the growth rate of Q for the six-firm group on the eve of the Civil War was 5·5 per cent per year, not a spectacularly high figure. Indeed, it was probably no more rapid than the rate of growth of cumulated output in the New England cotton textile industry as a whole *c.* 1858–60, and may well have been somewhat slower. The latter conclusions can be inferred from Davis and Stettler's ('The New England Textile Industry,' Table 4, p. 221) partial reconstruction of the historical record of textile production in New England.

[2] See discussion of the findings of other studies, footnote 1 on pp. 113 and 116 above, and note that the empirical range cited is established by those inquiries which used Q-measures of experience.

---

since in the two regressions from which the estimates are drawn the number of degrees of freedom (N) is identical. Testing R-III vs. R-II for the sample group, we have the difference $\hat{\lambda}_{III} - \hat{\lambda}_{II} = 0{\cdot}3845 - 0{\cdot}2085 = 0{\cdot}1760$, which is highly significant: $t = 7{\cdot}24$, with *df.* pooled $\simeq 32$. Alternatively, we can compare the difference between the estimates of λ_T and λ_Q provided by the constrained version of the regression model:

$$\hat{\lambda}_V - \hat{\lambda}_{IV} = 0{\cdot}3520 - 0{\cdot}2049 = 0{\cdot}1471,$$

which yields a still more strongly significant t-statistic, $t = 12{\cdot}55$.

coefficient λ' in the appropriately rewritten constant cost production function.[1]

$$Y_\tau = A_0^*\{(SH)_\tau e^{\beta_s \tau} G_\tau^{\lambda'}\}^{\alpha_s^*} \{(MH)_\tau G_\tau^{\lambda'}\}^{\alpha_L^*} \{(C)_\tau\}^{\alpha_c^*}. \qquad (13')$$

But if 'learning' enhanced the productivity of all the inputs, as (13) states, the underlying learning coefficient would really be only a fraction of the reduced-form parameter: from (14b) we find $\lambda^* = \lambda(1-\alpha_c^*)$. How large a fraction? To answer this question it is necessary to have at least a point-estimate of the elasticity of cotton textile output with respect to the input of purchased materials. If one is willing to obtain such a figure by assuming that elasticity would approximately equal the value of purchased materials as a proportion of total production costs, the United States Census of Manufactures for 1860 suggests $\alpha_c^* = 0{\cdot}50$ as a *maximum* estimate.[2] The

[1] Comparing (13′) with (13), it is apparent that λ^* in the latter would be equivalent to $\lambda^* = \lambda'(\alpha_s^* + \alpha_L^*)$. From (14b), however, one would find that $\lambda = \lambda^*/(1-\alpha_c^*) = \lambda'$, given the condition $\alpha_s^* + \alpha_L^* = 1 - \alpha_c^*$.

[2] Cf. footnote 1, p. 119, above. Actually the ratio of the value of materials to the value of shipments reported for the cotton textile industry by the Eighth Census was 0·51. But this gross materials share figure clearly overstates the (net) materials share that is appropriate in the present context. The reason it does so is simply that the gross value of materials purchased in the industry reflects not only (say) raw cotton purchases by spinning mills and integrated establishments, but purchases of yarn by weavers. It is not difficult to see that in general the industry's gross materials share will exceed the industry net materials share, which will also be characteristic of integrated establishments like those under examination here.

Suppose that Spinners buy cotton worth cX_s and ship yarn worth X_s to Weavers. The latter pay $X_s = yX_w$ for it, working it up into X_w of cloth. The value input coefficient for yarn is $y < 1$, while that for cotton is $cy < 1$. Integrated establishments ship X_I worth of cloth, for which they will require cyX_I worth of cotton, assuming the unit prices and technical coefficients are the same for each operation, whether it is conducted by integrated or by non-integrated establishments. Thus, the industry flows are as follows:

	Spinner	Weaver	Integrated
Value of purchases	cyX_w	$X_s = yX_w$	cyX_I
Value of shipments	$yX_w = X_s$	X_w	X_I

From this it is readily seen that the ratio of the industry's gross materials purchases to its gross shipments will be:

$$1 > \frac{yc(X_w + X_I) = yX_w}{(X_w + X_I) + yX_w} > cy.$$

But the (net) ratio of materials purchased from establishments outside the industry $(cyX_w + cyX_I)$ to shipments of output destined for use outside the industry $(X_w + X_I)$ is cy, identically equal to the materials share prevailing in the integrated plants alone.

The foregoing general proposition is confirmed by McGouldrick's (*New England Textiles*, pp. 144–5) fragmentary estimates of the share of the unit value of cloth output accounted for by the cost of raw cotton in 1839 and 1884. The estimates in question are derived from cloth and cotton price quotations, and technical coefficients – specifically, the weight of the

resulting minimum estimates established for λ^* by the regression models based on Q are thus very much below the customary range, for they cluster between 0·103 [= 0·205 (1 − 0·50)] and 0·115 [= 0·230 (1 − 0·50)].

By contrast, the use of cumulated time as the measure of experience leads to estimates of the reduced-form learning coefficient ($\lambda = \lambda'$) for the group which lies right at the center of the 0·2–0·5 range. Indeed, according to equation *R*-V^c, which places $\hat{\lambda}$ at 0·3266, we can say that primary factor costs per unit of output fell roughly 20 per cent with each doubling of the group's collective experience measured in years of firm operations. On this basis, using the (maximum) point-estimate of $\alpha_c^* = 0{\cdot}5$ previously employed, the corresponding minimum values for λ^* are found to lie immediately below 0·2 in the case of the group, while for Blackstone alone the estimated learning coefficient turns out to be $\hat{\lambda}^* = 0{\cdot}367$.[1]

Thus, it seems appropriate on two general counts to accord preference to formulations of the long-run learning model that involve T, the less conventional and less comprehensive of the two measures of production experience we have considered. First, such formulations suggest relative and absolute values of the input elasticities that are not historically implausible. Secondly, the learning coefficients estimated taking Q as the experience-index are significantly smaller than those based on T – so much so that they lie substantially beneath the low end of the customary range of estimates established by previous empirical inquiries into learning phenomena, studies which (successfully) relied on cumulated output measures of experience.

Learning, redundant entry, and the case for 'pilot' enterprises

Would-be apologists for the protection afforded United States cotton textile producers during the period 1824–60 have by this stage been left with little room for maneuver. Deprived of making any empirically substantiated appeal to the existence of reversible scale economies external

[1] Specifically, the group learning coefficients range from $\lambda^* = 0{\cdot}163$, implied by the estimate of λ from *R*-V^c, to $\hat{\lambda}^* = 0{\cdot}192$, implied by *R*-III. Only one estimate is available for the Blackstone learning coefficient, that provided by Zevin's equation *R*-V, shown in Table 3.

---

average yard of standard cloth, and the net weight loss in the conversion of a pound of raw cotton into cloth – obtained from the accounts of companies operating integrated mills in New England. According to these figures, we find that in 1839 outlays for all purchased materials were roughly 0·35 of total cloth production costs, with the share of cotton costs alone accounting for 0·31. The corresponding shares for 1884 – assuming fixed technical coefficients – would have been 0·41 for all purchased materials, and 0·37 for cotton taken alone. Compare these with the aggregate materials share (0·50) based on the U.S. Census of 1860.

to the firm, their case has to be grounded on the existence of irreversible scale effects such as learning by doing would generate. But if the ages of cotton textile firms, rather than their cumulated outputs, are accepted as the relevant measures of production experience from which, historically, significant cost-reducing modifications of practice were derived, the proponents of protection are pushed on to still more narrow grounds. The logic of the case presented by the preceding section would compel them to contend that the taxation of imported cotton-goods had a beneficial effect upon domestic productivity growth, if only because it encouraged the entry and subsequent survival of additional 'learning entities' in the industry – even though the tariff subsidy may have induced these firms to expand their volume of output beyond the point that would be justified by the entailed learning effects.

The last is already none too strong an argument. And an obvious line of rebuttal to it would be provided by indications that the acquisition of further experience in the industry through multiplication of the number of similarly situated firms proved less effective in reducing costs than an equivalent rate of addition in the years of experience of a small cadre of pioneer, or 'pilot' enterprises. In other words, it would be incumbent upon tariff advocates to maintain that the additional firm-experience elicited by subsidizing the entry of domestic enterprises would not turn out to be largely redundant.

Now there is ample reason to expect precisely such redundancies if much of the commercial and technical knowledge gained by pioneer enterprises were placed at the disposal of new entrants, either by the transfer of managerial and supervisory personnel, or by the ability of entrants simply to mimic observable business procedures established by the pioneers, without immediately grasping the underlying rationale. The evidence presented here in Table 3 does suggest that something close to this situation may well have prevailed in the New England textile industry after the 1820s. Notice that the reduced-form learning coefficient, estimated by Zevin for the Blackstone Manufacturing Co. using the firm's age as the measure of its experience (see equation *R*-V), is more than twice as large as the corresponding parameter estimate obtained (from either *R*-V or *R*-V^{c}) for the group of six leading textile companies. Whereas primary factor costs per unit of output – at constant prices of labor and capital services – appear to have fallen approximately 20 per cent with each doubling of the group's collective years of operating experience, they were reduced by 40 per cent with each doubling of the individual firm's operating age.[1]

[1] Recall that this interpretation can be justified on the basis of the production model represented in (13) – noting that the effect of learning on the rate of growth of the productivity of combined labor and capital inputs would, under the condition of constant returns to

Quite clearly it would take no longer to double the age of a single learning entity than to double the collective age of a simultaneously established group of six, ten, or one hundred firms, so long as the number of firms remained unchanged from the group's inception. Consequently, fiscal devices intended to promote ubiquitous acquisition of production experience instead of concentrating the learning process in one or two pilot enterprises, could achieve a faster rate of cost reduction only by expanding the membership of the learning population sufficiently rapidly to overcome the lower – and quite possibly diminishing – elasticity of efficiency with respect to experience. The latter elasticity is, of course, nothing other than what is meant here by 'the learning coefficient.' A comparison of the rates of accumulation of experience by the Blackstone Manufacturing Co. and the group of six leading firms is rather illuminating in this regard, since the former began producing power-loom cloth in 1818, three years later than the first date for which the production of cloth by a member of the group (the Boston Manufacturing Co.) is recorded.[1] In 1834–5, by which time the group's full complement of six firms were all engaged in production, its combined duration of operating experience was increasing at the annual rate of 13·6 per cent, about 2·3 times as rapidly as the contemporaneous growth rate of the same experience measure (T) for the Blackstone Co. The resulting primary factor productivity growth rate implied by the group learning coefficient ($\hat{\lambda} = 0{\cdot}327$) was thus roughly 4·5 per cent per annum, or only a shade above the rate implied by the much larger learning coefficient ($\hat{\lambda} = 0{\cdot}735$) applicable to Blackstone. Within another two years, moreover, the relationship between the group's and the Blackstone Co.'s respective rates of growth of experience (as given by the index, T, again) had been permanently reversed, and the annual rate of primary factor productivity growth due to learning for the group as a whole thereafter remained below the rate implied for the single firm.

However, before the line of argument just set forth can be fully credited as further weakening efforts to retrieve the case for the tariff protection

[1] Cf. Table 1, above, and footnote 2, p. 133; Zevin (unpublished work), p. 11.

---

scale, be given by $\lambda(\dot{T}/T) = \{\lambda^*/(1-\alpha_c^*)\}\,(\dot{T}/T)$. Alternatively, the statement would follow directly from the form of the production specified in (13′), since $\lambda' = \lambda$, where the estimates of the latter are obtained as the coefficient of $\ln(T)$ in the regressions of Table 3. For the comparison of the Blackstone learning coefficient based on T with that estimated for the group, the estimates provided by fitting the constrained form of the regression – i.e., *R*-V (rather than *R*-III) – are used, since Zevin presents only an estimate of that kind for Blackstone. Note that if $\lambda = 0{\cdot}735$, as in *R*-V: Blackstone, then the doubling of T between year 0 and year t will result in a 66·7 per cent rise in the productivity of labor and capital, or a 40 per cent fall in the level of unit primary factor costs between the two dates. See footnote 1, p. 119, for derivation of the relationship employed in making the computation.

accorded cotton textile manufacturers in the ante-bellum United States, it is necessary to remove any suspicions that the observed difference between the learning coefficients estimated for the individual firm and those for the group of (similar) firms might have simply resulted from an aggregation bias. The problem of aggregation effects in the present context can be posed as follows: Supposing a single firm, like Blackstone, was representative of each and every member of a collection of firms, would it still be possible for the coefficient of $\ln(T)$ estimated for the group to be below the coefficient estimated for the individual learning entity? Put most succinctly, the answer to this question is that while an aggregation effect of the sort described is possible, the conditions required for it to materialize were not met in the particular historical circumstances under consideration.

It is worth looking at the matter more explicitly. By making use of equation (11), the rate of growth of E_i, the level input efficiency attributable solely to learning in the i-th firm, can be reformulated for discrete time as:

$$\left(\frac{\Delta E_i}{E_i}\right)_\tau = \frac{\lambda_i}{T_i(\tau)}. \tag{17}$$

Suppose, then, that the learning coefficients of n firms comprising a (stable) group are identical, that is, $\lambda_i = \lambda$. The aggregated efficiency growth rate for this group will therefore be

$$\left(\frac{\Delta E}{E}\right)_\tau \equiv \sum_i^n (\Delta E_i/E_i)_\tau W_i = \lambda \sum_i^n W_{i\tau}/[T_i(\tau)], \tag{18}$$

where the weight W_i used in performing the aggregation represents the i-th firm's share in the output of the group at the base date $(\tau-1)$:[1]

$$W_{i\tau} = \{Y_i/\sum_i^n Y_i\}_{\tau-1}. \tag{19}$$

For convenience we may denote the ratio of the i-th firm's share of group output to its share of the collective experience of the group by θ_i; thus, dropping the time subscripts,

$$\theta_i = (W_i)/(T_i/\sum_i^n T_i). \tag{20}$$

Then, substituting for (W_i/T_i) in (18) we arrive at:

$$\left(\frac{\Delta E}{E}\right)_\tau = \lambda \sum_i^n \left\{\theta_{i\tau} \Big/ \sum_i^n T_{i\tau}\right\} = \lambda \left\{\sum_i^n \theta_{i\tau}\right\} \Big/ T, \tag{21}$$

where the aggregate 'experience' of the group is $\sum T_i = T$.

[1] To avoid unnecessary complications, this assumes there were no significant interfirm transactions among the members of the group; and further, that the output, being homogeneous, was marketed at a uniform price.

Consider now the estimated learning coefficient $\hat{\lambda}$ which one will obtain, say, from the exact-relationship between the measured productivity growth rate of the group and the growth rate of the latter's collective experience:

$$\left(\frac{\Delta E}{E}\right)_{\tau} = \hat{\lambda}\left(\frac{\Delta T}{T}\right)_{\tau}, \tag{22}$$

where the growth rate of the aggregate experience-index is simply

$$\frac{\Delta T}{T} = \left(\sum_{i}^{n} \Delta T_i\right) \Big/ \left(\sum_{i}^{n} T_i\right) = n/T. \tag{23}$$

Will the estimated aggregate coefficient $\hat{\lambda}_{\tau}$ coincide with the (uniform) learning coefficient of the group's members, λ? Substituting equation (23) into (22), and solving the resulting expression with (21), a simple relationship is found to exist between the two parameters at each point in time:

$$(\hat{\lambda}/\lambda)_{\tau} = \left(\sum_{i}^{n} \theta_{i\tau}\right) \Big/ n. \tag{24}$$

This immediately warrants the general assertion that there are indeed aggregation biases which need to be borne in mind when attempting to estimate firm learning coefficients from industry data, even under strict conditions of constant returns to scale throughout the industry. Further, equation (24) permits us to conclude that the effect of aggregation could hold the estimate of $\hat{\lambda}$ below the value of the uniform learning coefficient applicable to the underlying firms only when n exceeded the sum of $\left(\sum_{i}^{n} \theta_i\right)$.

It is not hard to see that this last condition would be fulfilled when those firms which had a disproportionately large share of the group's experience also happened to hold an even larger share of the group's aggregate product.[1] Yet, there is no general reason to expect that this would generally be the case, and it is just as readily seen that it was not in

[1] *Proposition:* For $\sum_{i}^{n} \theta_i < n$ it is sufficient that (a) $\left(\sum_{i}^{n} T_i\right) > 1$, and (b) $U > 0$, in the definitional expression

$$U\{[(T_i/\sum_i T_i)/(\sum_i T_i/n)] - 1\} = \{[(W_i)/(T_i/\sum_i T_i)] - 1\}.$$

Proof: If (a) and (b) hold, we can write

$$U(1 - \sum_i T_i) < 0, \tag{c}$$

fact true of the particular collection of firms that made up the sample group examined here. The information provided by Table 1 permits direct computation of the values of $\left(\sum_i^n \theta_{i\tau}\right)$ for two sample dates, 1839 and 1860. It is found that this sum stood at 7·85 on the earlier date, and had declined to 6·38 by the close of the period under review. Since the number of entities in the group in this case is $n = 6$, the aggregation bias alone could scarcely give rise to a finding that $\hat{\lambda} < \lambda$. Quite the contrary, the effect of aggregation by itself must have tended to *raise* the regression estimates of the learning coefficient for the group relative to the learning coefficient of the representative constituent firm.

We are thus left in the position of having to allow, perhaps, that the coefficient in the learning function of the Blackstone Co. was atypically high, in comparison with the corresponding coefficient representative of the other six leading firms we have studied. Alternatively, and rather more plausibly, it would seem there was in fact some considerable element of redundancy present in the group measure of experience, even when that is based on aggregated years of firm operations in the business of manufacturing power-loom cloth. But neither conclusion offers much support for the protective tariff as having been the appropriate instrument for effectively promoting – at minimum expense to American consumers – the acquisition of experience on the part of the firms that were best able to translate such knowledge into significant reductions of the costs of producing cotton textiles. Rather, the existence of important learning opportunities notwithstanding, the case for the retention of United States

---

and also

$$\sum T_i + U(1 - \sum_i T_i) < \sum T_i,$$

which, upon division by $\sum T_i$ and some rearrangement, becomes

$$[U + \sum_i T_i(1-U)]/(\sum_i T_i) < 1. \qquad \text{(d)}$$

We may now show, from the definition of U, that under conditions (a) and (b) the L.H.S. of (d) is identical to $\sum \theta_i/n$. First, recalling the definition of θ_i given by equation (20), and invoking condition (b) we have:

$$\theta_i = W_i/(T_i/\sum T_i) = UnT_i/(\sum T_i)^2 + 1 - U. \qquad \text{(e)}$$

Then, by summation over the index i, and division by n we find:

$$\frac{\sum_i \theta_i}{n} = \frac{1}{n}\{Un \sum_i T_i/(\sum T_i)^2 + n(1-U)\} = [U + \sum_i T_i(1-U)]/(\sum_i T_i).$$

Q.E.D.

tariffs on cotton goods after the 1820s would now seem to have to rest on imagining that a set of rather desperate circumstances prevailed. Had it been impossible – for political or institutional reasons – to arrange selective subsidies for a few pilot enterprises, the situation would have reduced to an all-or-nothing choice: one could opt either to protect all potential domestic manufacturers of cotton textiles indiscriminately, or to entirely forego command of the minimum experience needed to put the industry on an internationally competitive footing. There may indeed be instances in which such hard choices have to be faced, but at least from the present vantage point it is difficult to see why the particular historical case at hand should have been one of them.

LEARNING AND THE SOURCES OF PRODUCTIVITY GROWTH: SOME PROVISIONAL CONCLUSIONS

This article began by asking whether learning by doing was of importance to the growth of manufacturing efficiency during the early phases of American industrialization. A full reply will be some time in coming, as it must await the inception – to say nothing of the completion – of detailed, quantitative inquiries into the historical sources of productivity growth and technological advance in many different industrial pursuits. The econometric exercise undertaken here does enable us, however tentatively, to offer some explicit answers restricted to the case of the ante-bellum cotton textile industry. This is an opportunity well worth seizing, not only for its implications in the United States setting, but also in view of the cotton industry's rather special initial role in the outward spread of industrialization and the factory system from Britain during the nineteenth century.[1]

Considering the sample group of New England firms to represent the technologically leading element within the American cotton textile industry in the years (1834–60) following the era of major technical and organizational innovations, there seems little doubt that endogenous, experience-linked improvements in efficiency were a powerful force

[1] Cf., e.g., D. S. Landes, *The Unbound Prometheus* (Cambridge: Cambridge University Press, 1969), pp. 41–2, 64–6, 80–8 (on the industry in Britain), and pp. 158–69 (in Belgium, Holland, France, Germany, and Switzerland); P. I. Lyaschenko, *History of the National Economy of Russia to the 1917 Revolution* (New York: Macmillan, 1949), pp. 333–9, on the cotton textile mills of the Moscow region in period 1799–1861. On the Japanese experience, notably the subsidization of pilot enterprises by the Meiji State, cf. T. C. Smith, *Political Change and Industrial Development in Japan: Government Enterprise, 1868–1880* (Stanford: Stanford University Press, 1955), pp. 11, 26–7, 60–3; Y. Horie, 'Modern Entrepreneurship in Meiji Japan' (esp. pp. 183–6); and J. Hirshmeier, 'Shibusawa Eiichi: Industrial Pioneer' (esp. pp. 225–9), in *The State and Economic Enterprise in Japan*, W. W. Lockwood (ed.), (Princeton: Princeton University Press, 1965).

among the influences raising best-practice productivity. This is, at least, one salient feature of the story related by Table 6, which attempts to apportion the annual growth rate of cloth output per man-hour (observed for the sample group as a whole) among a number of contributory sources: increased inputs of the cooperating factors of production, exogenous improvements of the 'quality' of the cotton textile industry's labor force and its capital equipment, and the endogenously determined movement of efficiency along a long-run learning curve.

The computations underlying Table 6 run pretty much along the path first blazed by Robert Solow – a route so oft-trod nowadays that scarcely any discussion of the method is required.[1] By subtracting the weighted rates of increase of spindle-hours per man-hour, and of raw cotton per man-hour from the growth rate of cloth output per man-hour, an estimate of the rate of growth of the productivity of all these conventionally defined inputs is obtained simply as a residual: $\dot{A}/A$. The weights are, of course, supposed to represent the relevant elasticities of cloth output with respect to the non-labor inputs; but instead of following Solow in relying exclusively upon relative factor-shares to establish these values, use has been made of the information about the parameters of the constant cost Cobb–Douglas production function estimated for the group of leading firms in regression equation *R*-V^{c}.[2] Thus, fixing the sum of the elasticity coefficients at unity (in accord with the findings reported in Table 5), the elasticity of output with respect to capital service inputs has been determined from the estimated reduced-form parameter $\hat{\alpha}_s$ and an *a priori* estimate of α_c^* based upon the (0·5) industry-wide share of materials in the total cost of production, that is, *per* equation (14), $\hat{\alpha}_s^* = \hat{\alpha}_s(1-\alpha_c^*)$.

Although the resulting residual productivity growth rate and the estimated effect of the accumulation of production experience ($\lambda^*[\dot{T}/T]$) are both dependent upon the weight assigned to the inputs of raw materials in the underlying production function, notice that the relative importance of learning by doing versus 'exogenous' sources of productivity growth depends only upon the reduced-form parameter estimates ($\hat{\alpha}_s$ and $\hat{\lambda}$)

[1] The rationale for the calculation of Hicks-neutral efficiency growth is summarized in the definitions accompanying equation (2), above.

[2] Implicitly we are asking the production function (13), as estimated by *R*-V^{c} for 1834–60, to fit the observations exactly for each of the two brief periods selected in Table 6: 1833–9 and 1855–9. But, since the regression model is based on a stochastic reformulation of the production relationships, there is no reason to expect the fit to be exact. The last line of Table 6 shows the small fractions – about one-twentieth and one-tenth, respectively – of the 1833–9 and 1855–9 labor productivity growth rates which remain unaccounted for by regression equation *R*-V^{c}. This statistically unexplained portion, however, is allowed to reflect itself in the table as deriving from 'Exogenous Sources.'

TABLE 6 Sources of the growth of labor productivity in the group of leading cotton textile manufacturers

	Periods[a] 1833–9	1855–9	Basis for estimates[b]
Of the estimated annual growth rate of *cloth output per man-hour*:	0·0667	0·0320	$\dot{Y}/Y-\dot{L}/L-\psi$
we ascribe to the increase in *spindle-hours per man-hour*:	0·0074	0·0043	$\alpha_s^*\,[\dot{S}/S-\dot{L}/L]$
and to the increase in *raw cotton per man-hour*:	0·0333	0·0160	$\alpha_c^*\,[\dot{Y}/Y-\dot{L}/L-\psi]$
leaving a residual growth rate of *productivity of all inputs*:	0·0260	0·0117	$\dot{A}/A$
To this efficiency growth rate *learning by doing* contributed:	0·0202	0·0054	$\lambda^*\,[\dot{T}/T]$
while *exogenous sources* contributed:	0·0058	0·0063	$\xi^*-\{\alpha_s^*+\alpha_L^*\}\psi+\varepsilon$
which may be allocated between the *effect of a 0·010 annual rate of improvement of labor force 'quality'*:	0·0033	0·0033	$\alpha_L^*\,[0{\cdot}0100]$
and the effect of *capital-augmenting technical change at the annual rate of*:	0·0146	0·0175	$\beta_s=\dfrac{\{\alpha_s^*+\alpha_L^*\}\psi+\alpha_L^*\,(0{\cdot}01)-\varepsilon}{-\alpha_s^*}$
The portion of the total input productivity growth rate remaining *unexplained by the regression equation for 1834–60*[c]:	+0·0037	+0·0030	$\varepsilon=\dot{A}/A+\{\alpha_s^*+\alpha_L^*\}\psi-\lambda^*[\dot{T}/T]$

Notes and Sources:

[a] Growth rates computed as average compound rates between three-year averages centered on terminal years of the period, based on data in Table 8.

[b] Parameter estimates employed are those derived from Regression Model V[c] coefficients fitted to data for the six-firm sample. Growth rates of Y, L, S, and T (aggregated for all firms), are calculated from the same body of data.

Fixed parameter values: $\alpha_s^* = 0{\cdot}171$; $\alpha_L^* = 0{\cdot}329$; $\alpha_c^* = 0{\cdot}500$ (*a priori* estimate); $\lambda^* = 0{\cdot}163$.

Assumed rate of change of hours per day: $\psi = -0{\cdot}0042$ over the period 1833–9; $\psi = -0{\cdot}0067$ over the period 1855–9.

[c] Note that Regression Model V[c] imposes the constraint: $\xi^* = 0$. That means the regression model estimated for the entire period 1834–60 would 'explain' contributions to efficiency growth from exogenous sources amounting to 0·0021 in 1833–9 and 0·0033 in 1855–9, leaving +0·0037 and +0·0030 as statistical residuals for these periods, respectively. (The latter are shown in the last line of the table.)

In imputing the whole of the productivity growth rate ($\dot{A}/A$) not accounted for by learning effects to *exogenous sources*, we are implicitly including in the latter those portions of the productivity growth rate not explained by the regression; thus, 0·0021+0·0037 = 0·0058 for 1833–9. Therefore, when from the latter total we subtract our estimate of the effect of *labor quality improvement* (0·0033), we still find a positive remainder that can be attributed the *effect* of capital augmenting technical change (0·0058 − 0·0033 = 0·0025) in the period 1833–9. The entries in the next to last line of the table indicate the *annual rate of capital-augmenting technical change* that would, when weighted by the elasticity of capital inputs, have given rise to the difference between the 6th and 7th lines of the table.

provided by regression *R*-V.[1] The same holds true, perhaps more obviously so, of the comparative size of the learning effects *vis-à-vis* the contribution to the labor productivity growth rate made by the rise in 'spindles' per worker. It is to the findings emerging from Table 6, in regard to just these

[1] Making use of the notation employed in Table 6, note, first, the definitions:

$$\dot{A}/A = [\dot{Y}/Y - \dot{L}/L - \psi] - \hat{\alpha}_s(1-\alpha_c^*)[\dot{S}/S - \dot{L}/L] - \alpha_c^*[\dot{C}/C - \dot{L}/L - \psi], \quad \text{(i)}$$

and

$$\overset{*}{\lambda}[\dot{T}/T] = \hat{\lambda}(1-\alpha_c^*)[\dot{T}/T]. \quad \text{(ii)}$$

But, since we have argued for the validity of substituting $\dot{Y}/Y = \dot{C}/C$ in (i), it directly follows that the relative contribution of learning to total productivity growth is

$$\frac{\overset{*}{\lambda}[\dot{T}/T]}{\dot{A}/A} = \frac{\hat{\lambda}[\dot{T}/T]}{[\dot{Y}/Y - \dot{L}/L - \psi] - \hat{\alpha}_s[\dot{S}/S - \dot{L}/L]} \quad \text{(iii)}$$

and thus free of the influence of the value assumed for α_c^*. It may be left to the reader to prove to his own satisfaction that the following magnitudes are also quite independent of the parameter α_c^*, and therefore can be determined simply on the basis of the regression coefficients estimated in *R*-V[c]:

$$(\lambda^*[\dot{T}/T])/(\alpha_s^*[\dot{S}/S - \dot{L}/L]); \quad \text{(iv)}$$

$$(\lambda^*[\dot{T}/T])/(\alpha_L^*[\beta_L]); \quad \text{(v)}$$

$$(\lambda^*[\dot{T}/T])/(\alpha_s^*[\beta_s])' \quad \text{(vi)}$$

and

$$\beta_s \text{ itself.} \quad \text{(vii)}$$

relative magnitudes, that attention should be directed when considering the significance of endogenous efficiency improvements among the various sources of productivity advance. And on this point the evidence is strikingly clear. During the period 1833–9, learning by doing accounted for virtually eight-tenths (0·0202/0·0260) of the average annual growth rate of total productivity, and was nearly three times as important as the rise of capital intensity in maintaining the rather impressively rapid rate of increase in output per man-hour. Even at the close of the era under review, learning effects outweighed those derived from the substitution of capital for labor and remained equal in importance to the contributions made by 'exogenous' sources of productivity growth. Moreover, the secular change in the strength of these learning effects is seen from Table 6 to have exercised a dominant influence in altering the pace of overall productivity advance enjoyed by the group of leading firms: the marked retardation of the latter that occurred between the 1830s and the late 1850s is wholly attributable to the reduced rate at which effective additions were being made to the collective production experience of the group.[1]

A different aspect of this last observation presents itself in the theoretically satisfying temporal constancy of the remaining part of the productivity growth rate, which is thus attributable to secular developments largely exogenous to the manufacture of cotton goods. For the sake of completeness, Table 6 offers a somewhat arbitrary, and admittedly quite conjectural, apportionment of the 'exogenous contributions,' dividing it between the effects of improvements in average labor quality, on the one hand, and those of exogenous capital-augmenting technical changes, on the other. Because the total exogenous component of productivity growth was much the same during the latter half of the 1850s as it had been in the 1830s (0·0063 versus 0·0058 per annum, respectively), the fraction which is accounted for by positing a steady secular rate of labor quality improvement must remain essentially unchanged between 1833–9 and 1855–9.[2] From Table 6 the latter fraction is found to have been somewhat under six-tenths (0·52 to 0·57), at least if it is correct to visualize labor quality improving at something like a 1 per cent per annum trend rate.[3] High as that may seem, it would merely reflect the estimated impact of the replacement of women by male workers in Massachusetts' cotton textile labor force over the long period 1832–1900 – an effect to which direct investment in worker training may well have contributed by preventing the marginal

[1] As is easily found from the preceding footnote, the magnitude $\Delta(\lambda^*[\dot{T}/T])/\Delta(\dot{A}/A)$ = (0·0148/0·0143) is also independent of the value assigned to the parameter α_c^*, so long as the underlying production function was such that the latter remained constant over the period of the change (Δ) – from 1833–9 to 1855–9.

[2] Note that this fraction, $[\alpha_L^*\beta_L]/[\xi^*-(\alpha_s^*+\alpha_L^*)\psi+\varepsilon]$ is readily shown to be independent of the choice of an estimate for α_c^*.

[3] See Addendum to this chapter.

productivity of male textile workers from declining faster than it actually did *vis-à-vis* the marginal productivity of the female workers employed in the industry.

A further point of interest concerns what the remaining part of the contribution from exogenous sources of productivity growth implies about the pace of capital-augmenting technical change: β_s. Due to their residual derivation, the effects ascribed by Table 6 to this last source necessarily reflect the faster-than-trend rate of growth input per man-hour recorded for the sample group during the intervals 1833–9 and 1855–9.[1] It is therefore all the more striking to observe how modest are the annual rates of exogenous capital augmentation estimated for each of these sub-periods. Now a picture of utter technological stagnation in the design and operating capabilities of the industry's capital equipment is scarcely consistent with the 1·5–1·8 per cent per annum rates inferred for β_s. But on the other hand, these figures may reasonably be read as lending support for the view – widely accepted, but as a rule not quantitatively substantiated – that from the early 1830s onward the independent American textile machinery producers no longer were particularly innovative, and thus had ceased to provide significant impetus to further reductions in the cost of manufacturing cotton goods.[2]

The discussion until this point has focused on the relative importance of endogenous and exogenous developments that contributed to the growth of productive efficiency; it has thus skirted the question of the absolute size of the contribution that learning by doing made to lowering unit costs of production in the leading segment of the ante-bellum cotton textile industry. Any evaluation of the effectiveness of the protectionist policies pursued by the United States must, however, be concerned not with the former so much as with the latter issue. If taxing domestic consumers by levying duties on imported cotton goods is represented as the necessary price of securing the fruits of experience in the manufacture of such goods, the obvious thing to ask is by how much, and for how long a time, could domestic production costs continue to be reduced as a result of the lessons thereby learned.

More than usual caution is in order before the findings of Table 6 can safely be approached with such questions in mind. The absolute sizes of the estimated components of the labor productivity growth rate, unlike their comparative dimensions, are dependent upon acceptance of the

[1] Hence the rates computed for β_s exceed even the trend rates of capital augmentation which Table 4 would suggest as being consistent with regression model *R*-V (in which $\xi = 0$) under the assumption that $\beta_L = 0{\cdot}010$.

[2] For an exposition of this view of the U.S. textile machinery industry, cf. W. Paul Strassman, *Risk and Technological Innovation* (Ithaca, N.Y.: Cornell University Press, 1959), pp. 76–101.

approximate 1860 industry-wide shares of materials in total costs as an estimate of the elasticity parameter α_c^* in the underlying production function. But as we have already given grounds for suspecting that the indicated share, 0·5, is best regarded as a maximum estimate of α_c^*, it follows that Table 6 may accord rather too heavy a weight to the growth of materials inputs per man-hour, and inadequate weight to the more slowly rising capital-intensity of the production process. Hence, the bias of these computations is, if anything, toward understating the residual productivity growth rate ($\dot{A}/A$) and the absolute 'contributions' made to it – both those arising from learning effects and from the exogenous sources of growth in conventional input efficiency. Fortunately, such distortions as may be involved on this account do not appear to be serious enough to jeopardize the credibility of the general picture presented by Table 6.[1]

It appears that over the course of the boom of the 1830s best-practice cotton textile productivity was rising at an average rate in excess of 2·5 per cent per year, though probably not one exceeding 3 per cent per annum. This was certainly a respectably fast pace. Indeed, it is one that since World War II has been matched only by those comparatively young branches, like the chemical and electrical machinery industries among the United States' (two-digit SIC) manufacturing industries, whose overall efficiency has risen most rapidly.[2] And most of the increase, as

[1] To illustrate the point we may calculate the effects of a (large) 25 per cent overstatement of α_c^*. Assuming that $\alpha_c^* = 0{\cdot}4$ rather than 0·5, for 1833–9 one obtains: $\dot{A}/A = [0{\cdot}0667 - (6/5)(0{\cdot}0074) - (4/5)(0{\cdot}0333)] = 0{\cdot}0312$, instead of 0·0260, and

$$\lambda^*[\dot{T}/T] = (6/5)(0{\cdot}0202) = 0{\cdot}0242 = (0{\cdot}0202/0{\cdot}0260)(0{\cdot}0312).$$

On the same basis, for 1855–9 one obtains:

$$\dot{A}/A = [0{\cdot}0320 - (6/5)(0{\cdot}0043) - (4/5)(0{\cdot}0160)] = 0{\cdot}0140,$$

instead of 0·0117, and

$$\lambda^*[\dot{T}/T] = (6/5)[0{\cdot}0054] = 0{\cdot}0065 = (0{\cdot}0054/0{\cdot}0117)(0{\cdot}0140).$$

Note that all that is involved is a change of the levels of the growth rates: the rates of retardation of $\dot{A}/A$ and of the portion due to learning effects remain unaffected by the recomputation, as does the relationship between the total productivity growth rate and its endogenous component.

[2] This comparison is based on the findings of a forthcoming study of post-World War II U.S. manufacturing efficiency, which Moses Abramovitz and I have prepared in connexion with the Stanford S.S.R.C. Study of Economic Growth in the United States. Note that whereas the estimated productivity growth rates in Table 6 refer essentially to best-practice productivity within the cotton textiles branch, the efficiency growth rates cited for two-digit manufacturing industries pertain to average-practice productivity, and to much wider industrial segments. The comparison is favorable to the sample group of ante-bellum cotton textile companies on the latter count, but is stacked against them on the former – so long as it is possible in the short-run to raise average productivity rapidly by closing the gap between average-practice and best-practice.

we have already observed, arose from learning by doing: by raising the efficiency of the combined input of labor, capital, and materials at the average annual rate of 0·0202 established during 1833–9, learning effects might have brought about a 20 per cent cut in unit costs, *ceteris paribus*, within a span of approximately eleven years. In the event, it took longer; for in the very nature of the affair, the rate of progress along the group's learning curve could not be maintained. By the end of the period 1833–9 the estimated annual efficiency improvement due to 'learning' was only 1·3 per cent, already well below the 2 per cent average annual rate of the preceding six years; a decade after, the Compromise Tariff of 1833 found the rate cut to just under 1 per cent per year.[1] In the years immediately preceding the Civil War, as Table 6 shows, the endogenous contribution to productivity growth amounted to something only slightly above half a per cent per annum on the average. By then the rate of overall input efficiency advance recorded for the leading cotton textile firms had itself descended to a level that has been characteristic during the post-World War II era of the general run of thoroughly established manufacturing industries in the United States.

Without weighing the benefits of these implied rates of cost-reduction against the welfare burden that the ante-bellum textile tariffs imposed on American consumers, it would be improper to venture any final assessment. But on the face of the matter, it seems one would have a hard job making a case for the halting, very limited way in which lower tariffs were programmed for the 1830s under the terms of the 1833 Compromise Act – much less in defending the continuation of the duties on cotton goods in the period from 1842 to 1860. Inconclusive though it is on this point, Table 6 holds out scant encouragement for those who would argue that especially large benefits in the form of learning effects remained at stake.

Even this concedes too much; for only in the extreme, all-or-nothing situation in which public subsidization of pilot firms in the industry would prove institutionally unmanageable, if not technically ineffective, would it be appropriate to weigh the cost of the tariffs to consumers against the loss incurred by foregoing *all* experience in manufacturing cotton goods for the domestic market. Yet the burden of the econometric evidence examined in the foregoing pages is to raise serious doubts that the United States actually faced an all-or-nothing choice of that sort. At least, it suggests that the cost-reductions achieved in the cotton textile industry through learning by doing were not technically dependent upon the acquisition of widespread, repetitive experience of the kind measured by cumulated aggregate output. This conclusion in favor of the conception

[1] Cf. Appendix Table I for underlying data from which the annual rates of change in the experience-index T, and hence, the learning effects – $0{\cdot}163(\Delta T/T)$ – have been computed. The learning coefficient is naturally the same one used in Table 6.

of learning based upon the temporal duration of production experience, must be coupled with the evidence of constant returns to scale in the operation of multiple-mill firms. Together they imply that from a technical standpoint the blanket subsidies provided for all domestic entrants into cotton textile production were – from the Tariff of 1824 onward – a largely redundant set of measures; essentially the same benefits, such as they were, might have been secured under a Free Trade regime by policies which selectively subsidized the accumulation of transferable knowledge deriving from the commercial operation of pilot-learning enterprises.

Under such conditions it is hard to view United States tariff policy toward the industry as anything but a means of redistributing income in favor of the cotton textile producers – a policy for which, it should be recalled, no particular social justification can be found by taking into consideration the particular capital market imperfections that were present in the ante-bellum economy. The significance of learning by doing in the industrialization of nineteenth-century America undoubtedly deserves wider recognition and fuller investigation than it has hitherto received. And yet, in the important instance of the early cotton textile industry – harbinger of the factory system and the major branch of manufacturing developed in the United States before the Civil War – paying closer attention to the specific nature of the learning process overturns no great historical traditions, but instead effects a much-to-be-desired reconciliation of some old truths with some new. Perhaps we shall soon find ourselves once again taking up with similarly renewed convictions, Taussig's more sweeping critical stance towards the full range of the young nation's persistent departures from the righteous paths of Free Trade.

Addendum: Estimated rates of labor quality change in the cotton textile industry, 1830–60

The following tabular array brings together comparable wage rates and employment share figures for male and female cotton textile workers in 1830–2 and 1850.[1] These are the only dates in the ante-bellum era for which this complete set of price and quantity observations can be readily assembled.

TABLE 7 The structure of wages and employment in the cotton textile industry, 1830–50

	1830–2	1850
1. Daily money wages		
Males, w_m	\$1·01	\$0·76
Females, w_f	\$0·38	\$0·46
2. Female relative wage		
$\omega_f = w_f/w_m$	0·376	0·605
3. Share of employment		
Males, l_m	0·374	0·360
Females, l_f	0·626	0·640
4. Share of wage bill		
Females, $1-s_m$	0·387	0·518

Source:
(1.) S. Lebergott, 'Wage Trends, 1800–1900,' in *Trends in the American Economy in the Nineteenth Century*, N.B.E.R., Income and Wealth Conference, Vol. XXIV (Princeton, N.J.: Princeton University Press, 1960), Table 2, p. 462.
(3.) S. Lebergott, *Manpower in Economic Growth* (New York: McGraw-Hill, 1964), Table 2.7, p. 10.
(4.) Derived from lines 2 and 3, as indicated by equation (2) of the text of this Addendum.

It can be seen immediately that the evidence is at variance with the supposition that there was a unitary elasticity of substitution between males and females in cotton textile production, an impression created in the minds of some recent students of the industry by the constancy of the males' share of the cotton textile wage bill over the interval between 1832 and 1900.[2] Two coincident swallows do not make a Cobb–Douglas

[1] The material presented in the present Addendum has been drawn from David, 'Use and Abuse of Prior Information' (1972), pp. 721–4.
[2] Cf. J. G. Williamson, 'Embodiment, Disembodiment, Learning by Doing, and Returns to Scale in Nineteenth-Century Cotton Textiles,' *Journal of Economic History*, Vol. XXXII, 3 (September 1972), esp. pp. 700 *et seq.*

summer. Between 1830–2 and 1850, it turns out, there was a slight rise in the relative size of the female component of the factory work force, and a pronounced appreciation of the relative wage rate of female operatives. The result was *expansion* of the female's share of the wage bill from 0·39 to 0·52.

The apparent variation of the male–female wage shares during the ante-bellum era calls for an approach to the computation of rates of labor quality change that is different from, and rather more general than the method it was appropriate to apply in the special circumstances of the 1832–1900 comparison treated in footnote 2 on p. 123, above.

We may start with the usual definition of the average quality of labor, z, measured as the number of equivalent labor units per worker employed:

$$z = [\sum_j \omega_j L_j]/\sum_j L_j = \sum_j \omega_j l_j. \qquad (1)$$

In this expression ω_j represents the relative marginal productivity of the j-th type of worker, as compared with the m-th class of worker when the measure is understood to indicate the average labor input in equivalent m-worker units; l_j is implicitly defined as the proportion which the j-th class of workers comprise of the total labor force $L = \sum_j L_j$. Consider then the expression for the share of the total wage bill received by the j-th category of labor:

$$s_j = \frac{w_j L_j}{\sum_j w_j L_j} = \frac{(w_j/w_m)l_j}{\sum_j (w_j/w_m)l_j}. \qquad (2)$$

When the wage rate differentials are accepted as reflecting differentials in the marginal productivity of workers of different classes, that is,

$$\omega_j = w_j/w_m, \qquad \text{for all} \qquad j = 1, \ldots, m, \qquad (3)$$

we can simply substitute from equation (3) into equation (2), and solve the resulting expression with (1) to obtain a convenient general relationship between s_j, ω_j, l_j and the index of average quality

$$z = (\omega_j l_j)/(s_j). \qquad (4)$$

Differentiating the logarithmic transformation of equation (4) with respect of time provides us with an easily implemented general formula for reckoning the rate of change of the (Divisia) index

$$\overset{*}{z} = \overset{*}{\omega}_j + \overset{*}{l}_j - \overset{*}{s}_j, \qquad (5)$$

where the starred variables denote instantaneous rates of change, $(dz/dt)/z = \overset{*}{z}$, etc. It is now easy to see exactly how an unfounded supposition that $\overset{*}{s}_j = \overset{*}{s}_m = \overset{*}{s}_f = 0$ must lead the calculations astray: *were* the shares constant, it would also be true that

$$\overset{*}{w}_f - \overset{*}{w}_m = \overset{*}{\omega}_f = -(\overset{*}{l}_f - \overset{*}{l}_m), \tag{6}$$

which is the familiar condition characteristic of unitary elasticity of substitution between two inputs. Making use of this to rewrite (5) for the special case in which $\overset{*}{s}_f = 0$, we substitute $\overset{*}{\omega}_f$ from (6) for ω_j and arrive at

$$\overset{*}{z} = \overset{*}{l}_m = -0{\cdot}002,$$

essentially the erroneous estimate presented for the 1830–50 period by Jeffery Williamson.[1]

To obtain the correct answer, the male wage rate may be taken as the numeraire: setting $\omega_j = \omega_f$, the relevant average annual growth rates $\overset{*}{\omega}_f$, $\overset{*}{l}_f$, and $\overset{*}{s}_f$ are then readily computed from 1830 to 1850 from the information in the foregoing table. Upon inserting these rates in the formula provided by equation (5) it is discovered that the actual trend of *improvement* of average labor quality did not diverge from the long-term average rate found for the 1832–1900 interval:

$$\overset{*}{z} = 0{\cdot}0241 + 0{\cdot}0011 - 0{\cdot}0147 = 0{\cdot}0105.$$

We are not, alas, in a position to be so definite about the trend in the average quality of labor during the remaining ante-bellum decade, the 1850s. There was a decline in the female share in employment, from 0·640 to 0·614 between 1850 and 1860,[2] but it would be dangerous to treat this movement as tantamount to an improvement in labor quality. Indeed, there is *some* basis – albeit uncomfortably slender – for believing that the 1850s saw no net improvement whatsoever in the average quality of the cotton industry's labor force.

The only comparable wage rate figures for male and female cotton textile workers that Stanley Lebergott[3] has been able to assemble for the years 1850 and 1860 are those extracted from the original census reports of the great Merrimac Mill. The returns for that gargantuan establishment, which already employed upward of 1,400 workers in 1833, show the relative average wage of the female operatives declining from 0·576 to 0·562 over the course of the 1850s. It should be noted that the Merrimac relative wage (ω_f) for 1850 approximates, but does not agree exactly with the national relative wage estimate appearing in the table above; thus, using the national employment share figures in conjunction with the Merrimac relative wage gives rise to a slightly deviant 1850 female

[1] Cf. Williamson, 'Embodiment' (1972), p. 700.
[2] Cf. Lebergott, *Manpower*, Table 2.7, p. 70.
[3] Cf. Lebergott, 'Wage Trends' (1960), p. 465.

wage share figure $s_f = 0{\cdot}506$, which falls to $s_f = 0{\cdot}475$ by 1860. Accepting the set of average rates of change computed from these all too fragile estimates, the formula given by equation (5) tells us that if the 1850s witnessed any change in the quality index z worth mentioning, it was a deterioration and not an improvement:

$$\overset{*}{z} = \overset{*}{\omega}_f + \overset{*}{l}_f - \overset{*}{s}_f = -0{\cdot}0025 - 0{\cdot}0041 - (-0{\cdot}0063) = -0{\cdot}0003.$$

Treating this change as negligible ($\overset{*}{z} = 0$, for 1850–60), and taking it in conjunction with the more firmly grounded estimate of $\overset{*}{z}$ already presented for the preceding twenty-year period, one might conclude that the average rate of labor quality improvement during the entire interval from 1830 to 1860 lay in the range of 0·6–0·7 per cent per annum. Surely we shall be excused if we accord disproportionate weight to the firmer of the two components and so round the result up to 1 per cent. That is the heuristic figure which has been entered for the *secular* rate of exogenous labor quality improvement in the assignment of labor productivity growth among the several sources distinguished by Table 6.

TABLE 8 Variables used in estimating production functions for the sample group of cotton textile firms (BOSTON, HAMILTON, SUFFOLK, TREMONT, LAWRENCE, NASHUA)

Year	Yards per man-day Y/L	Average spindles in operation (in thousands) S	Man-days worked (in thousands) L	Cumulated output of cloth (in millions of yards) Q (end of year)	Cumulated firm-years of recorded production time T (end of year)
1834	38·59	103·5	630·0	101·255	44
1835	42·19	122·5	708·2	131·133	50
1836	40·07	131·4	852·5	165·291	56
1837	43·39	132·4	710·5	198·253	62
1838	45·94	130·4	814·8	235·683	63
1839	48·39	122·8	827·3	275·716	74
1840	45·88	123·1	842·8	314·382	80
1841	46·38	136·5	865·4	354·519	86
1842	50·72	131·7	734·6	391·780	92
1843	53·03	130·6	682·1	427·953	98
1844	56·62	133·7	691·6	467·114	104
1845	54·82	150·5	763·3	508·958	110
1846	50·47	155·3	892·5	554·000	116
1847	51·09	157·8	911·3	600·556	122
1848	59·33	166·8	868·6	652·089	128
1849	61·19	182·6	847·4	703·942	134
1850	66·00	165·0	722·4	751·620	140
1851	66·05	166·5	642·7	794·070	146
1852	66·62	200·7	792·3	846·850	152
1853	66·53	216·1	848·0	903·265	158
1854	64·80	181·2	883·5	960·508	164
1855	63·88	209·1	906·8	1,018·433	170
1856	61·14	234·3	1,003·9	1,079·811	176
1857	63·83	225·7	866·3	1,135·108	182
1858	74·82	231·8	774·5	1,193·059	188
1859	75·01	232·1	905·6	1,200·835	194
1860	60·12	228·6	1,190·3	1,332·397	200

Sources:

(Y/L), from Davis and Stettler, 'The New England Textile Industry, 1825–60: Trends and Fluctuations,' Table S, col. 1, p. 228.

(S), derived from estimates of (Y/S), *ibid.*; and (Y), obtained by aggregation of company output estimates given in Davis and Stettler, 'New England Textile Industry,' Table A.1, pp. 234–8.

(L), derived from the (Y/L) estimates by the same procedure used in obtaining S.

(Q), derived by cumulating the estimates of Y beginning with the earliest recorded output of the group's members, from Davis and Stettler, 'New England Textile Industry,' Table A.1, pp. 234–6.

(T), derived from the same source as Q.

3 The 'Horndal effect' in Lowell, 1834–56: a short-run learning curve for integrated cotton textile mills

For a period of fifteen years after the construction of the Swedish steel-works at Horndal, in 1835–6, no further investments were made in the facility. Only the bare minimum of maintenance was carried on, leaving the physical plant, and presumably the technical nature of the production process, essentially unchanged. And yet, over the course of this period output per man-hour at the mill rose at an average annual rate in the neighborhood of 2 per cent. This vivid example of pure productivity growth, deriving apparently from endogenous improvements in the performance of the work force and the operation of the steel-mill – rather than from technical progress embodied in new capital equipment – gave the name 'Horndal effect' to the phenomenon of learning by doing with fixed facilities.[1]

The pioneering empirical studies of the effects of the accumulation of production experience on unit costs, notably the early work of T. P. Wright, and the subsequent systematic studies by A. Alchian and H. Asher on the costs of airframe production, were also concerned with the fixed facilities case.[2] Whereas more recent work has generalized the idea of endogenous efficiency growth from the short-run to the long-run case of changing physical plant and equipment, where the firm rather than the specific factory becomes the logically relevant learning entity, the early studies focused on the reduction of unit labor input costs as a given plant's production run was extended.

Admittedly, the concept of a long-run learning curve is a powerful

* The research reported here was undertaken in connexion with the Stanford–S.S.R.C. Project on Economic Growth in the United States. I am grateful to Peter Temin for his comments on an earlier draft, and to the suggestions made by an anonymous referee of this journal. I retain full responsibility for any errors and deficiencies that remain.

[1] Cf. E. Lundberg, *Produktivitet öch Rantabilitet* (Stockholm: P. A. Norstedt and Söner, 1961), pp. 129–33. Lundberg's account is cited in the seminal theoretical work of K. J. Arrow, 'The Economic Implications of Learning by Doing,' *Review of Economic Studies*, XXIX, 2 (June 1962), pp. 155–73, who renders 'Järnverken' literally as 'iron works.' Lundberg, however, did not raise the question as to whether or not the temporal path of labor productivity conformed to a learning curve of the classic type.

[2] T. P. Wright, 'Factors Affecting the Cost of Airplanes,' *Journal of the Aeronautical Sciences*, Vol. 3 (1936), pp. 122–8; A. Alchian, 'Reliability of Progress Curves in Airframe Production,' *Econometrica*, XXXI, 4 (October 1963), pp. 679–92: H. Asher, *Cost Quality Relationships in the Airframe Industry* (Santa Monica: The Rand Corporation, July 1956), R-291.

one, and lends itself to empirical implementation in time series analyses of historical experience rather more readily than does the original, short-run concept – which requires us to be able to observe the inputs and output of an unchanging plant. But the long-run reformulation tends to blur some important distinctions. It would be nice to know, for example, how much of the endogenous reduction of labor and capital input requirements could have been achieved with essentially the same physical facilities. Equally, one would like to know whether over the long run the reduction of unit *capital* requirements made possible by 'learning' proceeded at the same pace, or at a rate that was slower than the rate of reduction *labor* input requirements per unit of output. Although most of the long-run studies[1] have explicitly assumed that learning by doing operated upon the production technology of the activity in a Hicks-neutral fashion, or have postulated a Cobb–Douglas function in which the difference between Hicks- and Harrod-neutrality vanishes, no test of the assumption has been contrived.

And yet, it seems that simple test is available where one is in a position to compare the magnitude of short-run learning coefficients for the same production activity. Should it be found that combined real unit costs of capital and labor declined by, say, 20 per cent when production experience (cumulated output) doubled, and that the fall in real unit labor costs accompanying the doubling of experience with *fixed* facilities was also 20 per cent, there would be a strong presumption in favor of maintaining that the long-run learning effects were Hicks neutral. Another way of looking at the same issue is to frame it in terms of a prior restriction on the acceptable magnitude of an estimated long-run learning coefficient: if the coefficient has been estimated employing the specification of Hicks neutrality, it should lie within the confidence interval that surrounds available estimates of short-run learning coefficients for the activity.[2]

These considerations make the appearance of the 'Horndal effect' rather more than a mere *curiosum*. From the study of those instances in which short-run learning effects can be isolated, we might hope to arrive not

[1] P. A. David, 'Learning By Doing and Tariff Protection: A Reconsideration of the Case of the Ante-Bellum United States Cotton Textile Industry,' *Journal of Economic History*, XXX, 3 (September 1970), pp. 521–601; R. B. Zevin, 'The Use of a "Long Run" Learning Function: With Applications to a Massachusetts Cotton Textile Firm, 1823–1860,' mimeographed for the University of Chicago Workshop in Economic History, 22 November 1968; E. Sheshinski, 'Tests of the Learning by Doing Hypothesis,' *Review of Economics and Statistics*, Vol. 49, 4 (November 1967), pp. 568–78; and L. Rapping, 'Learning and World War II Production Functions,' *Review of Economics and Statistics*, Vol. 47, 1 (February 1965), pp. 81–6. The first of these studies appears as Ch. 2, above.

[2] The argument is made more explicit in the concluding section below.

only at a fuller comprehension of the sources of historical productivity growth, but also at better understanding of the character of technical progress deriving from applying the fruits of experience to the design and operation of new production facilities.

THE LAWRENCE #2 MILL

Since such instances are likely to be known to us only rarely, and through the accidents of industrial history and historical record-keeping, so to speak, they are all the more to be treasured. The case of the #2 mill of the Lawrence cotton textile company, of Lowell, Massachusetts, is therefore worthy of more notice than it has hitherto received, inasmuch as it appears to provide an interesting instance of the 'Horndal effect' in the ante-bellum era of American industrial history. Moreover, the Lawrence #2 mill offers us a look at short-run learning effects in cotton cloth production in integrated spinning and weaving facilities, an important nineteenth-century manufacturing activity which recent research reveals benefited from *long-run* efficiency improvements derived from learning by doing.[1]

The Lawrence #2 mill was one of five company mills in operation in Lowell before the Civil War. It was built in 1834, and, according to P. F. McGouldrick who has examined the company records in the Baker collection:[2]

> Detailed inventories of machines at times of purchase and in place in 1867, show that this mill worked with just about the same stock of machines between 1835 (when it was new) and 1856. Other company records show no changes in power plant, or the mill itself, between these two dates.

Not only was there no new investment in expansion of this mill's capacity, for the annual rate of cloth production fluctuated below an essentially constant maximum level during the period from 1834 to 1857, but there appears to have been no significant machinery replacement before 1856. The following table, culled from McGouldrick's detailed survey of machine service lives, relates to the experience of the four Lawrence mills erected in the years 1834–6.

One can, in fact, be confident that the distributions around these mean retirement ages was not such as to admit the significant replacements of equipment originally installed when the #2 mill was built.[3] There were

[1] Cf. David, *op. cit.*

[2] P. F. McGouldrick, *New England Textiles in the Nineteenth Century, Profits and Investment* (Cambridge, Mass.: Harvard University Press, 1968).

[3] Had equipment retirements followed an exponential distribution as is often supposed by modern econometric and theoretical models of investment behavior, the median

no possible dates at which 1834 vintage openers could have been replaced before 1855–6. Of some 259 carding machines installed in Lawrence mills 1–4 between 1834 and 1836, 203 survived until 1867. The earliest replacements of either 'speeder' or 'stretcher' spindles were made in 1861, and *all* the Lawrence Co. installations of spinning frames dating from the years 1834–45 inclusive apparently survived through to the general inventory-taking of 1867. The company replaced its 1834–6 vintage

TABLE 9 Estimated mean retirement age of machinery newly installed by the Lawrence Co. in 1834–6

Machine type	Age at retirement
Openers and pickers	20
Carding machines	35
Drawing frames	23
Roving frames	31
Spinning frames (Throstle)	36
Warping machines	22
Dressing-process machines	38
Looms	37

Source: McGouldrick, *op. cit.*, pp. 223–30.

dressers in 1863, not before. Finally, in the important instance of weaving machinery McGouldrick's examination of information pertaining to the specifications of loom types, mill locations, and machinery orders led him to conclude that 'every single Lawrence company loom installed between 1834 and 1845 survived through 1867.'[1]

Now this must not be taken to mean that the equipment of the #2 mill was left physically undisturbed for over two decades after its installation. Quite the contrary. There was, of course, maintenance and repair work done in the mill, and it is to be expected that stronger parts were used to replace those which tended to break frequently, or quickly became badly worn; various attachments may well have been added to the 1835 vintage looms and spinning frames. Indeed, it is well known that although

[1] McGouldrick, *op. cit.*, p. 229.

would have lain *above* the mean in the distributions of retirement ages. The evidence presented by McGouldrick, *op. cit.*, pp. 223–30, however, makes it fairly clear that the exponential model is inappropriate in the case at hand, if only because for many important categories of machinery the age distribution of retirements was not continuous (or even approximately continuous) from installation onward. Instead, it was truncated at the lower end. Hence, the discussion in the text here focuses upon the temporal location of this lower truncation point, rather than on the direction and degree of the skewedness of the distribution above it.

textile machinery survived 'in place' to legendary ages, in many instances the equipment had been rebuilt in piecemeal fashion – sometimes more than once – so that by the end of its formal service life it contained scarcely a bit of metal dating from its debut on the mill floor. Many small modifications and improvements could be and doubtless were introduced in the course of this 'maintenance' work, but what is critical is that the machines as entities did not have to be scrapped and replaced in order to effect these incremental advances in operating efficiency. Further, the introduction of such modifications was intimately bound up with, and hence closely guided by, the accretions to the fund of technical knowledge gained in the actual running of the mill. This is precisely the sort of technical progress which may usefully be distinguished from innovations that *must be* 'embodied' in newly built, totally redesigned plant and equipment, and which therefore are more naturally the province of firms specializing in machinery production.[1]

Once the realities of maintenance and component repairs on machinery are admitted into the discussion, the sharp dichotomization of technical progress into 'embodied' as opposed to physically 'disembodied' innovations is seen to break down. Machines of a given vintage (installation) may be physically modified to incorporate innovations that only became available at subsequent dates, but those innovations can be said to require being 'physically embodied' nonetheless. The only sensible way to preserve the usefulness of the 'embodiment hypothesis' is to associate it exclusively with the proposition that (apart from the effects of physical wear and tear not rectified by maintenance) the operating characteristics, and costs, of machines of a given vintage remain fixed throughout their service life.

It is in this latter sense that we may say the Lawrence #2 mill provides an example of a fixed stock of capital which was subject only to disembodied technological improvement during the 22-year period in which it remained nominally unchanged. Throughout the period under consideration the specifications of the cloth (type 'C') produced by the mill were not varied, and the *peak* levels of annual production attained – in 1837–9 and 1854–5 – showed hardly any change. Thus, on all counts Lawrence #2 mill resembles the Horndal and airframe cases of fixed production with fixed facilities. And like those oft-cited cases, average real labor costs per unit exhibited an unmistakable secular decline. The unit labor cost measure in question is that developed by McGouldrick:

[1] Cf. R. M. Solow, 'Investment and Technical Change,' in *Mathematical Methods in the Social Sciences*, eds. K. J. Arrow, S. Karlin and P. Suppes (Stanford, California: Stanford University Press, 1960), pp. 89–104, for the seminal modern formulation of the view of technical progress as being 'embodied in,' and hence requiring the introduction of new *kinds* of machines.

total money labor costs – inclusive of wages of foremen, supervisors and workers in the carding, spinning, dressing and weaving divisions – per yard of cloth produced within the six-month intervals ending in February of each year, deflated by the Layer index of hourly wages of cotton textile operatives.[1]

The compound rate of increase of average labor productivity implied by the movement of the latter series (reproduced in Table 10) over the whole period from 1835 to 1856 was 1·95 per cent per annum. This quite closely approximated the 2·25 per cent per annum average rate of increase of cloth yardage per man-hour exhibited during the same period by a growing collection of integrated mills owned by a sample group of six leading New England textile firms, among which the Lawrence company was one.[2] The unit labor requirement figures for the Lawrence #2 mill do not, however, decline monotonically. Following the isolated initial observation for the six-month interval ending February 1835, which might well be discarded as pertaining to a period so close upon the plant's completion that one cannot be certain it reflected the labor requirements for normal operation of all divisions within the mill, the unit labor input series picks up on a more or less continuous basis at the end of the 1830s and displays a pronounced pattern of cyclical fluctuation which carried it to successively lower and lower troughs. These minimum unit labor requirements (peak labor productivity) levels were reached in 1839, 1846, 1851 and 1856 – as far as one can tell from the available data, from which observations relating to 1840 and 1854 are unfortunately missing.

It would be perfectly straightforward to fit a geometric trend to the unit labor requirements observations for the period 1839–56, and the rate labor productivity growth estimated on this basis would not differ greatly from the 2·15 per cent per annum average rate one may compute by considering just the 1839 and 1856 terminal points. To interpret the variations around the secular trend rate as deviations from a steady, presumably exogenous rate of technical progress would, however, be to commit a specification error. The timing of the fluctuations exhibited

[1] McGouldrick, *op. cit.*, Table 26, p. 147.

[2] L. E. Davis and H. L. Stettler, 'The New England Textile Industry, 1825–60: Trends and Fluctuations,' in *Output, Employment, and Productivity in the United States After 1800* (New York: Columbia University Press, for National Bureau of Economic Research Studies in Income and Wealth, XXX, 1966), pp. 213–38), Table 8, p. 228, for underlying data on output per worker for the group of firms, sometimes referred to as the 'Davis–Stettler sample' – which comprises a subset of the Baker sample of textile firms studied by McGouldrick. The same data are reproduced in David, *op. cit.*, Appendix Table 1, col. 1, where an estimate (0·55 per cent per year) is also provided for the 1830–60 trend rate of *decline* in the length of the average work day (*ibid.*, p. 549). The man-hours productivity growth rate cited in the text above is estimated as $0{\cdot}017+0{\cdot}0055 = 0{\cdot}0225$.

by the index of labor productivity suggest they were in large part connected with business-cycle-related changes in the intensity of utilization of fixed, and quasi-fixed inputs.[1] In other words, because we are unable to correct for the depressing effects on output per unit of labor input of varying degrees of less-than-full utilization of the fixed stock of plant and equipment, and of the 'overhead' elements of the labor force (e.g., supervisors, foremen, maintenance workers) as well, it would be misleading to accord equal weight to all the observations in drawing inferences about the influence that secular forces were exerting upon the level of labor productivity.

This problem may be skirted by focusing attention exclusively on the successive *minimum* levels recorded for real labor cost per yard of cloth, a much smaller number of observations. Direct computation of the average implied labor productivity growth rates yields the following pattern for the three peak-to-peak intervals: 1839–46, 3·37 per cent per annum; 1846–51, 2·35 per cent; 1851–6, 0·29 per cent. No matter how reasonable a geometric trend might appear to fit these four observations, we see that there was in fact a pattern of pronounced and continuous retardation.

Such retardation of the rate of growth of labor productivity, or of the rate of decline in real unit labor costs, is of course just what we should anticipate finding when the source of the changes was the *endogenous* improvement of efficiency based on the accumulation of production experience. And, as the following section shows, the classic formulation of the short-run learning function does indeed provide an excellent description of the *secular* path of full capacity utilization unit labor requirements recorded at the Lawrence Co.'s #2 mill.

A SHORT-RUN LEARNING CURVE FOR COTTON TEXTILES

The classic expression for a short-run learning curve, relating the real labor costs per unit of output (H/Y) to an index of production experience

[1] Cf. McGouldrick, *op. cit.*, Ch. 6 *passim*, pp. 213–15, and Davis and Stettler, *op. cit.*, pp. 221–5, on the relationship between fluctuations in textile production and business cycles before 1860. In 1839, 1846, 1856 the estimates of cotton textile output for New England exhibit peaks that coincide with, or fall within a year of the peaks indicated by the N.B.E.R. reference cycle chronology. The output of the Lawrence company's mills, *in toto*, shows peaks in 1839, 1846, 1853 and 1856 – but also in 1841 and 1848. Cf. *ibid.*, Table A.1, p. 234. The experience of the Lawrence #2 mill thus seems to have reflected only the stronger business cycle movements which impinged upon the industry: its (essentially constant) full-capacity level of production was attained, in 1837–9 and 1854–5, slightly in advance of the most pronounced general textile output peaks, and, except for 1851, the years of peak labor productivity coincide with episodes of high capacity utilization in the industry at large. For the application of the model presented in the following section, however, what is relevant is not the general state of the textile trade but the supposition that the four peak-labor productivity observations refer to episodes when the #2 mill's capacity was being fully utilized.

(G), takes the form:

$$H/Y = B_0 G^{-\lambda}, \qquad 0 < \lambda < 1. \tag{1}$$

In interpreting estimates of the learning coefficient λ, it is helpful to see that equation (1) may be obtained as a special case of a more general long-run production function

$$Y_t = A_0 e^{\mu^* t} G_t^{\lambda^*} K_t^{\alpha_K^*} H_t^{\alpha_L^*} C_t^{\alpha_C^*}, \tag{2}$$

which makes allowance for exogenous disembodied technical progress at the rate μ^*, as well as for total input efficiency growth due to learning by doing. In equation (2) K represents the flow of capital services, H the input of labor and C the volume of materials (essentially raw cotton) used in producing Y yards of standard-quality cloth. The raw cotton input variable, however, may be legitimately represented as a scalar transformation of the standard output measure.[1] Thus, employing the explicit transformation

$$C_t = mY_t, \tag{3}$$

and noting the following definitional identities,

$$\lambda \equiv \lambda^*/\alpha_L^*; \qquad \mu = \mu^*/l - \alpha_C^*;$$
$$\alpha_L = \alpha_L^*/l - \alpha_C^*; \qquad \alpha_K = \alpha_K^*/l - \alpha_C^*; \qquad \alpha_C = \alpha_C^*/l - \alpha_C^*; \tag{4}$$

we may rewrite the long-run production function in the form:

$$(H/Y)_t = \{A_0^{\alpha_C^* - 1} m^{-\alpha_C}\} e^{-\mu t} K_t^{-\alpha_K} H_t^{1-\alpha_L} G_t^{-\lambda \alpha_L}. \tag{5}$$

This last expression is particularly suitable as a basis for a regression model to be employed in time series analysis, since the choice of (H/Y) as the dependent variable mitigates the usual problems of serial correlation among Y, H and K. But when the relevant set of time series observations pertain to an activity which maintained essentially the same rate of output, say, $\bar{Y}$, a constant level of cloth yardage per month or year, equation (5) becomes rather unsatisfactory in that the same variable $(H/\bar{Y})$ would in effect appear both on the right-hand side and as the dependent variable in a regression equation. Because (H/Y) is likely to be subject to errors of observation, it is in any event desirable to remove it from the company of the 'explanatory' variables; that is readily accomplished in a general way by multiplying the R.H.S. of (5) by $(Y^{1-\alpha_L}\ Y^{\alpha_L - 1})$, and rewriting the resulting expression so that it reads

$$(H/Y)_t = \{A_0^{\alpha_0^* - 1} m^{-\alpha_C}\}\{Y_t\}^{1-1/\alpha_L}\{K_t\}^{-\alpha_K/\alpha_L} e^{-(\mu/\alpha_L)t} G_t^{-\lambda}. \tag{6}$$

[1] Cf. David, *op. cit.*, pp. 546–7 for a more detailed discussion and supporting evidence.

Equation (6) still expresses a general long-run production relationship with both exogenous and endogenous 'technical progress.' When, as in the present case, attention is confined to the labor inputs used in producing (a) an essentially constant volume of output, $Y_t = \bar{Y}$, which called for (b) full utilization of the original fixed physical facilities, $K_t = \bar{K}$, the first three bracketed terms on the R.H.S. of (6) can be represented by a constant, B_0. If it is then further posited that (c) exogenously generated improvements in efficiency contributed nothing to the reduction of real unit labor costs, i.e., $\mu = 0$, the long-run relationship degenerates to the pure short-run learning function given by equation (1).[1]

To secure an estimate of the learning coefficient applicable in the case of the Lawrence #2 mill, it remains for us only to settle upon a suitable index of G, the measure of production experience in the short-run learning function described by equation (1). In principle two candidates present themselves for consideration: the cumulated output of the mill, Q, and the length of the time during which the mill had been in production, T.[2]

Although the choice between these alternative specifications may carry significant policy implications, and therefore becomes a matter of some concern in discussing the nature of learning phenomena at the level of industries, or other aggregations of production units, such issues need hardly detain us in the case at hand.[3] Indeed, in the special circumstances of a fixed production facility which is being operated at full capacity, or even at a constant output rate, the distinction between the measures Q and T loses any practical significance. For, when $Y(t) = \bar{Y}$, a constant,

$$Q_T = \sum_0^T Y(t)dt = \bar{Y}\sum_0^T 1\,dt = \bar{Y}[T]. \tag{7}$$

[1] It is easily shown that the imposition of these three conditions, (a), (b) and (c), is sufficient to collapse the more general CES long-run production function with exclusively *labor-augmenting* learning effects, i.e.,

$$Y = \{a(LG^{\lambda})^{-\rho} + b(K)^{-\rho}\}^{-\delta/\rho}\, m'$$

into the short-run function given by equation (1). However, the presently available estimates of long-run learning functions for the ante-bellum U.S. cotton textile industry [David, *op. cit.*] are based on the Cobb–Douglas specification ($\rho \to 0$) described by equation (2), which renders the correspondence demonstration in the text more appropriate as an aid in comparing long-run and short-run estimates of the learning coefficient λ.

[2] Cf. David, *op. cit.*, pp. 542–5, and W. Fellner, 'Specific Interpretations of Learning by Doing,' *Journal of Economic Theory*, I, 2 (August 1969), pp. 119–40. P. 124, for discussion of these alternative specifications. The measure of experience proposed by Arrow, *op. cit.* – namely, the integral of gross investment – is obviously inappropriate here.

[3] The implications for the choice between the use of tariffs and other devices to subsidize learning in 'infant industries' are examined in David, *op. cit.*, pp. 527–39, 585–91, and are discussed further in P. A. David, 'The Use and Abuse of Prior Information in Econometric History: A Rejoinder to Professor Williamson on the Antebellum Cotton Textile Industry,' *The Journal of Economic History*, XXII, 3 (September 1972).

In the instance of the Lawrence #2 mill, however, a difference does arise between the integral of actual cloth output and the establishment's operating life: although we shall confine attention to the peak levels of labor productivity observed when the mill was producing close to its fixed capacity output rate, in the course of each of the intervening periods of less intensive utilization, cumulated actual output must have fallen progressively farther and farther below the path of cumulated potential output, $\bar{Y}[T]$. It might be possible, therefore, to discriminate between the two formulations; by comparison with T, the index Q would tend to rise in a less regular, and secularly less rapid fashion. Of course, if the minimum rate of production sufficient for 'learning' $\bar{y} \leq Y(t) \leq \bar{Y}$, even that possibility would vanish; and, in any event, having but four observations at our disposal, it would seem rather pointless to place much credence in the outcome of such a test of comparative goodness-of-fit.

Instead of seeking to decide the issue here, I have allowed myself to be influenced by my previous findings for the ante-bellum cotton textile industry, which favored the designation of T – the cumulated years of production experience – as the more generally satisfactory index of G.[1] There is some practical virtue in making this choice as well, inasmuch as the annual production series for the Lawrence #2 mill (needed to construct the measure Q) is not immediately available. It should be clear, however, that even were annual observations for Q available, we should still be restricted to making use only of those which related to the peak labor productivity dates (1839, 1846, 1851 and 1856). A meaningful test of the comparative goodness-of-fit of the alternative specifications (Q vs. T) could not be provided by fitting the learning curves to all the observations of (H/Y), for the reasons that already have been stated; the likelihood that in a purely statistical sense Q would 'perform' better that T as the explanatory variable in such a regression exercise, cannot be permitted to influence the ultimate choice between them. It is only to be expected that in the fixed facility case the rate of growth of cumulated output would register fluctuations (around a declining trend rate) which were positively correlated with cyclical movements in the utilization of the mill's capacity, and hence would exhibit an inverse correlation with the cyclical movements in the rate of decline of unit labor requirements. The specification error involved in regressing $\ln(H/Y)$ on $\ln(Q)$ using *all* (uncorrected) observations would thus tend to generate over-estimates of the elasticity coefficient λ.

For the Lawrence #2 mill the 'elapsed time' measure is constructed by reckoning the start of the experience acquisition process as having occurred in August 1834, so that by the end of the production period to which the observation for '1839' refers (the six months ending in February

[1] Cf. David, 'Learning by Doing,' pp. 57–85.

1839), the value of $T = 4{\cdot}5$ years (cf. Table 10). Despite the small number of data points, the least-squares fit of the regression model derived from equation (1) proves most satisfactory indeed:

$$\ln(H/Y)_t = \underset{(0{\cdot}1208)}{5{\cdot}1061} - \underset{(0{\cdot}0477)}{0{\cdot}2278} \ln(T)_t; \; \bar{R}^2_{df=2} = 0{\cdot}8787.$$

This equation leaves unexplained a mere 8 per cent of the variance in the peak labor productivity observations, and, even when one corrects for the 2 degrees of freedom remaining after fitting the two parameters of the learning curve, the value of $\bar{R}^2 = 0{\cdot}88$ is statistically significant at the conventional 5 per cent level. The estimated elasticity coefficient, $\hat{\lambda} = 0{\cdot}228$, implies that real unit labor costs declined by approximately 15 per cent with each doubling of the mill's operating life.[1] This fulfils such expectations as we might have formed by considering the many studies of short-run learning curves for more recent industrial processes; these have obtained estimates of the learning coefficient in the 0·2–0·5 range, typically clustered around 0·33, as in the well-known instance of airframe production. The 95 per cent confidence interval which the foregoing regression estimates enables us to construct for the cotton textile learning coefficient –

$$0{\cdot}02257 < \lambda < 0{\cdot}43307$$

– clearly places the case of the Lawrence Co. mill within this larger universe of manufacturing experience.

Surely the evidence presented here provides sufficient cause for American economic historians to insist that Horndal share with Lowell the honor – and, indeed, even yield precedence to accord with the historical priorities – in giving its name to the productivity effects of learning by doing in the context of a fixed industrial facility.

LEARNING EFFECTS IN THE SHORT RUN AND THE LONG: A COMPARISON FOR THE ANTE-BELLUM COTTON TEXTILE INDUSTRY

The short-run learning curve estimated from the data for the Lawrence #2 mill can be put to more interesting uses, however, than the launching of a campaign to establish universal reference henceforth to the 'Lowell–Horndal effect.' One set of issues on which it may shed some light concerns the specific manner in which endogenous 'technical progress' contributed to the long-run growth of productivity in the U.S. cotton textile industry before the Civil War. There were significant long-run learning effects

[1] This is computed as $1-(0{\cdot}5)^{0{\cdot}23} = 0{\cdot}15$. For derivation of the formula, cf. equation (1) and let the relative standing of the *G*-index be 2. Similarly derived arc-elasticity estimates are presented below.

during that phase of the industry's history: in an earlier study of the mills owned by six leading New England firms engaged in manufacturing low count cloth, for the years 1834–60 the long-run learning coefficient estimated was $\hat{\lambda}'_{LR} = 0{\cdot}327$, with a standard error of 0·0565. The latter coefficient indicates the elasticity of the efficiency of the primary factor (labor and capital) inputs with respect to T_{LR}, defined as the total cumulated years of *firm* experience held by the collection of enterprises that made up the sample group.[1] This group, it should be emphasized, does form a proper subject for a study of *long-run* learning effects: while the firms that comprised it remained the same, within the period of observation new mills were being added, and the machinery of some among the existing mills was rearranged.

The short-run learning parameter estimated for the Lawrence Co.'s mill is thus seen to have been rather smaller than the long-run coefficient characteristic of the leading firms in the industry during the whole period 1834–60. And even if the difference between them (0·23 vs. 0·33) were not statistically significant – although in fact it is – there would be no doubt but that the effect of a given proportional increase of the stock of experience on the level of unit real costs of capital and labor inputs combined was appreciably weaker in the fixed facility case than it was where firms were free to adjust existing plants and build new ones.

Does this mean the actual contribution that learning by doing made to the growth of labor productivity at the Lawrence #2 mill was smaller – either absolutely or in relative terms – than the contribution being made by endogenous efficiency improvements in the industry at large? To answer the question it seems desirable to attempt to arrive at a comparison of the sources of productivity growth within a common time frame, and to select for this purpose '1839–55' as an interval between occasions when both the single Lawrence mill and the larger aggregation of plants were producing at comparable high rates of capacity utilization.

Table 10 offers such a comparison of the sources of the rise in cloth output per man-hour over this sixteen-year period, by utilizing the production function parameter estimates obtained in my study of the collective experience of six leading textile firms during 1834–60.[2] But rather than taking the estimate of the long-run learning coefficient

[1] Cf. David, *op. cit.*, Regression V^c, p. 569. λ'_{LR} thus corresponds to the learning coefficient obtained from a logarithmic regression model based on equation (5), above:

$$\lambda'_{LR} = \lambda^*/(1-\alpha^*_c) = \lambda\alpha_L.$$

[2] The format of Table 10 is derived from the analysis of the sources of labor productivity growth presented by David, *op. cit.*, and the selection of 1839 and 1855 as terminal observations has been dictated largely by consideration of the suitability of this long interval for measuring the growth of the full-capacity utilization level of productivity recorded for the six-firm sample.

TABLE 10 The contribution of short-run and long-run learning, cotton textiles 1839–55

Average annual growth rates	Short-run learning: Lawrence #2 mill Version A	Short-run learning: Lawrence #2 mill Version B	Long-run learning: six-firm sample group
1. Standard yards per man-hour	0·0227		0·0237
2. Contribution of labor input	0·0116		−0·0003
3. Contribution of capital input	0	−0·0017	0·0068
4. Residual, factor productivity	0·0111	0·0128	0·0172
5. Exogenous efficiency improvement	0	0·0051	0·0051
6. Contribution of endogenous improvements to labor and capital productivity growth rate	0·0111	0·0077	0·0121
7. 'Learning effect' on labor productivity (yards per man-hour)	0·0227	0·0157	0·0247
8. 'Elapsed time' experienced-index	0·0994		0·0538
9. Labor-augmenting learning coefficient, $\lambda =$	0·2272	0·1579	0·4494
10. Primary factor-augmenting learning coefficient, $\alpha_L\lambda \equiv \lambda' =$	0·1114	0·0774	0·2204

Notes and Sources:

The underlying observations from which growth rates have been computed for the period 1839–55 are drawn from Table 11, below, in the case of the Lawrence #2 mill; and from David, *op. cit.*, Appendix Table 1, in the case of the sample group. For the former, the terminal (H/Y) observations are for the six months ending in February of 1839 and 1855, the latter being linearly interpolated from observations for 1851 and 1856. For the six-firm sample the terminal figures relate to calendar 1839 and the three-year average of observations centered on calendar 1855.

Computations were made using the following formulae:

line 1: $\dot{Y}/Y - \dot{H}/H \equiv \dot{Y}/Y - \dot{L}/L - \psi$

line 2: $-(1-\alpha_L)\,[\dot{H}/H]$

line 3: $\alpha_K[\dot{K}/K] \equiv \alpha_K[\dot{S}/S + \psi]$

line 4: $\dot{A}/A \equiv [\dot{Y}/Y - \dot{H}/H] + (1-\alpha_L)\,[\dot{H}/H] - \alpha_K[\dot{K}/K]$

line 5: $[\alpha_L\beta_L + \alpha_K\beta_K] = -(\alpha_L + \alpha_K)\psi$, since α_L and α_K were estimated under this constraint;

line 6: $\dot{D}/D \equiv \dot{A}/A - [\alpha_L\beta_L + \alpha_K\beta_K] = \alpha_L[\dot{T}/T]$.

The input elasticity parameters used are those fitted to the sample group data for 1834–60 in David, *ibid.*, *R*-V^c: $\hat{\alpha}_K = 0{\cdot}2552$ and $\hat{\alpha}_L = 0{\cdot}4904$. Note that this gives mild decreasing returns to labor and capital.

Line 7 is obtained by dividing the entries of line 6 by $\hat{\alpha}_L$. Lines 9 and 10, respectively are obtained by dividing lines 7 and 6 by $(\dot{T}/T)$ from line 8.

The growth rate of the average length of the workday $\psi = -0{\cdot}0068$, was computed for the period 1839–60 from data presented by David, *ibid.*, pp. 548–9, interpolating the average hours figure for 1839 from the available observations relating to 1830 and 1844.

provided by the regression analysis of the data, the procedure adopted in Table 10 is to employ the estimates of the capital and labor input elasticity coefficients, and of the implicit rate of exogenous productivity growth, to make an exact residual computation of the endogenous component of the rate of primary factor productivity growth for the shorter period between 1839 and 1855.[1]

As may be seen from the first line of the table, the 1839–55 trend rate of growth of labor productivity was essentially the same in the Lawrence Co. #2 mill as that experienced by the entire group of mills: 2·3–2·4 per cent per annum on the average, when one reckons labor productivity in terms of cloth yardage per man-hour. The growth of labor and capital productivity combined, however, was somewhat faster (by 0·4 to 0·6 percentage points per year) in the case of the sample firms than for the single, fixed facility (cf. Table 10, line 4). This is so even were we to undertake an adjustment for the likely trend rate of reduction of the length of the workday in the Lawrence mill, paralleling the downward adjustment of the operating spindles measure of capital input growth which had been made on the same account for the multi-mill sample. Version B in the table entertains this adjustment, which raises the calculated factor productivity growth rate for the Lawrence #2 mill – because it pushes the rate of growth of actual capital inputs below zero.

The contribution that endogenous improvements (learning) made to raising the joint productivity of the primary factors was likewise rather smaller in the case of the single mill. Line 6 indicates the residual productivity growth rate obtained after due allowance is made for the effects of exogenous sources of efficiency increase, which – in the case of the sample group – are assumed to have offset the effect on output of the secular shortening of the workday.[2] On the other hand, when it is maintained that *all* the productivity gain enjoyed in the Lawrence Co.'s mill was endogenous in origin (version A) the annual average growth rate is not

[1] As indicated in the Notes and Sources to Table 10, the H/Y figure used in making these calculations in the case of the Lawrence #2 mill in 1855 is not the datum that appears in Table 11 for that date. The latter cannot be presumed to lie on the short-run learning curve estimated for conditions of full-capacity utilization, and hence minimum H/Y. Instead, an adjusted 1855 estimate has been (linearly) interpolated from the minimum H/Y observations for 1851 and 1856. The close agreement between the computed value of λ which appears in Table 10 (line 9, version *A*) for the Lawrence #2 mill, and the regression estimate of $\hat{\lambda} = 0{\cdot}228$ reported in the text above, shows that the interpolation succeeds in providing a satisfactory estimate of the hypothetical full-capacity utilization level of H/Y in 1855.

[2] Cf. David, *ibid.*, pp. 550–6 and 594–5. In effect, referring to equation (5) above, K was replaced by S, and H by L, the difference being a common exponential trend factor due to the changing length of the workday, $e^{\psi t}$. Omitting any time trend from the regression equation in fitting the long-run version of the model suggested by equation (5) is equivalent to setting $\mu + (\alpha_L + \alpha_K)\psi = 0$.

significantly lower than that found for the multi-firm sample – after deducting the contribution of exogenous efficiency growth.

Now it is evident that for a fixed facility the effect of learning on labor productivity growth is – according to the classic version (A) – simply the recorded labor productivity growth rate itself, and this was only slightly less than the long-run effect of learning on output per man-hour in the case of the sample group. The calculations offered as Version B, however, indicate that admitting a role for exogenous sources of improvements of labor's efficiency in the Lawrence #2 mill, and assuming the same rate as that estimated for the multi-firm sample, leads to the conclusion that the contribution of learning by doing in a short-run, fixed plant setting was only two-thirds of that derived under the long-run, variable plant conditions.

But over the 1839–55 interval, the production experience of the single mill – reckoned in years of operation – was mounting at an average rate almost twice that applicable in the case of the group of firms comprising the sample. Hence, in line 9 of the table it is found that the learning coefficient λ was actually almost twice as large in the long-run as in the short-run situation.

The latter coefficient indicates the elasticity of unit real labor input costs with respect to experience, and it is to be noted that for the Lawrence #2 mill the value of λ ($= 0{\cdot}227$) computed in Table 10 from the 1839 and 1855 extremes is essentially identical to the estimate previously secured by least-squares regression. As the final line of the table shows, if we suppose the same labor input elasticity parameter (α_L) is appropriate to the single mill as to the multi-mill production function, the learning coefficient (λ') which gauges the effect on combined real unit costs of labor and capital is found, like the coefficient relating to the behavior of unit real labor costs alone, to have been greater under long-run conditions. Thus, in the period 1839–55 a doubling of production experience, T, would result in the reduction of real primary factor costs per yard by 7·5 per cent in the short run, but by 15 per cent when adjustments could be made in the (economic) long run.

From the foregoing comparison of findings pertaining to long- and short-run learning effects in the manufacture of low count cotton cloth, during an identical temporal interval, it appears that freedom to adjust the physical plant did enhance the benefits which these early American manufacturing enterprises derived from the acquisition of production experience. Putting the point slightly differently, some of the modifications made between successive *vintages* of textile plant in the ante-bellum era most probably stemmed from endogenously generated technical advances.

Were improvements of this sort purely capital-augmenting, as embodied technical progress is widely presumed to be? As a factual answer to this question can hardly be supplied on the basis of the evidence presently available, perhaps the best course is to follow conventional, and theoretically convenient presumptions, i.e., to assume that the exclusive fruits of long-run learning, which required harvesting through the installation of new plant and equipment, had the effect of augmenting capital and not labor.[1] Under this assumption it becomes natural to try to see what comparison of long- and short-run learning coefficients may imply about the factor-augmenting bias of *endogenous* technical progress.

Inasmuch as the estimates considered here are derived on the assumption that the production function was of the Cobb–Douglas form, we are patently in no position to *test* whether over the long run the effect of learning was relatively capital-augmenting, relatively labor-augmenting, or on balance Hicks-neutral.[2] It is possible, nonetheless, to read the estimates of the learning coefficients in a fashion that does tell us something on this score, if we are willing to accept the initial formulation – in equation (1) – of the true 'Lowell–Horndal effect' as being purely labor-augmenting. Proceeding on this premise, the learning coefficient estimated for the Lawrence #2 mill would reflect the elasticity of labor efficiency with respect to experience, whereas the long-run coefficient obtained for the sample group of mills could be treated as representing the weighted sum of the elasticities of labor and capital efficiency with respect to experience:

$$ {}_{SR}\lambda' = \lambda_L \alpha_L \quad \text{and} \quad {}_{LR}\lambda' = \lambda_K \alpha_K + \lambda_L \alpha_L. $$

Taking the labor and capital input elasticity estimates used in the construction of Table 10, and employing the short-run and long-run learning coefficients derived for 1839–55 in the same place, the effect of the imputed capital-augmenting portion of the long-run learning effects can readily be determined: it is simply the difference between the long-run and the short-run coefficients appearing in the last line of the table, i.e., $\lambda_K \alpha_K = {}_{LR}\lambda' - {}_{SR}\lambda' = 0{\cdot}109$. From this, making use of the relative

[1] If embodied technical progress is not purely capital-augmenting it cannot be wholly represented as an improvement in the quality of 'machines,' and equipment of different vintage cannot be aggregated into a single measure of 'equivalent capital,' *J*, the so-called 'Solovian jelly.' On this point cf. F. M. Fisher, 'Embodied Technical Change and the Existence of an Aggregate Capital Stock,' *Review of Economic Studies*, Vol. 32 (October 1965), pp. 263–88.

[2] On the equivalence of the factor-augmenting classification of technical change with the Hicksian classification, which holds for production functions in which the elasticity of factor substitution is less than unity, cf. e.g., P. A. David and Th. van de Klundert, 'Biased Efficiency Growth and Capital–Labor Substitution in the U.S.,' *American Economic Review*, Vol. 55, 3 (June 1965), pp. 362–3.

elasticity estimate for the capital inputs, $\hat{\alpha}_K = 0{\cdot}255$, the implied capital-augmenting learning coefficient is found to be

$$\hat{\lambda}_K = ({}_{LR}\lambda' - \alpha_L \lambda_L)/\hat{\alpha}_K = \frac{0{\cdot}109}{0{\cdot}255} = 0{\cdot}427.$$

As $\hat{\lambda}_K > \hat{\lambda}_L = 0{\cdot}227$, the inference to be drawn from this exercise is clear: the long-run effects of learning by doing tended to be biased toward capital augmentation in the case of the ante-bellum cotton textile industry. But along with this 'finding' one should remember that there really is no empirical warrant at present for supposing that the labor-augmenting effects of learning by doing in the long run were identical in magnitude to the endogenously derived labor productivity gains recorded at the

TABLE 11 Labor inputs per yard of standard 'C' cloth produced in #2 mill, Lawrence Co., 1835–56

Period six months ending in February	Index of unit labor inputs, (H/Y) 1844–6 = 100	Cumulated years of production experience (T), at end of period
1835	127·2	0·5
1836		1·5
1837		2·5
1838		3·5
1839*	119·4	4·5
1840		5·5
1841	129·4	6·5
1842	123·1	7·5
1843	110·3	8·5
1844	104·4	9·5
1845	100·9	10·5
1846*	94·7	11·5
1847	104·4	12·5
1848	119·2	13·5
1849	93·7	14·5
1850	90·5	15·5
1851*	84·4	16·5
1852	109·9	17·5
1853	88·7	18·5
1854		19·5
1855	87·0	20·5
1856*	83·1	21·5

Source: McGouldrick, *op. cit.*, Table 26, p. 147, for computation of the index of (H/Y) from mill records.

Lawrence #2 mill. Consequently it would be merely an act of faith for us to insist that the observed differences between the long-run and short-run real cost reductions attributable to learning is assignable wholly to the added effects of capital-augmenting technical improvements embodied in new investment. One suspects that only a patient, detailed study of the technical information preserved in textile company records will make it possible to offer a more satisfying, definitive response to the questions posed here regarding the nature and bias of the efficiency improvements to which the acquisition of production experience gave rise.

Diffusion

4 The mechanization of reaping in the ante-bellum Midwest

I

The widespread adoption of reaping machines by Midwestern farmers during the years immediately preceding the Civil War provides a striking instance of the way that the United States' nineteenth-century industrial development was bound up with *concurrent* transformations occurring in the country's agricultural sector. On the record of historical experience, as Alexander Gerschenkron has cogently observed, 'the hope that industry in a very backward country can unfold from its agriculture is hardly realistic.'[1] Indeed, even when one considers countries that are not very backward it is unusual for agricultural activities to escape an uncomplimentary evaluation of their efficacy in creating inducements for the growth and continuing proliferation of industrial pursuits. As Albert Hirschman puts it, 'agriculture certainly stands convicted on the count of its lack of direct stimulus to the setting up of new activities through linkage effects: the superiority of manufacturing in this respect is crushing.'[2] But having conceded that much regarding the general state of the world, the student of economic development in nineteenth-century America is compelled to stress the anomalous character of his subject, to insist that in a resource-abundant setting, highly market-oriented, vigorously expanding, and technologically innovative agriculture did provide crucial support for the process of industrialization.

Such support in the form of sufficiently large demands for manufactures and supplies of raw material suitable for industrial processing would,

* I wish to acknowledge my gratitude to Peter Temin for stimulating criticism and helpful suggestions offered when this paper was first being drafted. The present version has benefited from the comments of my colleagues in the Economics Department, the members of the Graduate Seminar in Economic History at Stanford University, 1964–65, and many participants in the Purdue Conference on Quantitative and Theoretical Research Methods in Economic History (4–6 February 1965). My debts on this account are so numerous that those who hold them must perforce remain anonymous. Errors or deficiencies that have survived all this counsel are assuredly mine alone.

[1] A. Gerschenkron, *Economic Backwardness in Historical Perspective* (Cambridge, Mass.: Harvard University Press, 1962), p. 215.

[2] A. O. Hirschman, *The Strategy of Economic Development* (New Haven, Conn.: Yale University Press, 1958), pp. 109–10. On the now familiar concepts of 'forward' and 'backward' linkages between a sector of the economy (or an industry) and other sectors (or industries) that buy its output and supply it with inputs, respectively, see *ibid.*, Ch. 6, *passim.*

undoubtedly, have been less readily forthcoming from a small, or economically backward agrarian community. It is precisely in this regard that United States industrialization may be seen as having diverged most markedly from the historical experience of continental European countries, where backward agriculture militated against gradual industrial growth, and the successful pattern of modernization of the economy tended to be characterized by an initial disengagement of manufacturing from the agrarian environment.[1]

However, to treat the generation of demand for manufactures during the process of industrialization as taking place within a framework of static, pre-existing intersectoral relations, summarized by a set of input–output coefficients, does not prove to be an entirely satisfactory way of looking at the connexions between the character of agriculture and the growth of industrial activities in the United States. Adherence to such an approach leads one, *inter alia*, to gloss over the problems of accounting for alterations in the structure of intersectoral dependences, although those alterations often constitute a vital aspect of the process of industrialization. It is not wholly surprising that pursuit of a static 'linkage' approach has tended to promote the misleading notion that the expansion and modernization of the agrarian sector constituted a temporal precondition for rapid industrial development in the United States,[2] whereas in many crucial respects it is far more useful to regard the two processes as having gone hand in hand. As a small contribution to the study of the interrelationship between agricultural development and industrialization in the American setting, this essay ventures to inquire into the way that – with the adoption of mechanized reaping – an important element was added to the set of linkages joining these two sectors of the mid-nineteenth-century economy.

II

The spread of manufacturing from the eastern seaboard into the transmontane region of the United States during the 1850s derived significant impetus from the rise of a new demand for farm equipment in the states of the Old Northwest Territory. That impetus was at least partially reflected by the important position which activities supplying agricultural investment goods came to occupy in the early structure of Midwestern industry. In the still predominantly agrarian American economy of the time it is not unexpected that a substantial segment of the total income generated by industrial activities was directly attributable to the manufacture of durable producers' goods specifically identified with the farmer's

[1] See Gerschenkron, *op. cit.*, pp. 215, 354, 107–8, 125–6.

[2] See, e.g., W. W. Rostow, *The Stages of Economic Growth* (Cambridge: Cambridge University Press, 1960), pp. 17–18, 25–6.

needs – leaving aside the lumber and related building materials flowing into construction of farm dwellings, barns, sheds and fences. If, in addition to value added in the production of agricultural implements and machinery in 1859–60, one were also to include half the value added by the manufacture of wagons and carts, saddles and harnesses, and the variety of items turned out by blacksmiths' shops, the resulting aggregate would represent over 4 per cent of the value added by the nation's entire industrial sector. That is, rather more than the proportion contributed by the manufacture of machine shop and foundry products, which at the date in question ranked as the country's seventh largest industry in terms of current value added.[1] However, on the eve of the Civil War the production of agricultural implements and machinery *alone* generated just as large a proportion of total industrial value added in the preponderantly agrarian Western states; in Illinois, this single branch of manufacturing accounted for fully 8 per cent of the total value added by the state's industries in 1859–60.[2]

To appreciate the importance of the position that the agricultural implements and machinery industry assumed in the structure of Illinois' early manufacturing sector, it must be realized that at the time there was no single branch of industry which in the nation as a whole contributed so large a portion of aggregate value added in manufacturing. Cotton goods production, America's largest industry in 1859–60, accounted for only 6·6 per cent of the national aggregate.

When one looks at a rapidly developing center of industrial activity in the Midwest such as was Chicago during the 1850s, the manufacture of agricultural implements and machinery is found to have had still greater relative importance as a generator of income. The growth of agricultural commodity-processing industries, especially meat-packing in Chicago during the latter half of the century, suggests that the Garden City's meteoric rise to the status of second manufacturing center in the nation by 1890 might be taken as a demonstration of the strength of *forward* linkages from commercial agriculture. It is not an object of the present essay to assess the validity of that impression. Nevertheless, it should be remarked that during Chicago's first major spurt of industrial development, a movement which saw manufacturing employment in the city rise from less than 2,000 in 1850 to approximately 10,600 by 1856, the forward-linked processing industries were less significant to the industrial life of the city than was an activity based on *backward* linkage from agriculture. The branch of manufacturing in question was the farm implements and

[1] See *U.S. Eighth Census* (1860), 'Manufactures,' pp. 733–42.

[2] See *ibid.*, pp. ccxcii, 725, 729. The 'Western' states here are: Ohio, Indiana, Michigan, Illinois, Wisconsin, Minnesota, Iowa, Missouri and Kansas. The share of the agricultural implements industry in aggregate value added by U.S. manufacturing was 1·4 per cent, according to the Census of 1860.

machinery industry: in 1856 it accounted for 10·8 per cent of total value added by Chicago's industrial sector, compared with 6·3 per cent contributed by the principal processing industries, meat-packing, flour- and grist-milling, and distilling combined.[1]

Among the salient characteristics of the agricultural scene in the ante-bellum Midwest, two appear as having been crucial to the emergence during the 1850s of a substantial regional manufacturing sector bound by demand-links reaching backward from commercial agriculture. First, the settlement of the region and the extension of its agricultural capacity during that decade proceeded with great rapidity, encouraged by favorable terms of trade and improvements in transportation facilities providing interior farmers with access to distant markets in the deficit foodstuff areas to the east. Between the Seventh and the Eighth Censuses of Agriculture over a quarter of a million farming units came into existence, and about 19 million acres of improved farm land were added in Illinois, Indiana, Michigan, Iowa and Wisconsin. This represented a rate of increase in the number of farms of 7 per cent per annum, and a 9 per cent annual rate of expansion in improved acreage.[2]

Secondly, agricultural practice in this region of recent settlement was not the static crystallization of long experience typical of stable agrarian societies. Far from being a closed issue, choices among alternative production techniques were rapidly being altered and Western farming was thereby being carried in the direction of greater capital-intensity and higher labor productivity. On the eve of the Civil War this burgeoning farm community was in the midst of a hectic process of transition from hand methods to machine methods of production, from the use of rudimentary implements to reliance on increasingly sophisticated machinery. Among the items of farm equipment being introduced on a large scale in the Midwest during the 1850s were steel breakers and plows, seed drills and seed boxes, reapers and mowers, threshers and grain separators.[3]

[1] See P. A. David, *Factories At the Prairies' Edge: A Study of Industrialization in Chicago 1848–1893* (Manuscript), Appendix C, Table C.2, for annual estimates of manufacturing employment in Chicago; Appendix A.III and Ch. 3 for industrial value added estimates cited. Estimates of value added in Chicago industries are based on local census statistics for gross value of product (in 1856) and the ratios of value added to gross product in the corresponding industries reported by the U.S. Eighth Census (1860) for Cook County, Illinois. In 1859–60, according to the latter source, meat-packing, milling, and distilling together accounted for 8·7 per cent of manufacturing value added in Cook Co., compared with 7·9 per cent contributed by the agricultural implements and machinery industry.

[2] See *U.S. Eighth Census* (1860), 'Agriculture,' p. 222.

[3] See Leo Rogin, *The Introduction of Farm Machinery in Its Relation to the Productivity of Labor in the Agriculture of the United States during the Nineteenth Century* (Berkeley, Calif.: University of California Press, 1931), Publications in Economics, Vol. IX, esp. pp. 33–4, 47, 72–80, 165–6, 196, 201.

An editorial pronouncement appearing in the *Scientific American* during 1857 suggests the extremes to which the mechanization of farming had proceeded:

every farmer who has a hundred acres of land should have at least the following: a combined reaper and mower, a horse rake, a seed planter, and mower: a thresher and grain cleaner, a portable grist mill, a corn-sheller, a horse power, three harrows, a roller, two cultivators and three plows.[1]

The importance that the newly introduced reaping and mowing machines (especially the former) had assumed among the products of the agricultural implements and machinery industry of the Midwest by the end of the 1850s provides some indication of the direct impact of the shift to more capital-intensive farming techniques upon the expansion of an agrarian market for industrial products.[2] According to the Census of 1860, reapers and mowers accounted for 42 per cent of the gross value of output of all agricultural implements and machinery in Illinois and for 78 per cent of the gross value of output of the corresponding industrial group in Chicago. A few years earlier, in 1856, when the Midwestern boom was still in full swing, reaper and mower production in Chicago had dominated that center's farm equipment output-mix to an even greater degree.[3]

Despite the fact that the history of commercial production of mechanical reaping machines in the United States stretched back without interruption to the early 1830s, this industry was one that only began to flourish in the 1850s. From 1833, the date of the first sale of Obed Hussey's reaping machine, to the closing year of that decade, a total of 45 such

[1] Quoted in C. Danhof, 'Agriculture,' in H. F. Williamson (ed.), *The Growth of the American Economy*, Second Edition, New York, N.Y.: Prentice-Hall, Inc., 1951), p. 150.

[2] The term 'direct impact' is used here with two considerations in mind. First, this neglects the indirect (input–output) effects of expanded reaper and mower production on the production of intermediate inputs used by the industry, e.g., pig iron, bar iron, malleable castings, sheet zinc, leather, oils, paint, turpentine, physical input coefficients for each of which are available. (See David, *op. cit.*, Ch. 3.) Secondly, we here neglect the favorable indirect impact on the growth of agricultural demand in general, arising from the fact that substitution of machinery for labor on the farms raised labor productivity and facilitated faster expansion of agriculture during this, and subsequent periods. The latter point is discussed further below.

[3] See *U.S. Eighth Census* (1860), 'Manufactures,' Tables 3, pp. 11, 86. The enumeration of establishment output given in the (Chicago) *Daily Democratic Press*, 'Review of 1856,' shows that 5,860 reapers and mowers contributed 87 per cent of the gross value of all agricultural implements and machinery produced in Chicago in that year. (Separate mowers, in contrast to reapers and combined reaper-mowers, accounted for less than 32 per cent of the value of reaper and mower production in Chicago in 1856.) The balance of Chicago's production for 1856 consisted of a miscellany of 541 threshers, 200 separators and horse powers, 1,000 plows, and an unknown number of corn-shellers and cob-crushers.

machines had been purchased by American farmers. At the end of the 1846 harvest season Cyrus H. McCormick determined to abandon his efforts of the previous six years at manufacturing his reaping machine on the family farm in Rockbridge County, Virginia, and set about transferring the center of his activities to a more promising location, Chicago. The known previous sales of all reaping machines at that time aggregated to a mere 793, but by 1850 some 3,373 reapers in all had been produced and marketed in the United States since 1833. A scant eight years later it was reckoned that roughly 73,200 reapers had been sold since 1845, fully 69,700 of them since 1850. And most of that increase had resulted from the burst of production enjoyed by the industry during the five years following 1853![1]

The major portion of this production had taken place in the interior of the country, and it is apparent that in the absence of farmers' readiness to substitute machinery for labor during the 1850s, an equally rapid pace of agricultural expansion – had such in fact been feasible – would have provided a considerably weaker set of demand stimuli for concurrent industrial development in the region. The latter facet of the late ante-bellum agrarian scene must, therefore, be the prime focus of our interest; it cannot be taken as a given, but must be explained. That should not, however, be regarded as a dismissal of the first-mentioned aspect of Midwestern agricultural development in this period. As shall be seen when we come to grips with the problem of explaining the movement of mechanization, the speed of agricultural expansion and the substitution of machines for farm labor were intimately connected developments between which causal influences flowed in both directions.

III

In view of the consequences for agricultural and industrial development that followed from the mechanization of reaping during the 1850s, it

[1] See Rogin, *op. cit.*, pp. 72–8, for the record of reaper production before 1860. The figures given above are cumulated from yearly sales data, save for the estimate of 73,200 reapers sold between 1845 and 1858. The latter figure can be traced to pro testimony in the litigation over extension of the McCormick Patent of 1845, not the original 1834 Patent. At that time it was asserted to represent the number of machines sold that had made use of principles patented by McCormick; the claim was sweepingly inclusive, as it covered the 23,000 machines sold by McCormick (directly and under license) and all other machines sold since 1845. See William T. Hutchinson, *Cyrus Hall McCormick*, Vol. 1: *Seed-Time, 1809–1856* (New York, N.Y.: The Century Company, 1930), p. 470. Leo Rogin (*op. cit.*, pp. 78–9) erroneously accepts this figure as an estimate of the stock of reaping machines in operation on farms west of the Alleghenies in 1858, evidently following a misinterpretation perpetrated by *Country Gentleman*, Vol. 13 (1859), pp. 259–60, the proximate source cited by Rogin for the number in question.

might be supposed that this episode in the modernization of American farming and the formation of backward linkages between the enterprises of field and factory would have been thoroughly explored by economic historians. To be sure, virtually all the standard accounts of the development of agriculture in the United States up to 1860 mention the introduction of the machines that Obed Hussey and Cyrus H. McCormick had invented in the 1830s. Yet, the literature remains surprisingly vague about the specific technical and economic considerations touching the adoption of these devices by American farmers. We have called attention to the fact that although the twenty years prior to 1853 had witnessed a slow, limited diffusion of the new technique, the first major wave of popular acceptance of the reaper was concentrated in the mid-1850s. Thus, the intriguing question to which an answer must be given is: why only at that time were large numbers of farmers suddenly led to abandon an old, labor-intensive method of cutting their grain, and to switch to the use of a machine that had been available since its invention two decades earlier?

In this inquiry, the impact of the mechanization of small grain harvesting upon U.S. agriculture is not the prime subject of concern.[1] Nevertheless, it would hardly be possible to account for the upsurge of demand for reaping machinery without considering the economic implications of the new harvesting technology and the specific circumstances surrounding its introduction. The traditional story of the ante-bellum adoption of mechanical reaping, a version to be found in any number of places,[2] is set out along the following lines.

During the first half of the nineteenth century arable land was abundant in the United States, but the amount of small grains (especially the amount of wheat) that an individual farmer could raise was limited by the acreage that could be harvested soon after the ripening of the crop. Labor was scarce, and harvest labor notably dear as well as unreliable in supply. Compared with the method of harvesting using the grain cradle – an improvement on the sickle that had come into quite general use even in

[1] See William N. Parker and J. L. V. Klein, 'Productivity Growth in Grain Production,' in *Output, Employment and Productivity in the United States after 1800* (New York: Columbia University Press for the N.B.E.R., 1966), for a recent quantitative study which attributes to mechanization the major part of the increase of labor productivity in U.S. small grain production during the nineteenth century. Wheat, oats, and rye are the small grain crops considered in the present paper.

[2] Monographic and textbook treatments of the subject include, e.g., Percy W. Bidwell and John I. Falconer, *History of Agriculture in the Northern United States* (Publication No. 358), (Washington, D.C.: Carnegie Institute of Washington, 1925), pp. 281–94; Clarence Danhof's essay, 'Agriculture,' Ch. 8, in Harold F. Williamson (ed.), *op. cit.*, pp. 144–6; Paul W. Gates, *The Farmer's Age: Agriculture 1815–1860* (*The Economic History of the United States*, Vol. III) (Holt, Rinehart, Winston, 1960), pp. 258–88; Ross M. Robertson, *History of the American Economy*, Second Edition (New York: Harcourt, Brace, and World, Inc., 1964), pp. 257–58.

the transmontane wheat regions by the middle of the century – the new mechanical reapers effected a saving in labor. When Midwestern farmers were led to increase production as a result of the rise in wheat prices during the 1850s (a rise augmented by the impact of the Crimean War upon world grain markets), the demand for reaping machines rose, and their adoption went forward at an accelerated pace. The movement thus initiated received renewed impetus from the extreme shortage of agricultural manpower occasioned by the Civil War. By saving labor, and therein relaxing the constraint on cultivated acreage imposed by hand methods of harvesting, the introduction of the reaper made possible the rapid expansion of small grain production that occurred during the latter half of the nineteenth century.

This account may vary in some details from any particular historian's version of the events in question, but it contains all the generally accepted elements of the story. It specifically follows the historiographic tradition of ascribing to the rise in wheat prices during the 1850s a causal role in bringing about the transition from cradling to mechanical reaping prior to the outbreak of the Civil War.[1] Upon a moment's reflection, the latter

[1] Although it is generally accepted that the Civil War provided an important stimulus to the widespread use of agricultural machinery in Northern agriculture, the extensive use made of reapers and mowers in the West during the 1850s is now regarded as a well-established point. See Leo Rogin, *op. cit.*, p. 79; and Emerson D. Fite, *Social and Industrial Conditions in the North during the Civil War* (New York: Macmillan Company, 1910), p. 7. Rogin estimates that about 70 per cent of the wheat harvested west of the Alleghenies was cut by mechanical reapers on the eve of the War. There are grounds for questioning the validity of that figure, although they are not such as to lead us to doubt that the reaper had won general acceptance, if not universal adoption, in Midwestern agriculture by 1860.

Rogin's estimate is based on his acceptance of a figure (73,200), giving the number of reaping machines sold in the U.S. between 1845 and 1858 as an estimate of *the stock of reapers in operation on Western farms* in 1859. Since the average life of a reaper was roughly ten years (see Addendum, section 2b), a calculation of the stock of machines net of replacements (but gross of depreciation, i.e., the gross stock, assuming reapers simply fell apart after ten years in the manner of the 'one horse shay') would require disregarding at least the sales of machines prior to 1848. This would merely lower the stock estimate from 73,200 to approximately 72,300. (See annual sales data given by Rogin, *op. cit.*, pp. 72–8.) However, if one allows for continuous straight-line decay of the machine at the rate of 0·1 per annum, and takes the time pattern of known sales of machines by the McCormick Co. during 1848–58 (from data in William T. Hutchinson, *op. cit.*, p. 369, n. 60) as a representation of the time distribution of all reaper sales, it is found that the net stock figure for 1858 should be only 56 per cent of the cumulated number of machines sold during 1848–58, or the equivalent of 43,400 full-capacity machines. This probably understates the true net stock, since the time distribution of McCormick's production was less skewed toward the latter half of the period of 1848–58 than was the time distribution of aggregate reaper production in the country: 50,000 full-capacity machines might be accepted as not too high a figure for the net reaper stock in 1858.

But, it is necessary to compare both the net and the gross stock estimates with the

is seen to be the analytically unexpected aspect of this tale of a change in technology; it is far more usual for discussions of the choice of technique to be couched, implicitly or explicitly, in terms of the relative prices of the substitutable factors of production (grain cradlers and reaping machines in this instance) and to say nothing about the price of the commodity being produced.

Yet, precisely why this departure from the classical (or, properly speaking, neoclassical) treatment of the choice between labor-intensive and capital-intensive factor proportions is called for in the case of the adoption of the mechanical reaper, is not revealed by the statement. That it remains rather ambiguous about the lines of causation linking dear labor, high grain prices, expanded production acreage, and the spreading use of the reaper must, with all diffidence, be attributed to the ambiguities of the literature from which the statement itself has been constructed.

Vide Bidwell and Falconer's classic work on Northern agriculture prior to the Civil War:

During the early fifties the reaper was gradually supplanting the cradle in the wheat fields of the country, but as yet the acreage in grain in the Western States was largely limited to the capacity of the cradle.... Moreover, in a large part of the West there was little incentive to produce large amounts of wheat on account of the lack of markets and low prices. Rising prices of wheat caused a 'boom' in agriculture from 1854 to 1857 and caused almost universal demand for reapers in the wheat-growing regions.[1]

P. W. Gates's recent study of ante-bellum agriculture proceeds in much the same general vein:

With wheat well above the dollar mark from 1853 to 1858, Illinois, Wisconsin, Iowa, and Minnesota farmers enjoyed real prosperity and were in a position to buy and pay for reapers.... Since the amount of wheat a man could sow was limited by the

[1] Bidwell and Falconer, *op. cit.*, p. 293.

national wheat harvest, rather than with the Western wheat harvest as Rogin does. Taking the average national wheat yield per acre as 11·4 bushels in 1859 (a figure computed by weighting regional yield estimates given in Table 5 of Parker, *op. cit.*), and following Rogin's procedure of comparing the wheat acreage harvested, as estimated from the Census of 1860 output figure and the yield per acre, with that acreage which could be cut by the stock of reapers if each machine were worked at the normal seasonal capacity rate of 100 acres, one obtains the following estimates as alternatives to that given by Rogin. At a maximum, if we accept the net stock figure, 48 per cent of the 15 million acres of wheat harvested in the U.S. in 1859 was cut by machines; at a minimum, if we accept the net stock figure, 33 per cent was thus harvested. Rogin's figure of 70 per cent seems somewhat exaggerated for the West, since even were the entire stock of reapers (implausibly) thought to have been located west of the Alleghenies, the above revisions of the reaper stock figure would suggest that the proportion of wheat acreage in the West cut by horse power at the end of the 1850s lay somewhere between 70 per cent and 50 per cent.

short period in which it had to be harvested and by the man-days of labor required to cut it, it can readily be seen how much the reaper expanded the possibilities of wheat growing.[1]

Comparable passages of other contributions to American economic history could be examined without finding any clear views as to whether it became profitable for farmers to adopt the mechanical reaper only when they found it profitable to increase the acreage of wheat sown per farm, or whether it was the expansion of grain cultivation in Midwestern agriculture as a whole that led to a general substitution of machinery for labor in harvesting operations. The literature is no less ambiguous in the answers it offers to two closely related questions. Did the adoption of mechanical reapers make rapid expansion of grain cultivation in the Midwest possible by raising the scale on which it could be profitably grown (and harvested) by individual farming units? Or, was it simply that the widespread substitution of the reaper for the grain cradle alleviated the scarcity of agricultural labor which otherwise would have restricted wheat production in the newly developing Western regions to appreciably lower levels?

There is no question that mechanical reaping effected a saving in harvest labor requirements; the evidence marshaled in Leo Rogin's path-breaking work, *The Introduction of Farm Machinery in Its Relation to the Productivity of Labor*, is nothing if not conclusive on that point.[2] Yet, to the writer's knowledge, no systematic attempt has been made to compare the magnitude of the savings in wage costs with the capital costs of a reaper to Western farmers during the first decade of the machine's widespread adoption. It is therefore not surprising that the accounts cited fail to divulge whether (or not) the new harvesting technique proved more profitable than grain cradling under all plausible relative factor-prices, or whether (or not) it was economically superior to cradling for all scales of farm operations.

These are, indeed, crucial questions. If the answers are in the affirmative, then the rate at which the reaper replaced the cradle in Western grain fields during the 1850s depended solely upon the capacity of the agricultural machinery industry; Bidwell and Falconer's assertion that 'reapers were introduced as fast as they could be manufactured' (*op. cit.*, p. 293), would be more than a mere figure of speech. It would be literally true and would carry the implication that the replacement of hand-harvesting methods would have occurred much earlier in American history were it not for technically unsolvable problems of designing and manufacturing a reaping machine. One would then have to find more convincing technical

[1] Gates, *op. cit.*, p. 287.

[2] See Rogin, *op. cit.*, pp. 125–37, and the discussion of Rogin's conclusions in the Addendum.

obstacles than are discussed in the authoritative works on the reaper to account for the lag between the first sale of Hussey's machine in 1833, the filing of the original McCormick patent in 1834, or even the first sale of McCormick's machine in 1840, and the eventual adoption of the innovation in the 1850s.[1] If the mechanical reaping technique actually was superior to hand-harvesting with the cradle, regardless of relative factor-prices or scale, this would also pose something of a problem for those writers who, like H. J. Habakkuk, regard the mechanization of agriculture in the United States as an 'obvious' illustration of the labor-saving bias of American technology fostered by nineteenth-century conditions of relative labor scarcity.[2]

It is quite clear, however, that the traditional accounts of the introduction of the reaper do not entertain such notions. By placing emphasis on the effects of rising grain prices and the extension of wheat production, they imply that altered demand for agricultural products was of fundamental significance in determining the rapid rate at which the innovation supplanted hand methods of harvesting small grains in the Midwest during the 1850s. This line of explanation, taken broadly, would suggest that the sudden growth of the market for the reaper – coming nearly two decades after the machine first began to be sold – was a consequence of the specific conditions surrounding Midwestern agricultural development towards the close of the late ante-bellum era. Even had the rise of the market for farm machinery not provided significant impetus to the initial industrialization of the region, the implications of this hypothesis for our general view of the process of the diffusion of technology would make it important to try to formulate the economics of the traditional account in a fashion sufficiently precise to permit its re-examination in the light of pertinent evidence.

Suppose, for the moment, that it is justifiable to assert that the saving of labor achieved with the mechanical reaper was not so great as to render cradling an inferior technique in all relevant factor-price situations. It may then be argued that mechanization of reaping spread through the agricultural sector as a result of an alteration in factor-prices which accompanied the expansion of grain cultivation in the West.[3] In other

[1] See Hutchinson, *op. cit.*, Chs. 5–10, *passim*; and Rogin, *op. cit.*, pp. 72–5, 85–91. It is not, however, necessary to depend upon a completely supply-determined explanation of this lag.

[2] H. J. Habakkuk, *American and British Technology in the Nineteenth Century* (Cambridge: Cambridge University Press, 1962), pp. 100–2 especially.

[3] It is sometimes argued that the rate of growth of an industry is an important determinant of the extent to which it adopts new techniques of production (see, e.g., P. Temin, 'The Relative Decline of the British Iron and Steel Industry' in *Industrialization in Two Systems: Essays in Honor of Alexander Gerschenkron*, ed. Henry Rosovsky (New York: John Wiley and Sons, 1966), Ch. 5), because with equipment of given durability the rate of growth of the industry will affect the equilibrium age of the capital stock and, therefore,

words, the standard versions can be read as saying that the 'agricultural boom' set in motion by rising grain prices in the mid-1850s added to already existing pressures upon the available harvest labor force in the region, drove up the farm wage rate relative to the cost of harvesting machinery, and, thereby, created a situation in which it became profitable for farmers to substitute machinery for labor in harvesting small grain. This argument requires the not unreasonable assumption that the supply schedule of harvest labor facing the farm sector in the Midwest was less elastic than the supply schedule for agricultural machinery; otherwise, the outward shift of the demand schedules would not have resulted in the relative price of harvest labor being raised to a level at which continuing substitution of machines for cradlers would take place. Granting that assumption, the argument may be completed by noting that as the availability of the new method of harvesting rendered the demand schedule for labor more elastic than would otherwise have been the case, substitution itself tended to check the extent of the actual rise in relative wages caused by the expansion of aggregate grain production. In this manner, the use of the reaper throughout the grain regions held down the total cost of production although it could not prevent some rise in costs, and made possible a large volume of total output at any given level of grain prices.

For this analysis, in which mechanization appears as a change 'imposed' upon grain farmers by the general expansion of Midwestern agriculture, the relative inelasticity of the farm labor supply schedule is crucial. The greater the emphasis that is placed upon the role of related competitive demands for labor, such as regional railroad construction, to cite but one significant source, the less thoroughly tied to exogenous events (e.g., the Crimean War) affecting world grain prices is the explanation offered for the timing of the adoption of mechanical reaping.[1]

[1] See, David, *op. cit.*, Ch. 5, for discussion of the competing sources of demand in the Midwestern labor market during the 1850s, in which it is argued that the effects of urban construction and internal improvements activity in creating a relative scarcity of unskilled labor in the region should be accorded more importance than they usually receive.

---

the extent to which the capital stock embodies the newest techniques. This line of reasoning, which would connect the rising demand for wheat and the rapid growth of the Midwestern agricultural sector in the 1850s with the adoption of the reaper on a large scale in that part of the country – even if mechanical reaping was an unambiguously superior technique – does not carry much force in the present context. The reason is simply that the 'older' technique (i.e., cradling) of cutting grain was not embodied in any fixed capital on farms; since they had virtually no sunk costs connected with the old method, established farmers would not *on that account* find it more expensive to switch to mechanical reaping than would new entrants to the industry or farmers who were making significant increases in their productive capacity. In explaining the adoption of the reaper in Midwestern agriculture it therefore does not seem important to concentrate on any distinction between new and old farms in the industry.

In the picture just presented, *the individual farmer's* desire to increase his acreage under wheat does not appear as influencing his decision to purchase a reaper and dispense with the services of cradlers. Nor can the personalization of the collective market process described be justified with any plausibility. Since there is no reason to suppose that the labor supply schedule facing the individual farmer was less elastic than the supply curve for agricultural machinery that confronted him, why should there be any connexion between *the individual farmer's* decision to sow more wheat and his choice of the new reaping technique? Yet, the literature is replete with statements suggesting such a connexion: 'When the wheat from an acre of land would sell for more than the price of the land, it was considered a safe investment to sow more land in wheat and buy a reaper';[1] '. . . Americans also had a very strong incentive to develop machines which would enable farmers to cultivate a larger area. The alternative was to leave land uncultivated.'[2] If such statements represent something more than illustrations of the ease with which efforts to write economic history as the intended outcome of purposive individual actions, rather than their interplay in impersonal markets, can lead to what may be called 'fallacies of decomposition,' their authors must have in mind a set of considerations influencing the introduction of mechanical reaping which is quite distinct from the process of market-imposed adjustments already set forth.

To put it most simply, these statements may be taken to imply either that there were significant economies of scale associated with the use of the reaper, or that diseconomies of scale existed in the use of labor for cradling grain that were not encountered with the mechanized technique in the range of farm size relevant to the ante-bellum Midwest.[3] Both situations would arise from the presence of indivisibilities among the inputs of the microproduction function for harvesting small grains.

[1] Bidwell and Falconer, *op. cit.*, p. 293.

[2] Habakkuk, *op. cit.*, p. 101.

[3] It must be confessed that what Habakkuk has in mind in this connexion is less than completely clear, and that some doubt remains whether the rationalization put foward in the text is appropriate. The last sentence of the quotation – i.e., 'The alternative was to leave land uncultivated' – does suggest that the American farmer had a passion for cultivating all available land without consideration of profitability and that a machine permitting greater acreage to be cultivated with a given amount of labor would be adopted by the farmer under any product and factor-price conditions. The former part of this view is perhaps shared by other writers. Of the American farmer, Hutchinson, *op. cit.*, pp. 50–1, says, '. . . environment compelled him to be acquisitive, and he was prone to add more acres to his freehold than he could well keep under cultivation.' Yet, to proceed along these lines is to turn the question of the choice of harvesting techniques into one more properly dealt with by psychologists and sociologists, whereas, as will be seen, a satisfactory explanation can be provided in traditional economic terms.

In the apparent absence of feasible cooperative arrangements for sharing the use of harvesting machinery among farms, at this time the reaping machine itself constituted an indivisible input for the farmer.[1] Since he typically had to purchase it, rather than rent it when it was needed, the relevant cost of using a reaper in harvesting was the average annual cost over the life of the machine. Within a particular season, however, the cost of a reaper per acre harvested would fall as the acreage was increased. It would continue falling to the point at which the cutting capacity of a single machine during the feasible harvest interval was reached. By contrast, given a perfectly elastic supply of labor and no diseconomies of scale in its use, the saving in wage costs obtained by substituting the mechanical reaper for cradlers would remain constant per acre harvested.

It is possible, therefore, that below some level of acreage to be harvested – which we shall call the 'threshold' farm size – the total capital cost of a reaper (or of more than one reaper) exceeded the potential reduction

[1] See Fred A. Shannon, *The Farmer's Last Frontier: Agriculture 1860–1897* (*The Economic History of the United States*, Vol. v) (New York: Farrar and Rinehart, 1945), pp. 329–48, for farm groups' efforts at cooperative manufacture and large-scale purchase of agricultural machinery after the Civil War. Such cooperative ventures did not emerge prior to the War, nor are they equivalent to the cooperative use of farm machinery. There is little evidence of commercial renting of reaping and mowing machines, or commercial grain harvesting by horsepower, such as developed in connexion with the use of the large breaking-plows on the prairies during the 1850s. Neighbors may well have shared the use of a reaping machine owned by one farmer, compensating the owner on an informal basis, but, some inquiry into contemporary accounts, farm newspapers, and such sources does not suggest that this was common practice. It certainly does not appear to have been as characteristic a feature of inter-farm relations as was the joint use of corn shellers, threshers, and other equipment used in *post-harvest* tasks. The foregoing impressions are, perhaps, not sufficiently firmly established to demand a hypothesis which would account for the absence of commercial renting of early reapers and the lack of arrangements for sharing the use of jointly (or singly) owned machines. Nor are we able at this point to offer more than a possible line of explanation. The fact that the maximum cutting capacity of the early machines was not very large, especially when the time constraint on harvesting in any given locale is taken into consideration, coupled with the high costs of overland transport for a bulky machine weighing upward of half a ton, would appear to have militated against operating a profitable itinerant commercial reaping enterprise during the ante-bellum era. The time constraint seems the crucial factor, since it was not present to the same degree either in prairie-breaking operations or in post-harvest tasks such as threshing and corn shelling; the former became established on a commercial basis, while sharing of equipment among neighbors was not unusual in the latter cases. Furthermore, as a consequence of the time constraint, the problems of deciding who was to have priority of use of a jointly owned reaper would have required the owners (users) to form some compensation arrangement, equivalent to an output pool or a profit pool. The economic, not to mention the political and sociological consequences of a reorganization of independent small farms into such pooling arrangements would have profoundly altered the course of agrarian history in the United States.

in wage costs, making its adoption unprofitable in comparison with the method of cradling.

Exactly where the threshold point was located in the spectrum of farm acreage devoted to the small grains was determined by relative factor-prices. The saving of labor achieved with the reaper being essentially technologically fixed per acre harvested, a doubling of the total yearly reaper cost relative to the money wage cost of harvesting an acre with the cradle would double the number of acres that would have to be harvested before the costs per acre would be the same with either technique.[1] While it is conceivable that so great a saving of labor was effected by the reaper that the costs per acre harvested by machine were lower for any finite total acreage, so long as both the money wage rate and the cost of a reaper were positive and finite, the existence of *significant* economies of scale associated with mechanical harvesting cannot be taken to have been a purely technical matter; relative wage rates must not have been so high that it was profitable for the farmer to adopt the new method at any level of grain production.

In principle, the existence of diseconomies of scale in the use of labor for harvesting grain with cradles would operate in the same manner as economies of scale associated with the mechanical reaper. Harvest workers required a certain amount of supervision to maintain their efficiency, and the addition of hands required to cut the grain on a larger acreage presumably taxed the farmer's supervisory capacities. The harvest had to be carried out in a limited number of days lest the ripe grain be lost through shattering or spoilage. It was therefore not feasible simply to employ the optimum number of hired hands that could be supervised at any one time for as long a period as it would take to cut the grain. Assuming that the average productivity of cradlers would begin to decline when the amount of supervision they received fell below some minimum, we may say that the manpower requirements per acre would have been greater for larger acreages. Consequently, savings in labor obtained by switching to mechanical reaping would tend to rise as the acreage to be harvested increased. Even if the capital costs of a reaper were constant per acre, this could define a threshold size beyond which farmers would find it advantageous to mechanize.

Although there is evidence of considerable contemporary dissatisfaction about the quality of hired help on farms during the first half of the nineteenth century, and notwithstanding the comments of American farmers on the need to supervise temporary help in order to keep them on the job

[1] The relationship between the relative price of labor, *vis-à-vis* the reaper, and the threshold size is developed formally in the Addendum. Even without that derivation it is readily seen that if the labor cost per acre harvested by cradle is constant, threshold size must vary in direct proportion to the *relative* cost of capital.

and careful in their work,[1] it is difficult to gauge the extent to which the inferior quality of hired help failed to reflect itself in the general level of farm wages. Moreover, among the complaints registered by farmers are those specifically citing the carelessness of hired hands with machinery. If there were diseconomies of scale in the use of harvest labor, we do not know that these were restricted to the employment of cradlers rather than labor in general, and that they did not simply place a limit on the scale of farm operations in the free states.[2] It therefore seems justifiable to restrict the discussion of scale effects to consideration of those which arose from spreading the fixed cost of harvesting machinery over large grain acreage.

Figure 8 depicts the hypothesized situation in terms of alternative long-run unit cost curves for hand methods and machine methods of grain harvesting on independent farms. The supposed existence of some fixed costs with either process causes both the hand-method cost curve,

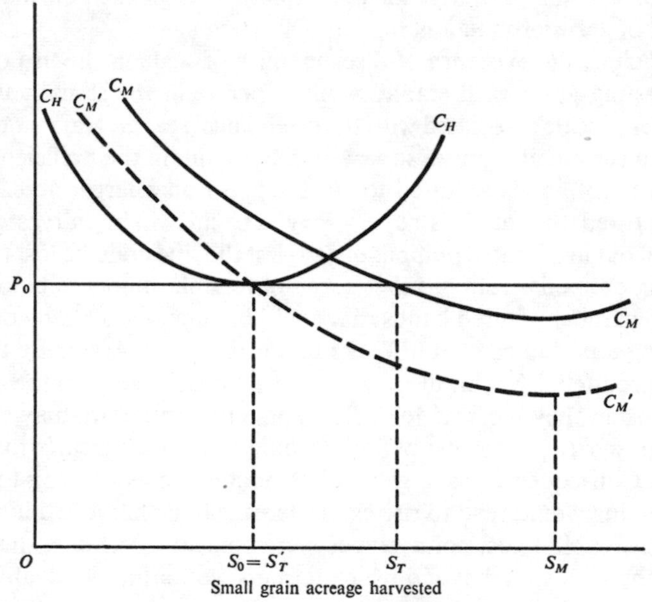

Figure 8 Hypothetical long-run average cost curves for harvesting

[1] See, e.g., Gates, *op. cit.*, pp. 272, 275 and also the passage cited from Edmund Ruffin's essay on 'Management of Wheat Harvest,' *American Farmer*, n.s., 6 (June 1851) in Rogin, *op. cit.*, p. 131. One of the state sales agents employed by the McCormicks in the 1850s took it as the object of his work to 'place the farmer beyond the power of a set of drinking Harvest Hands with which we have been greatly annoyed,' Hutchinson, *op. cit.*, pp. 355–6.

[2] Monographic studies of the effects of mechanization in agriculture, such as Rogin's, make no mention of increases in labor productivity associated with the reaper having been influenced by the size of the farm on which the machines were used.

$C_H C_H$, and the machine-method curve, $C_M C_M$, to decline over a range of total harvested acreage, but, because of the *additional* fixed cost entailed by the reaper, the curve $C_M C_M$ goes on falling after the hand-method unit costs begin rising. The rise in unit costs results from the limitation on the total supervisory capacity of the farmer, a restraint which eventually also causes the $C_M C_M$ curve to turn upward at a larger scale of operations. In the situation shown, the representative farm operating with the hand method of harvesting is taken to be in equilibrium at size S_0, with the market price of grain (per acre harvested, assuming constant yield per acre) equal to minimum units costs at P_0. Beyond S_0 lies S_T – the threshold size at which unit costs with the hand method are equal to those with the machine method – whose location is determined by the factors influencing the relative positions of the two curves.

The argument that the individual farmer's decision to adopt mechanical reaping was tied up with a simultaneous decision to expand his grain acreage, the latter being prompted by a rise in the relative price of grain, may now be quickly restated in terms of Figure 8. If there were no initial costs involved in increasing grain acreage under cultivation, with the market price at P_0 it would clearly pay to abandon the hand method of harvesting and expand the representative farm from S_0 to S_M. However, the costs of acquiring, clearing, and fencing new land, or simply of preparing land already held, were hardly insignificant even on the open prairies. If these costs, relative to the market price of grain, were sufficiently large to prevent expansion beyond S_T, they would have effectively blocked the concomitant adoption of the mechanical reaper.[1] The significance of the rise in grain prices during the 1850s in bringing about widespread introduction of the reaper accordingly was that by lowering the relative unit costs of expansion, higher prices induced the typical farmer to increase his grain acreage beyond the previous threshold size. In so doing, the farmer would take advantage of the $C_M C_M$ cost curve.

The presence of scale considerations in the choice between hand method and machine method of harvesting grain was not explicitly recognized in setting forth our earlier argument, in which the change in technique was depicted as an adjustment imposed by the relative inelasticity of the aggregate supply schedule for farm labor. That hypothesis may, nonetheless, be

[1] On clearing and fencing costs, see Allan G. Bogue, 'Farming in the Prairie Peninsula, 1830–1880,' *Journal of Economic History* (March 1963); Clarence Danhof, 'Farm-Making Costs and the Safety-Valve: 1850–1860,' *Journal of Political Economy* (June 1941); Gates, *op. cit.*, pp. 34, 186–8; M. Primack, 'Land Clearing under Nineteenth Century Techniques: Some Preliminary Calculations,' *Journal of Economic History*, Vol. XXI (1962), pp. 484–97. In terms of Figure 8 we can say that, spread over the total acreage to be harvested, the unit costs of expanding farm acreage beyond S_T was at least equal to the difference between P_0 and the $C_M C_M$ cost curve at S_M.

quite readily stated within the framework of the micro-analysis summarized by Figure 8. The contention is simply that because the threshold point will move inversely with changes in the price of labor relative to the cost of the reaper, the relative rise in farm wage rates (resulting from the collective response to higher grain prices in the mid-1850s) drove the threshold size downward to the point at which adoption of the reaper became profitable even on farms that had not enlarged their cultivated acreage. This is shown in Figure 8 as a downward shift in the position of the long-run machine-method cost curve relative to $C_H C_H$, lowering the threshold size (S_T) to the optimum acreage (S_0) established under the regime of the older harvesting technique.[1]

IV

On formal grounds our two versions of the switch to mechanical reaping thus turn out to be entirely compatible; they merely stress what may have been different aspects of a single story – one directing attention to forces working to push farms across the old threshold point, and the other emphasizing that an alteration in relative factor-prices may have brought the threshold down to the vicinity of previously established farm sizes.

The empirical requirements of these two hypotheses are equally apparent. To credit either version we must at least have some evidence that at the beginning of the 1850s the threshold size for adoption of the reaper lay above the average small grain acreage on Midwestern farms, not only in the region as a whole, but in those areas especially devoted to these crops. Once that is established, further evidence of a substantial decline in the threshold size *as the result of an alteration in relative factor-prices* during the decade would lend credence to the view that the adjustment in technique was imposed by the inelasticity of the labor supply, whereas acceptance of the pure individual farm-size adjustment view would hinge on a finding that over the course of the 1850s average grain acreage on Midwestern farms rose to the neighborhood of the threshold size that had existed at the opening of the period.

The Addendum provides a detailed discussion of the evidence assembled and the way it has been used in calculating the threshold acreages for adoption of mechanical reapers by grain farmers in the Western states during the period under study. The computations are readily made by linearizing the cost functions for the hand and machine methods of harvesting. It is sufficient here to consider the results of those calculations in conjunction with such information as is available regarding actual

[1] Note that the downward shift in $C_M C_M$ is equivalent to a rise in the market price of grain being accompanied by an upward shift in the $C_H C_H$ cost curve which is not matched by a rise in unit costs with the machine technique.

small grain acreage on the average Midwestern farm.[1] This may be done with the aid of Figure 9, which depicts the relationship between threshold size and relative costs of labor and capital for the basic hand-rake reaping machine on the assumption of linear cost functions.

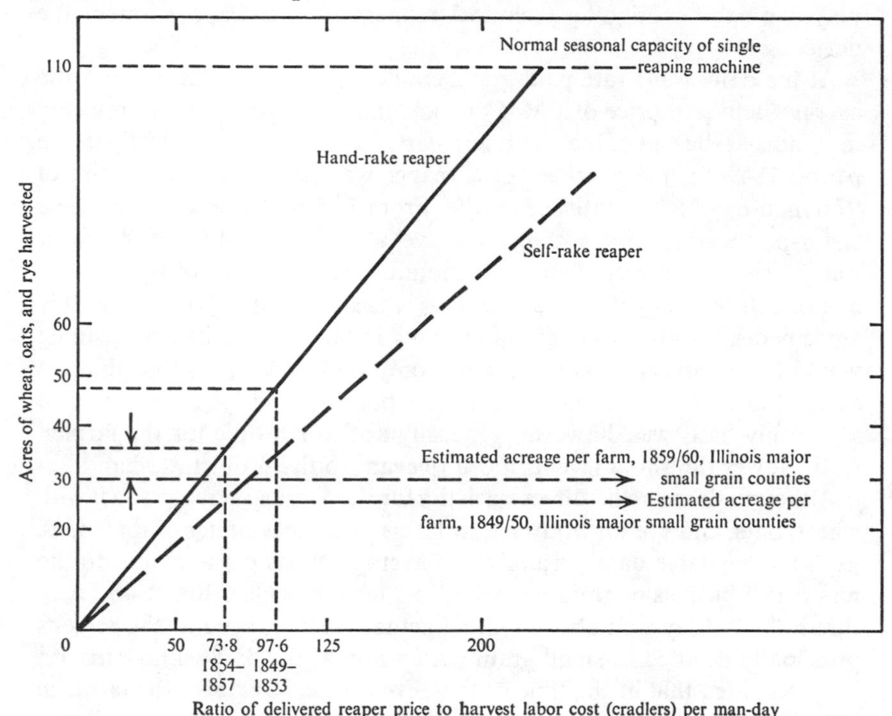

Figure 9 Threshold functions for adoption of the reaper. *Source:* See Addendum, Tables 12, 13

Figure 9 also shows the threshold function for the self-raking type of reaper which, by mechanically delivering the cut grain to the ground beside the machine either in gavels or in swath, dispensed with the need for a man to sweep the grain from the platform of the hand-rake machine. However, since self-rakers did not win popularity in the Midwest until the latter half of the 1850s, well after the initial acceptance of the basic

[1] It should be noted that although the traditional accounts consider the introduction of the reaper only in connexion with the harvesting of wheat, the Midwest's principal cereal and leading commercial crop at the time, the discussion here has been phrased in terms of the small grains – wheat, oats, and rye, all of which were harvested with the cradle and could be cut with the early reaping machines. (See Bidwell and Falconer, *op. cit.*, p. 353 and Hutchinson, *op. cit.*, p. 310.) This means that calculated threshold sizes for adoption of the reaper must be compared to estimates of average small grain acreage, rather than average wheat acreage alone. See Addendum, Section 3.

hand-raking machines manufactured by the McCormick Co., the discussion will focus on the pioneering hand-rake model.[1] Our conclusions therefore, relate to the influence of alterations in market forces on the adoption of the basic reaper during the 1850s, rather than to the role played by the continuing technical refinement and elaboration of the device.

At the daily wage rate paid grain cradlers during the harvest and the average delivered price of a McCormick hand-rake reaper that prevailed in Illinois at the end of the 1840s and early in the 1850s, specifically in the period 1849–53, the purchase of a reaper was equivalent to the hire of 97·6 man-days' labor with the cradle. From Figure 9 it is seen that these factor-prices established a threshold level at 46·5 acres of grain. Where it was possible to hire cradlers on a monthly basis, instead of by the day, and therefore to pay the lower *per diem* wages implied in typical monthly agreements, the abandonment of cradling in favor of mechanical reaping would have reduced harvesting costs only on farms with more than 74 acres of grain to be cut. Hiring all the labor required for the harvest on a monthly basis was, however, generally not worthwhile for the farmer, so the lower threshold level is more relevant to the problem at hand.

Although there are no direct statistics for the average acreage sown with wheat, oats, and rye on Midwestern farms at the beginning of the 1850s, from the available data pertaining to average yields per acre and to the number of bushels of grain harvested per farm it is clear that a 46·5 acre threshold still lay well above typical actual acreage, even in the regions principally devoted to small grain cultivation at the mid-century mark.[2] It is estimated that at the time of the Seventh Census (1850) the farms in Illinois averaged from 15 to 16 acres of wheat, oats, and rye. In the sixteen leading grain counties (of ninety-nine counties in Illinois) which as a group produced half of the state's principal cereal crop at that time, the average farm land under wheat, oats, and rye ran to approximately 25 acres. Among these major small grain counties, Winnebago County in the northernmost part of the state is representative of those with the highest average small grain acreage per farm, while Cook County was one of those having the lowest average acreage. Yet, on the 919 farms in Winnebago County the average worked out to 37·2 acres of small grain, still 10 acres under the threshold level, and in Cook County the 1,857 farms averaged but 18·6 acres apiece.

[1] The McCormicks did not manufacture a self-rake model until the introduction of the 'Advance' in the post-Civil War period. See the Addendum for further discussion of the self-rakes and of the combined reaper-mowers which came in use in the Western states before the war.

[2] See Addendum, Section 3, for discussion of this data and the acreage estimates in Table 12 on which the present statements are based.

Two closely related points thus emerge quite clearly. First, in the years immediately surrounding the initiation of reaper production in the Mid-west (1847) and the establishment of the McCormick Factory at Chicago (1848), the combination of existing average farm size and prevailing factor-prices militated against widespread adoption of the innovation. The admonitions appearing in Western agricultural journals during 1846 and 1847 against purchase of the new reaping machine by the farmer 'who has not at least fifty acres of grain,' would appear in the light of the considerations presented here to have been quite sound advice.[1] Secondly, observations of the sort made by a reliable contemporary witness, Lord Robert Russell, who traveled in the prairie country in 1853, that 'the cereals are nearly all cut by horsepower on the *larger farms* in the prairies,'[2] become understandable simply on the grounds of the scale considerations affecting the comparative profitability of the reaper. It is not necessary to explain them by contending that the larger farms were run by men more receptive to the new methods of scientific farming or less restrained by the limitations of their financial resources and the imperfections of the capital market, however correct such assertions might prove to be.

An initial empirical foundation for the plausibility of both our hypotheses having thus been established, we turn now to consider the evidence relating to the character of the adjustments themselves. During the mid-1850s, as the aggregate labor supply constraint hypothesis suggests, the price paid for harvest labor in Illinois did rise more rapidly than the average delivered price of a hand-rake reaper; in the period 1854–7 a McCormick reaper cost the farmer, on average, the equivalent of only 73·8 man-days of hired cradler's labor, compared with 97·6 man-days in the preceding period 1849–53. In consequence, as Figure 9 shows, the threshold point dropped from over 46 to roughly 35 acres. By the middle of the decade, then, the average small grain acreage above which it paid the farmer to abandon cradling had fallen below the average acreage that had existed in a leading grain-producing area like Winnebago County, Illinois, at the beginning of the decade, and lay only 10 acres above the average on the 21,634 farms in the state's leading grain counties in 1849–50.

[1] See quotations from the *Ohio Cultivator*, 1 October 1846, p. 147; 15 April 1847, p. 64; and *Chicago Daily Journal*, 2, 24, 28, 30 July and 15 August 1846, abridged in Hutchinson, *op. cit.*, p. 234, n. 15. Note that, as one might expect in view of the upward trend of relative farm wages from the trough of the depression of the 1840s, the 46·5 acre threshold estimated here to be appropriate for the period 1849–53 lies very close to, but slightly *below* the 50-acre threshold level implied by the advice to farmers in the years 1846–7.

[2] *North America, Its Agriculture and Climate*, London, 1857, p. 114, quoted in Rogin, *op. cit.*, p. 79. Italics added. Rogin says he regards this statement as 'more in accord with the facts' than are claims that in the early 1850s cereals as a rule were cut by machinery but, he offers no further evidence or reasoning to support this judgement.

At the same time, there is some evidence pointing to a rise in the average grain acreage harvested per farm, such as is proposed by the individual farm-size adjustment version. Just how large an increase in average acreage occurred during the 1850s in the specialized small grain regions of the Midwest is difficult at present to say, for the simple reason that the Census of Agriculture in 1860, unlike the previous Census, neglected to publish the statistics of the number of farms on a county-by-county basis. In Illinois as a whole, however, the number of acres of wheat, oats and rye harvested per farm is estimated to have been roughly 19 per cent higher in 1859–60 than it had been ten years earlier. Of course, it is possible that in the transition from cereal to corn and livestock production that was under way in the state during this decade, specialization in small grain cultivation became more concentrated and, therefore, that the increase in the typical farm acreage sown with those crops in the leading regions of their cultivation was considerably greater than 19 per cent. But, such evidence as can be brought to bear on the matter does not point in that direction. Broadly speaking, small grain cultivation was spatially no more concentrated in Illinois at the end of the 1850s than at the beginning, and wheat production became, if anything, geographically more dispersed.[1] It is therefore not wholly unreasonable to assume that small grain acreage per farm in the areas especially devoted to those crops increased at the same rate as it did in Illinois as a whole. On that basis one may conjecture that the number of acres under wheat, oats and rye on a typical farm in the state's leading small grain regions increased from 25 to 30 in the course of the 1850s.

The story of the adoption of the mechanical reaper in the years immediately before the Civil War should thus be told in terms of the effects of both an expansion of grain acreage sown on individual farms and the downward movement of the threshold size as a result of the rising relative cost of harvest labor. But, of the two types of adjustment taking place during the 1850s, the former must properly be accorded lesser emphasis. As Figure 9 reveals, on farms in the leading grain regions of Illinois the estimated increase in average small grain acreage was responsible for less than a third of the subsequent reduction of the gap existing between threshold size and average acreage at the opening of the decade. Moreover, among the Midwestern states experiencing rapid settlement during the 1850s Illinois was singular in the magnitude of the expansion of its average

[1] Whereas only 16 counties made up the group producing half the total wheat crop, and 19 counties accounted for half the total number of bushels of wheat, oats, and rye harvested in Illinois in 1860, it took 21 counties to account for 50 per cent of the wheat and 19 counties to account for 50 per cent of the three small grain crops harvested at the end of the decade. See J. D. B. DeBow, *A Statistical view... Compendium of the Seventh Census of the United States* (1854), pp. 21–7; *U.S. Eighth Census* (1860), 'Agriculture,' pp. 31–5.

farm size.[1] Elsewhere in the Midwest, the relative rise in farm wage rates is likely to have played a still greater role in bringing the basic reaping machine into general use during the decade preceding the Civil War.

V

Although the questions considered in the preceding pages are very specific, we have arrived at answers with rather broader implications. Historians of United States agriculture have maintained that during the nineteenth century the transfer of grain farming to new regions lying beyond the Appalachian barrier played a significant part in raising labor productivity in agriculture for the country as a whole. The connexion between the spatial redistribution of grain production and the progress of farm mechanization figures prominently among the reasons that have been advanced to support this contention. Some writers suggest that inasmuch as heavier reliance was placed on the use of farm machinery in the states of the Old Northwest before the Civil War, and, similarly, in the Great Plains and Pacific Coast states during the last quarter of the century, the geographical transfer of agriculture into these areas was tantamount to a progressive shift of grain farming toward the relatively capital-intensive region of the technological spectrum.[2] But, the mechanism of this putative interaction between spatial and technological change has not been fully clarified, and as a result, important aspects of the interrelationship between the historical course of industrialization and the settlement of new regions in the United States remain only imperfectly perceived.

To make some headway in this direction it is necessary to distinguish two possible modes of interaction between spatial and technological changes in United States agriculture: one involves adjustments of

[1] Between 1850 and 1860 improved acreage per farm increased from 66·2 to 91·3 in Illinois, from 53·4 to 62·4 in Indiana, from 55·6 to 62·1 in Iowa, and from 51·8 to 54·0 in Wisconsin. Illinois average farm size was thus not only growing more rapidly than that in the surrounding states, but was initially larger. (See *U.S. Eighth Census* (1860), 'Agriculture,' p. 222.) On these grounds alone, Cyrus McCormick's decision to move his base of operations from Rockbridge County, Virginia, in 1847 and embark upon manufacturing the reaper at Chicago during the following year proved extremely cogent.

[2] As an indication of the extent of the geographical redistribution of grain farming prior to the Civil War, it may be noted that Pennsylvania, Ohio, and New York were leading wheat growing states in 1849–50, but, a decade later roughly half the total U.S. wheat crop was raised in Ohio, Indiana, Michigan, Illinois, and Wisconsin – i.e., in the Old Northwest Territory. See Everett E. Edwards, 'American Agriculture – The First 300 Years,' *Yearbook of Agriculture 1940* (G.P.O., Washington, 1940), p. 203. Among recent explicit treatments of this question, Parker's (*op. cit.*) offers a set of calculations designed to gauge the impact of regional shifts in the pattern of production on the productivity of labor in small grain farming and attributes much of the 'region effect' (on national productivity) to the interaction between location and the degree of mechanization.

production methods in response to alterations of relative prices that were associated, either causally or consequentially, with the geographical relocation of farming; the other turns on purely technological considerations through which regional location influenced choices among available alternative techniques. Now, the general statement that the conditions under which farmers located in the country's interior carried on grain production especially favored the spread of mechanization is sufficiently imprecise to embrace both interaction mechanisms, the influences of market conditions as well as those of technological factors peculiar to farming in the different regions. One may well ask whether such ambiguity is justified. Without establishing the dominance of purely technical considerations it would be unwarranted to suggest that shifts of small grain farming away from the Eastern seaboard automatically, in and of themselves, accounted for increases in the extent to which that branch of United States agriculture became mechanized.

In the case of reaping operations, it is certainly true that there were technical features of Midwestern farming which in contrast with those characteristic of the Eastern grain regions proved inherently more congenial to the general introduction of ante-bellum reaping machines. On the comparatively level, stone-free terrain of the Midwest, the cumbersome early models of the reaper were less difficult for a team to pull, less subject to malalignment and actual breakage; because the fields were unridged and crops typically were not so heavy as those on Eastern farms, the reapers cut the grain close to the ground more satisfactorily, and the knives of the simple mechanism were not so given to repeated clogging.

Yet, despite the relatively favorable technical environment (and larger average small grain acreages on farms) in the Midwest, we have seen that the prevailing factor and product market conditions during the 1840s and early 1850s militated against extensive adoption of mechanical reaping equipment even in that region. Against such a background the fact that a large-scale transfer of small grain production to the Old Northwest Territory took place during the 1850s does not appear so crucial a consideration in explaining the sudden rise in the proportion of the total American wheat crop cut by horse power between 1850 and 1860.[1] Instead, it seems appropriate to emphasize that during the Midwestern development boom that marked the decade of the 1850s the price of labor – as well as the prices of small grains – rose relative to the price of reaping machines, and that the pressure on the region's labor supply

[1] This proportion rose from a negligible level at the beginning of the decade to somewhere between 33 and 50 per cent by the close of the 1850s. At the latter date the proportion of the trans-Appalachian wheat crop cut by horse power was appreciably higher than that for the nation as a whole. See footnote 1, p. 202, for the sources and some discussion of these estimates.

reflected not only the expanded demand for farm workers, but also the demand for labor to build railroads and urban centers throughout the region – undertakings ultimately predicated on the current wave of new farm settlement and the expected growth of the Midwest's agricultural capacity. If one is to argue the case for the existence of an important causal relationship between the relocation of grain production and the widespread acceptance of mechanical reaping during the 1850s, the altered market environment, especially the new labor market conditions created directly and indirectly by the quickening growth of Midwestern agriculture, must be accorded greater recognition, and the purely technical considerations be given rather less weight than they usually receive in this connexion.

There is, however, a sense in which the decline in the cost of reaping machines relative to the farm labor wage rate may be held to have reflected the interaction of the technical factors favoring adoption of the early reapers in the Midwest with that region's emergence as the nation's granary during the 1850s. The rising share of the United States wheat crop being grown in the interior did mean that, *ceteris paribus*, a larger proportion of the national crop could be harvested by horse power without requiring the building of machines designed to function as well under the terrain and crop conditions of the older grain regions as the early reapers did on the prairies. For the country as a whole, as well as for the Midwest, this afforded economies of scale in the production of a simpler, more standardized line of reaping machines. It thereby contributed to maintaining a situation in which the long-run aggregate supply schedule for harvesting machinery was more elastic than the farm labor supply. Thus it may be said, somewhat paradoxically, that the movement toward regional specialization in small grain farming directly made possible greater efficiency in manufacturing and thereby promoted the simultaneous advance of mechanized agriculture and industrial development in the ante-bellum Midwest.

Addendum: Threshold farm size

The element of fixed cost present with the mechanical reaping process for harvesting grain makes it necessary to take account of the scale of harvesting operations in cost comparisons of hand and machine methods. One means of doing this would be to stipulate the acreage to be harvested and then proceed to ask how the profitability of mechanical reaping compared to cradling was affected by the prevailing level of factor-prices. This addendum tackles essentially the same question by posing it in a slightly different way. The question can be put as follows: given alternative sets of factor-prices, at what alternative scales of harvesting operations would it be a matter of indifference (on cost grounds) to the farmer of the 1850s whether he adopted the reaper or continued to rely on the cradle? The answer is to be found from a computation of that acreage, called the threshold size, at which the total costs of the two processes were just equal and beyond which abandoning the cradle would become profitable, other things remaining unchanged.

1. THE FORM OF THE CALCULATION

It is not, however, necessary to calculate the total cost of harvesting an acre of grain at different scales of operation, as depicted by the long-run unit cost curves of Figure 8. In the first place, as will be seen, the activities of the harvest other than the actual cutting of grain – raking, binding, and shocking – may be omitted from consideration as not significantly influencing the choice between machine and hand reaping. Secondly, all that is required is the computation of the total *saving* of money wages effected by adoption of the mechanical reaper and the *additional* fixed capital charge that the farmer would incur in order to have the machine at his disposal during the harvest season. What must be known, therefore, is:

- L_s: the number of man-days of labor dispensed with by mechanizing the cutting operation, per acre harvested;
- w: the money cost to the farmer of a man-day of harvest labor;
- c: the fixed annual money cost of a reaper to the farmer.

From this information the threshold size, S_T, in acres to be harvested, can be determined, since, *ceteris paribus*, total costs of cutting the grain are the same for both processes when

$$c = \sum_{i=1}^{S_T} L_{si} w, \tag{1}$$

where the index i designates the acre in the sequence of acres ($i = 1, \ldots, S_T$) harvested.

Actually, the problem can be further simplified. From the available information it appears justifiable to assume that within the range of normal daily cutting capacity of the reaper there were no economies or diseconomies of scale in the use of cradling, and that the saving of labor effected by the mechanical reaper per acre harvested was a technically determined constant. In a word, the cost functions for the two processes may be taken as linear over the relevant range of operations. Consequently, we may replace equation (1) with the much simpler expression,

$$c = S_T L_s w. \tag{1a}$$

The average annual capital cost, or effective rental rate, on a piece of durable equipment may be reckoned as the sum of the annual depreciation of the equipment and the annual interest cost on the capital invested in it. One can think of the latter as an opportunity cost, since for half the year, on average, the owner's funds are tied up in the machine instead of being lodged elsewhere at interest. Alternatively, the interest cost is to be thought of as the actual charge made for a loan of the purchase price of the equipment. Strictly speaking, in calculating the interest cost of a mechanical reaper, allowance should be made for the fact that funds were locked up in these machines for periods longer than a year; it is known that Midwestern farmers purchased reapers on credit during the 1850s and paid them off only over an extended period of time.[1] Yet, within the range of accuracy we can hope to attain in the present calculations, the niceties of compound interest may be foregone and the interest cost computed on a simple basis. The half life of reapers was not, in any case, so long as to render this a serious omission.

An equivalently liberal attitude is warranted regarding the question of depreciation charges. Rather than play with formulae that attempt to take into account the actual time pattern of loss of value through wear and tear and obsolescence, straight-line depreciation over the physical life of the machine will be assumed.

As a result of the foregoing simplifications, the average annual gross money rental charge is quite straightforwardly given by

$$c = [d + 0{\cdot}5(r)]C,$$

[1] See, e.g., Hutchinson, *op. cit.*, pp. 368–9.

where d = the straight-line rate of depreciation,
r = the annual rate of interest,
C = the purchase price of a reaper.

Putting this together with equation (1a), we have the relation defining the threshold size in terms of the prices of harvest labor and reapers, the rate of interest, and the 'technical' coefficients L_s and d:

$$S_T = \left(\frac{d+0{\cdot}5r}{L_s}\right)\left(\frac{C}{w}\right). \tag{3}$$

From the form of equation (3) it is apparent that, given the rate of interest, the threshold for harvesting machines of specified durability and labor-saving characteristics is directly proportional to the relative price of the reaper (C/w). Thus, the threshold functions shown in Figure 9 of the text appear as positively sloped rays from the origin of a graph of acreages against the ratios of reaper prices to wage rates. In the following section we proceed first to consider the evidence that establishes the slope of the threshold function, and then to take up the question of the relevant range of variation of factor-prices in the Midwest during the period preceding the Civil War.

2. PARAMETERS AND VARIABLES

Labor savings

The reduction in harvest labor requirements achieved by the introduction of the mechanical reaper is perhaps the single most widely cited instance of the improvement of agricultural technology in small grain production during the ante-bellum era. Various estimates of the magnitude of the saving in labor appear in the secondary literature and the economic history texts, but virtually all of them derive from the evidence and conclusions presented in Leo Rogin's pioneering study.[1] A crucial issue that arises in this connexion is one that is frequently overlooked, namely, the amount of grain that the mechanical reaper could cut in a normal day's work. Notwithstanding the fact that during the 1850s the McCormick reapers were warranted to cut 15 acres of wheat in a twelve-hour day and that frequently mentioned records set in reaper trials cite 15 as the acreage cut in a day, Rogin's survey of the evidence leads him to conclude that 10 to 12 acres per day was closer to normal practice even on the broad prairies of Illinois:

[1] See, e.g., Bidwell and Falconer, *op. cit.*, pp. 293–4; Gates, *op. cit.*, p. 237; Robertson, *op. cit.*, p. 258; and Danhof, in Williamson, *op. cit.*, pp. 145–56.

. . . the foregoing rate is the one most frequently mentioned in other contemporary accounts and may be taken to represent the average performance with the hand-rake reaper after it came into prominence, as well as with the self-rake which superseded it.[1]

Following Rogin, then, 11 acres per day may be taken as the average cutting rate with the reaper during its first decade of widespread use.

The significance of establishing the normal daily rate of harvesting with the reaper lies in the fact that the information available on the labor requirements of the hand method of cutting grain comes in the form of statements about the number of acres that it was common for a man working with a cradle to harvest in a day. In Rogin's judgement, 'there appears . . . to have been a norm of performance for the country as a whole which approximates two acres' per man-day,[2] despite regional and local variation in speed caused by differences in the heaviness of the grain. If an allowance of a 10 per cent difference in speed above the national average is made in recognition of the lighter yields in small grains typical in the Midwest,[3] we are led to conclude that during the 1850s a mechanical reaper cutting 11 acres accomplished the work of 5 man-days' labor with cradles. Two men were required to operate the hand-rake reapers, whereas the self-rakers, first marketed commercially in 1854, needed but a driver. Therefore, in the cutting operation itself the use of the hand-rake reaper effected a saving of 3 man-days in cutting 11 acres, while the self-rakers saved 4 man-days.[4]

Although two other sources of economy are sometimes cited as associated with the mechanization of reaping, their inclusion does not seem warranted in the presented connexion. The reduction of losses due to the shattering of overripe grain by cradlers cannot be considered as anything more than a consequence of the saving in labor already mentioned: if this source of losses had proved sufficiently costly in the era prior to the introduction of the reaper, farmers would have found it worthwhile to hire enough cradlers to ensure that their grain was harvested before

[1] Rogin, *op. cit.*, pp. 134–5. The Census of 1860 also maintained that 'a common reaper will cut from ten to twelve acres in a day of twelve hours' (*U.S. Eighth Census* (1860), 'Agriculture,' p. xxiii). Bidwell and Falconer (*op. cit.*, pp. 293–4), who base their discussion on the results of the Geneva, N.Y. Trials of 1852, where mechanical reapers were matched in competition with cradlers cutting 15 acres, point out that the latter 'was a large area for an early reaper to cut in one day.' See Hutchinson, *op. cit.*, p. 336 for the McCormick warranties, and *ibid.*, p. 73, for a characteristic example of McCormick advertising giving the savings achieved with the reaper, based on an assumed daily cutting capacity of 16 acres.

[2] *Ibid.*, p. 128.

[3] See Parker, *op. cit.*, Table 5, 14. Wheat and oats yields were higher in the Northeast and Middle Atlantic states.

[4] Rogin, *op. cit.*, p. 135, gives the same savings in cutting 10 acres a day, but in some places says 10 to 12 acres per day.

it became especially prone to shattering.[1] The second point is somewhat more complex. Both the hand-rake and the self-rake models left the grain swept into gavels, or lying in swath, for the binders who followed behind the machine. Rogin[2] notes that binding behind machines was done at a faster rate than when the binders followed cradlers through the fields, and allows an additional saving of labor (amounting to two binders per day in harvesting 10 to 12 acres) on this account. This leads him to state that the total saving in labor connected with the use of the reaper amounted to 5 man-days, instead of just 3 in the cutting operation itself. However, it is pointed out that the reduction in the labor requirements for binding that accompanied the mechanization of grain-cutting may be attributed to the increased pressure put on the binders to get the grain out of the way of the machine, and the consequent adoption of the system of 'binding stations' in place of the traditional practice of permitting binders to range over the field at their (comparative) leisure.[3] Since it would appear that the 'saving' of the hire of two binders per 10 to 12 acres was achieved by having those employed behind the reaping machines work at a faster pace, the binders thus engaged would ask a higher daily wage, and, so long as there was work available for those who preferred the less demanding task of binding for cradlers, the differential wage would have to be paid. Thus, in terms of the cost of labor to the individual farmer, the greater efficiency exacted in binding by the introduction of the reaper cannot be considered to have resulted in any further economies. Indeed, on this account the reaper may have entailed additional monetary costs for the farmer, since when the system of binding stations was employed the use of women and children was precluded by the pace of the work.[4]

[1] Note that there is no evidence of smaller losses from shattering resulting from machine cutting as contrasted with hand cutting of *equally* ripe grain. Yet the tradition has grown up in the secondary literature of treating the reduction of shattering losses owing to more rapid harvesting as distinct and supplementary to the reduction in labor requirements with the mechanical reaper. See e.g., Bidwell and Falconer, *op. cit.*, p. 294; and Robertson, *op. cit.*, p. 258, where specific mention is made of the reduction of output losses due to damage from the elements as a further benefit bestowed on the farmer by the reaper. This would be justified if, for purely technical reasons, expected post-harvest yields as well as harvest labor requirements per acre were different with the two methods.

[2] See Rogin, *op. cit.*, p. 136.

[3] See *ibid.*, pp. 103 and 136, n. 339.

[4] *Ibid.*, p. 103, n. 179. Even if the reduction in manpower requirements for binding were due to purely technical considerations, e.g., the way the grain was delivered from the machines, it would be necessary to take account of the lower daily wage rates typically received by binders, compared to cradlers. In Ohio during the harvest of 1857 binders were paid roughly two-thirds the rate received by cradlers. (See Ohio State Board of Agriculture Report, 1857, pp. 181–2.) With this as a basis for weighting the reductions in labor requirements by relative wage rates, Rogin's procedure should give a total saving in labor with the hand-rake machine equivalent to 4·33 cradler man-days, on 10 to 12 acres, rather than the 5 (heterogeneous) man-days saving frequently mentioned in the literature.

In summary, for a normal day's work cutting 11 acres, the saving in labor per acre (L_s) with the hand-rake reaper is estimated at 0·273 man-days hire of cradlers' services, whereas with the self-raker introduced in the latter half of the 1850s the saving amounted to 0·364 man-days.

Durability and the interest rate

Information as to the average length of life of the reaping machines produced in the Midwest prior to the Civil War is very scanty. Yet those statements regarding durability that have come to light agree in placing the useful life of a reaper at close to ten years.[1] This figure apparently assumed good care and normal use of the machine. Just what constituted 'good care,' and how closely such a standard was approached by the notoriously casual practices of American farmers during the nineteenth century in the maintenance and storage of equipment, is extremely difficult to say.[2] What 'normal use' entailed is somewhat clearer; for, as a rule, the rate of ripening of the grain confined the feasible period for the harvest of small grain crops to roughly a two-week interval, with ten full working days.[3] Thus, approximately 110 acres of wheat would represent the annual normal use cutting capacity of a reaper. On the assumption that the life of the machine would not be prolonged beyond ten years by utilization at rates below the capacity level imposed by custom, the available hours of daylight, and the length of the period within which the grain could be safely gathered, we shall take the annual straight-line depreciation rate as $d = 0{\cdot}10$.

Selection of an appropriate rate of interest proves to be quite simple. It appears that the McCormick Co. charged farmers a standard rate of 6 per cent on the unpaid balance of their reaper notes throughout the 1850s.[4] Even if the farmer was able to secure a higher rate of return by placing his funds elsewhere than in a reaper, the 6 per cent is nonetheless appropriate from an opportunity cost viewpoint; so long as the McCormick Co. was willing to encourage sales by foregoing the opportunity to earn more than 6 per cent on its funds, the farmer who purchased a reaper would find it profitable to borrow at that rate and free such capital as he had for investment at higher rates of return.

Combining the 10 per cent depreciation rate and the simple interest cost, according to formula (2), we estimate the annual average rental rate of a reaper at 13 per cent of its purchase price.

[1] See Hutchinson, *op. cit.*, pp. 73 and 311, and Rogin, *op. cit.*, p. 95.

[2] See Harry J. Carman, 'English Views of Middle Western Agriculture, 1850–1870,' *Agricultural History*, 13 (January 1934), p. 13; and Hutchinson, *op. cit.*, p. 365.

[3] See Rogin, *op. cit.*, p. 95.

[4] See Hutchinson, *op. cit.*, pp. 337, 369, n. 31.

Threshold functions and relative factor-prices

The data assembled in the two preceding sections define a pair of functions relating threshold size to the relative price of reapers – one for the choice between cradling and hand-rake reaping machines, the other for the choice between cradling and self-rakers. With the parameter values substituted for d, r, and L_s in equation (3), the functions are found to be,

$$\text{hand-raker versus cradle:} \quad S_T = 0{\cdot}4765 \frac{C_{HR}}{w}$$

and

$$\text{self-raker versus cradle:} \quad S_T = 0{\cdot}3572 \frac{C_{SR}}{w}$$

Both functions are shown in Figure 9, but, since the self-raker was introduced in the Midwest only after 1854 and the reapers for which reliable price information is readily available are those of the hand-rake type manufactured by C. H. McCormick in Chicago, rather than the Wright–Atkins self-raking variety, the present discussion has been restricted to consideration of the threshold levels for adoption of the hand-rake machine.

We have also chosen to avoid the complications that would arise in attempting to work out the threshold function for the basic McCormick reaper with the mower attachment (perfected in 1855) that enabled the machine to cut grasses as well as grain. The technical advance embodied in the reaper-mowers obviously contributed to the abandonment of the cradle by permitting the fixed cost of the machine, which was sold either with or without the mower attachment, to be spread over a greater total (grain and grass) acreage harvest. It thus became profitable to use the reaper on farms with small grain acreages too low to justify the expense of a machine that could not also be converted to cut grass crops.[1] Similarly, since the advent of the self-raker did not drive the McCormick hand-rake machines out of the market, it must be inferred that the former were not less expensive than the hand-rake model on which McCormick continued to rely until after the Civil War. Instead, they must have been priced low enough to drop the threshold grain acreage at which their introduction as a replacement for the cradle was profitable to a level under the hand-raker threshold.[2]

[1] On the mower and the reaper-mower, see Hutchinson, *op. cit.*, pp. 309–16 and Rogin, *op. cit.*, pp. 96–102.

[2] From the threshold functions given above, one would surmise that, at a maximum, the self-raking machine could have carried a delivered price as much as a third higher than that of the hand-rake reaper, and still have been competitive with the latter, i.e., $(C_{SR}/C_{HR}) = (0{\cdot}4765/0{\cdot}3572) = 1{\cdot}33$.

The price of a reaper

From 1849 through the harvest of 1853 the price of the McCormick hand-rake reapers actually sold to farmers averaged 113 dollars plus freight charges, whereas the average unit price, also f.o.b. Chicago, for the period 1854–8 was 133 dollars.[1] Comparable average figures for the transport charges obviously present a greater problem, for such charges varied from place to place according to the distance from Chicago and the mode of transportation available.[2] An element of arbitrariness is thus unavoidable in fixing upon any particular figure to represent the transport cost component of the total delivered price of a reaper. For the years before 1854 we shall take the 11 dollar freight charge paid by the farmers in Peoria, Illinois, to get their reapers by canal and steamboat from the McCormick Factory wharf in Chicago; in 1854, by which time Chicago's tributary railroad connexions in the West totaled 814 miles of line, the printed McCormick sales forms stipulated a maximum charge of 5 dollars for freight, which we shall accept as representing the average transport cost during the period 1854–8.[3] Therefore, in the absence of any additional warehousing charges – sometimes imposed on the farmer by local sales agents – the delivered price of a hand-rake reaper averages out at roughly 124 dollars for 1849–53, and 138 dollars in years immediately thereafter. The estimate for 1849–53 is in close agreement with the figure of 125 dollars cited in 1851 by an Illinois periodical, the *Prairie Farmer*, as the amount the farmer would have to spend for a reaper.[4]

The cost of labor

Obtaining appropriate figures for the average daily money wage paid to grain cradlers during the harvest season poses a problem no less difficult than that of settling upon representative delivered prices of reaping machines during the 1850s. It must be noted, first, that daily wages for

[1] Average unit prices, reflecting the proper mixture of cash and credit prices on actual transactions have been computed from financial data of the McCormick Co. presented in Hutchinson, *op. cit.*, p. 369, n. 60. The figures given above agree fairly closely with the announced f.o.b. prices of 115 dollars prior to 1854 and 130 dollars for the harvest of 1854 (*ibid.*, p. 323).

[2] Strictly speaking this was not true: the McCormick Co., when competing for sales with its rivals in the Eastern market – outside of McCormick's home territory in the region west of Cincinnati – absorbed the total freight charges and sold reapers on the seaboard at Chicago prices on at least one occasion during the 1850s (see *ibid.*, p. 324).

[3] On transport costs, see *ibid.*, pp. 322–3. The Western tributary railroad mileage figure is for 1 January 1854. At the beginning of the previous year, Western tributary rail mileage amounted to only 157 miles. (See David, *op. cit.*, Appendix C, Table C.5.)

[4] See *Prairie Farmer*, Vol. XI (November 1851), p. 482.

farm labor in the Midwest during this period appears to have been roughly 37 per cent higher than the *per diem* rate received by workers hired on a monthly basis, even after allowing for the money equivalent of board usually provided workers hired by the month.[1] The mass of wage rate quotations for farm labor, however, reports monthly wages in addition to board and not the wage cost to the farmer. To get an idea of movements in the general level of daily wage rates for farm labor one must, therefore, rely on the movements in the level of daily rates for common labor during the decade. Finally, the existence of skill differentials within the structure of farm wages, often overlooked in studies that deal with problems of industrial skill differences and skill margins, must be taken into account. Cradlers tended to receive higher money wages than those who followed in their wake through the fields during the harvest; the differential was as much as 50 per cent over the general average daily wage paid farm workers during the harvest of 1857 in Ohio.[2]

To obtain estimates of cradlers' daily wage rates in Illinois we shall start with two quotations for the common day labor wage in the state: the rate of 85 cents per day reported as the state average by the Census of 1850 can be taken as roughly representative of rates prevailing in the period 1849–53, while from a number of contemporary sources it appears that $1.25 per day is an appropriate average rate for the 'boom' years 1854–7.[3] Adjusting these figures on the assumptions that they were equal to the general level of money rates in the agricultural sector and that cradlers received a constant 50 per cent differential over the general farm labor rate, one arrives at $1.27 and $1.87 as estimates for 1849–53 and 1854–7, respectively. In view of the report at the end of the 1850s (after the collapse of the Midwestern construction boom) that Wisconsin farmers were still paying harvest workers wages ranging from 1 dollar to 3 dollars per day,[4] the estimate of $1.87 per day for the boom does not appear too generous. If the level of the estimated rate for the latter period is accepted, that for

[1] See the data given for Indiana, and discussion of farm wages in Stanley Lebergott, *Manpower in Economic Growth: The United States Record since 1800* (New York: McGraw-Hill, 1940), pp. 244–5.

[2] See Ohio State Board of Agriculture, *Report* (1857), pp. 181–2. At the end of the 1840s, Wm. Marshall of Cordova, Illinois, testified that many farm laborers seeking work in the harvest did not know how to use a grain cradle. It is quite likely that they were recently arrived immigrants, probably Irish but possibly Germans. See 'McCormick Extension Case, Patent of 1845,' pro testimony, pp. 48–9, cited in Hutchinson, *op. cit.*, p. 206, n. 8.

[3] See J. D. B. DeBow, *Compendium of the Seventh Census*, p. 164; quotations for the mid-fifties cited in Gates, *op. cit.*, p. 278 and *The Illinois Central Railroad and Its Colonization Work* (Harvard Economic Studies, 42) (Cambridge, Mass.: Harvard University Press, 1934), pp. 94–8; also, David, *op. cit.*, Ch. 5, p. 35.

[4] See *Chicago Daily Democrat*, 19 July 1860, cited in Bessie L. Pierce, *A History of Chicago*, Vol. II: *From Town to City, 1848–1871* (New York: Alfred A. Knopf, 1940), p. 156, n. 31.

the earlier years follows from the assumption that over short periods the whole structure of farm wages moved with the regional common labor wage rate. Since the most plausible objection that could be raised against this assumption is that the adoption of mechanical reaping may have prevented cradlers' wages from rising as rapidly as the general level of farm wages (and the common labor rate) it appears that $1.27 cannot be regarded as *too high* an estimate of the daily wage received by cradlers at the close of the 1840s. Actually, that estimate comes remarkably close to the figure of '$1.25 or more' per day mentioned in contemporary sources as being paid for 'extra laborers' taken on during the 1847 harvest in Illinois.[1]

As we have already noted, farmers paid considerably more for a day's work when labor was hired by the day than when a monthly agreement was made. With a 37 per cent differential required to compensate farm hands for the insecurity of day labor, the implied *per diem* wage for cradlers hired by the month would have been 80 cents instead of $1.27 for 1849–53, and $1.18 instead of $1.87 during 1854–7. Of course, given that differential, it would be cheaper for the farmer to hire a worker by the month only if his services were needed for 17 or more out of the 26 working days in a month, that is, for a period considerably longer than the average 10 working days within which small grain crops like wheat could be safely harvested on a single farm. Farmers undoubtedly hired some hands on that basis, and we shall therefore make use of the lower rates to compute the relative price of a reaper when labor was hired by the month. Nevertheless, it is more reasonable to suppose that the marginal workers replaced in the harvest by the mechanical reaper were those that would have been taken on at the higher daily wage rates just for the duration of the harvest.

3. ESTIMATES OF THRESHOLD SIZE AND ACTUAL SMALL GRAIN ACREAGE

In Table 12 the delivered prices of reapers and the alternative wage rate data developed in the preceding section have been brought together to provide a set of estimates of the relative price of a hand-rake reaper for the early and middle years of the 1850s. Columns 3 and 6 of the table present the corresponding threshold acreages at which the costs of harvesting with cradles and with the reaper would have been exactly equal. The latter are computed from the threshold function for the hand-rake reaper given on p. 226, and provide the basis for the discussion of Figure 9 in the text.

[1] *Prairie Farmer*, October 1847, p. 324, and January 1848, pp. 26–7.

The acreage figures in Table 12 are discussed in conjunction with estimates of the actual average acreage sown with small grain crops on farms in Illinois during the 1850s, which are presented in summary form by Table 13. A few words about the sources of the latter figures are therefore in order.

The estimates for the harvest of 1849–50, in panel I of Table 13, are based on the statistics given in the Census of Agriculture for 1850 on the volume of wheat, oats, and rye harvested and the number of farms in each county of Illinois, as well as in the state as a whole.[1] From this information, the bushel harvest per farm of wheat, and of oats and rye (combined) were obtained and then converted into acreage estimates per farm for these crops by employing coefficients of the average grain per acre. For the harvest of 1859–60 (in panel II) only the estimate of the state average is given since the Census of 1860 does not supply statistics of the numbers of farms on a county basis.[2]

Fortunately, the choice of appropriate average grain yield coefficients for Illinois at the beginning and end of the 1850s is immensely simplified by the availability of William Parker's recent exhaustive research on the problem of wheat and oat yields per acre in nineteenth-century America. In comparison with oats, the rye crop of the Western states was insignificant prior to the Civil War; application of the yield estimates for oats to the combined oats and rye output statistics – figures for the harvest of these crops not being given separately on a county basis by the Census of 1850 – should not, therefore, introduce serious error into the acreage estimates. In constructing the acreage figures shown in Table 13, the yield of wheat in Illinois from the harvests of 1849–50 *and* 1859–60 was taken to have been 11·3 bushels per acre. The yield of oats we take to have been 30·5 bushels per acre. The following evidence, drawn from Parker's work, may be adduced in justification of these coefficients:

1. Between 1849 and 1859 the average yields of wheat and oats in the region comprising Ohio, Indiana, Illinois, Missouri, and Iowa were virtually unchanged; the wheat yield dropped from 12·4 to 12·1 bushels per acre, and the oats yield rose from 29·3 to 30·4 bushels per acre. (Parker, *op. cit.*, Table 10.)

2. From the U.S. Department of Agriculture revised data on acreage yields in the period 1866–75 it is found that the wheat yield in Illinois was slightly below the typical yields in the Midwestern region, whereas the yield of oats was virtually the same as that for the region as a whole. The Illinois yield of wheat per acre was 11·3 bushels, compared with the mid-range figure of 12·2 bushels per acre for the group of states mentioned

[1] J. D. B. DeBow, *Compendium of the Seventh Census*, pp. 21–7.

[2] See *U.S. Eighth Census* (1860), 'Agriculture,' pp. 31–5.

TABLE 12 Threshold grain acreage for adoption of hand-rake reapers in Illinois

	Period: 1849–53			Period: 1854–7		
	(1)	(2)	(3)	(4)	(5)	(6)
	w	$\frac{C_{HR}}{w}$	S_T	w	$\frac{C_{HR}}{w}$	S_T
Terms of hire of farm labor	cradlers, in current dollars	in man-days per reaper	acres harvested	cradlers, in current dollars	in man-days per reaper	acres harvested
Day	1·27	97·6	46·5	1·87	73·8	35·1
Month	0·80	155·0	73·8	1·18	117·0	55·6

above, while the Illinois oats yield was 30·5 bushels per acre compared with the mid-range figure of 30·3 for the region. (Parker, *op. cit.*, Table 1.)

Since the Department of Agriculture regional yield data for 1866–75 agree so well with the regional data for 1849 and 1859, the U.S.D.A. data on Illinois were accepted as appropriate for the decade of the 1850s.

The average yield coefficients employed fcr Illinois may be, if anything,

TABLE 13 Average acreage per farm in Illinois

Region	Number of farms	Wheat acreage per farm	Total wheat, oats, and rye acreage per farm
I. Harvest of 1849–50			
State	76,208	11·0	15·4
Major wheat counties[a]	21,634	19·3	25·2
Winnebago Co.	919	30·4	37·2
Cook Co.	1,857	11·4	18·6
II. Harvest of 1859–60			
State	143,310	14·6	18·3

[a] The group of leading wheat-growing counties which accounted for 50 per cent of the total wheat production in the state.

somewhat low for the major grain regions of the state.[1] Consequently, if the derived acreage estimates do contain a bias, it is certainly not one which would favor our conclusion that a considerable gap existed between the threshold size for adoption of the hand-rake reaper and the actual average small grain acreage per farm in the early 1850s.

[1] A sample of Illinois county estimates drawn from different dates in the period 1843–55 gives a median wheat yield of 16 bushels per acre, and a median yield for oats of 40 bushels per acre. See Parker, *op. cit.*, Table 3, p. 8.

5 The landscape and the machine: technical interrelatedness, land tenure and the mechanization of the corn harvest in Victorian Britain

The observation that the mid-nineteenth-century farming landscape in Britain was not congenial to the introduction of agricultural machinery forms the point of departure in this inquiry. There is certainly little novelty attached to this observation. Contemporary writers in the agricultural press frequently discussed the problems of using horse-drawn and steam-powered equipment in field operations where terrain conditions were 'unsuitable,' and references to these difficulties dot the major scholarly works on the modernization of British agriculture. The admirable account of the 1846–1914 era provided by Christabel S. Orwin and Edith H. Whetham is, perhaps, rather unusual in going beyond mere mention of the subject to suggest that terrain conditions exerted a powerful influence upon the extent and the spatial pattern of farm mechanization. In reference to the situation that existed in Britain during the period immediately following Repeal of the Corn Laws, Orwin and Whetham have written:[1]

> Broadcast sowing, hand weeding, the sickle and the scythe were still generally used on many farms, especially where rough or stony soils shook to pieces the new-fangled inventions of the agricultural engineers.... To be effective and economic, the new implements required a new system of farming – large level fields with straight hedges and wide gateways, and with no boggy patches and land-fast stones. It is not surprising therefore that the new implements and the new type of farming were to be found mainly on the eastern side of the country, from Lincolnshire through Northumberland and the Lothians to Aberdeenshire, on the big fields newly created from marsh and stones, and farmed in large units.

Yet, it must be agreed that the influence of the landscape inherited from long periods of agricultural settlement has still to be acknowledged as one of the dominant themes in the agrarian history of nineteenth-century Europe. Indeed, the point that the physical state of the land and the arrangement of existing fields, fences and buildings may have been of some consequence to the reception farmers accorded the new techniques of production, generally fails to be properly understood and appreciated. For one thing, it has tended to be all but submerged in

[1] C. S. Orwin and E. H. Whetham, *History of British Agriculture 1846–1914* (London: Archon Books, 1964), p. 10.

the flow of words currently lavished upon the putative role of rural 'labor abundance' as the crucial condition inhibiting farm mechanization throughout western Europe. This prevailing theme, which could be illustrated with innumerable passages from the recent literature dealing specifically with the British experience, is concisely conveyed in Folke Dovring's comments on the slow European adoption of agricultural machinery:[1]

> Machinery primarily designed to save labour could not be expected to spread very much until labour could be said to be scarce in one sense or another. Such was the case in America through most of the nineteenth century; but in fully settled western Europe, farm labour could become scarce only after the agricultural population had ceased to increase in absolute numbers and, to a higher degree, after it had begun to fall.

A re-evaluation of the validity of the 'American labor scarcity thesis' – and that of its obverse, the hoary 'British labor abundance doctrine' – as it has been applied to agricultural technology, is certainly long overdue.[2] But I must beg leave to defer treatment of this larger subject to another occasion. It is not my purpose here to adduce arguments and detailed evidence controverting the oft-stated, but sometimes only implicit proposition that comparative labor abundance *vis-à-vis* the United States bore principal responsibility for delaying or retarding the adoption of mech-

[1] F. Dovring, 'The Transformation of European Agriculture,' *Cambridge Economic History of Europe*, eds. M. M. Postan and H. J. Habakkuk (Cambridge: Cambridge University Press, 1965), Vol. 6, Part II, p. 645. Dovring's statement is not unrepresentative in its lack of precision regarding 'scarcity'; since all economic goods are scarce, something more must be meant. But if increased scarcity relative to other goods, or a rise in relative price compared with the relative price of the commodity in another country or region, is what we have in mind in speaking of farm labor 'becoming scarce,' then the required conditions cited by Dovring are palpably not necessary. For parallel statements relating specifically to Britain, among recent contributions to the literature cf., e.g., John Saville, *Rural Depopulation in England and Wales 1851–1951* (London: Routledge and Kegan Paul, 1957), pp. 140–1; E. L. Jones, 'The Agricultural Labour Market in England, 1793–1872,' *Economic History Review*, 2nd Series, Vol. 17 (1964), pp. 322–38; W. Harwood Long, 'The Development of Mechanization in English Farming,' *Agricultural History Review*, Vol. 11 (1963), p. 22; J. D. Chambers and G. E. Mingay, *The Agricultural Revolution 1750–1880* (London: B. T. Batsford, 1966), pp. 186–90.

[2] Although the choice of techniques in both industry and agriculture received attention in H. J. Habakkuk's stimulating book *American and British Technology in the Nineteenth Century* (Cambridge: Cambridge University Press, 1962), the discussion provoked subsequently in the journals has dealt almost exclusively with industrial technology and the influence exerted upon it by national factor endowment. Cf., e.g., Peter Temin, 'Labor Scarcity and the Problem of American Industrial Efficiency in the 1850s,' *Journal of Economic History*, XXVI (September 1966), pp. 277–98; R. W. Fogel, 'The Specification Problem in Economic History,' *Journal of Economic History*, XXVII (September 1967), pp. 298–308.

anized techniques in British (and western European) farming during the nineteenth century. I do mean, nevertheless, to advance the independent claim for considering the state of the farming landscape of mid-Victorian Britain as a significant impediment in the way of rapid and widespread mechanization of field operations during the second half of the century. Although I shall not be able to refrain from remarking upon the differences separating Britain and America in this respect, and in others closely related to it, the focus of the discussion must remain almost exclusively upon the British side of the story; an explicit assessment and explanation of Anglo-American differences in the extent and rate of diffusion of agricultural innovations would pose complications demanding treatment at much greater length.

Even so, from the study of one such innovation – the mechanical reaper – it will become apparent that the substitution of horse-powered machines for manual labor on Britain's farms was a more complex affair than naïve neoclassical models of factor substitution allow; it transcends the over-simplified general theoretical terms in which the choice of techniques in Britain versus that in America has lately come to be discussed; it forces an encounter with technical and institutional conditions whose significance has for too long passed without adequate consideration from either economists or historians. Limited as the case of the reaping machine is, the complexities it raises do seem well worth confronting and trying to analyze in rigorous fashion.[1]

THE LANDSCAPE AND THE REAPING MACHINE

When the Crystal Palace was opened in 1851, the reaping of grain crops with machines as a practical matter was essentially unknown anywhere in the British Isles, and the appearance of the 'American Reapers' sent by Obed Hussey and Cyrus H. McCormick to the Great Exhibition created a considerable stir of interest. Given the prevailing structure of harvest

[1] The present paper forms part of a comparative, monograph-length work now in preparation: *American Reapers and British Fields, The Mechanization of Grain Harvesting in Britain and America Before 1875.* I have made use here of some findings contained in this larger study, without presenting explicit documentation on points which are ancilliary to the central argument of the present paper.

In the course of my research on harvest mechanization in Britain I have had the benefit of comment and criticism from many friends and scholars – on both sides of the Atlantic. Proper acknowledgement of these debts must wait until the study as a whole is published. Meanwhile, for having extended their financial support at one time or another to this undertaking (as part of an ongoing project on the Trans-Atlantic Diffusion of Technology in the Nineteenth Century), the Ford Foundation, the Warden and Fellows of All Souls College, Oxford, and the Stanford University Committee on Research in International Studies, all have my expressed gratitude.

wage rates, the overdraft rate of interest at which equipment purchases might be financed, and the prices at which these machines could be purchased in the years around 1851, manual sheaf-delivery reapers like those exhibited at Hyde Park would have represented a quite profitable investment for a British farmer whose acreage under cereals matched the average of the farms situated on 'the corn side of the island' – provided his fields offered suitable conditions under which to operate the device.[1] Such field conditions were to be found in the Lothians, for example; and it is notable that by the early 1860s virtually the whole of the corn-harvest of that region was being cut by machine.

The progressive Lothians formed a British topographical counterpart to the American Midwest, but – unlike the latter region – carried little quantitative weight on the national farming scene. Although the corn-harvest of this lowland district might have been entirely mechanized, my estimates indicate that by 1863 rather less than 6 per cent of Great Britain's wheat, oats, rye, and barley acreage was being cut with reaping machines. The corresponding measure of the extent of the mechanical reaper's adoption in the United States had already passed the 15 per cent mark on the eve of the Civil War, and climbed above 78 per cent by 1870, whereas in 1874 probably more than 53 per cent of the British corn-harvest was still being cut by the sickle and the scythe.[2]

[1] The basis and full import of this assertion cannot be adequately discussed here. Skeptics may, however, consult the data presented below (in Table 14) which pertains to the threshold wheat, rye, and oats acreage for 'simple mechanization' in 1851. Note, then, that the average amount of arable devoted to wheat, rye, and oats on *all* farms in England and Wales at the time was probably close to 23 acres. On what James Caird called 'the corn side of the island,' the average would have stood well above the 25 acre break-even point indicated by Table 14.

[2] The details of the estimates of the overall extent of diffusion of mechanical reaping are too lengthy to be reproduced here, but a preliminary draft of the chapter of the larger study in which they are developed can be made privately available to the impatient. It should be noted that the figures cited in the text for the United States relate to the proportion of national acreage under wheat, oats, rye, *and barley*, and supersede the diffusion measures offered in P. A. David, 'The Mechanization of Reaping in the Ante-Bellum Midwest,' Ch. 1 in *Industrialization in Two Systems*, ed. H. Rosovsky (New York: Wiley and Sons, 1966), p. 10, n. 16. Keen-eyed readers will also note that the figures given here for Great Britain differ from those attributed to me by E. J. T. Collins, 'Labour Supply and Demand in European Agriculture 1800–1880,' Ch. 3 in *Agrarian Change and Economic Development, The Historical Problems*, eds. E. L. Jones and S. J. Woolf (London: Methuen, 1969), p. 75. The preliminary estimates cited by Collins, though in error, are nevertheless sufficiently accurate for the comparative use which Collins makes of them. As for the extent of harvest mechanization in the Lothians, cf. the data for the East Lothian harvest of 1860 from the *North British Agriculturalist*, cited by Jacob Wilson, 'Reaping Machines,' *Transactions of the Highland and Agricultural Society of Scotland*, 3rd Series, XI (January 1864), p. 144; R. Scot Skirving, 'Ten Years of East Lothian Farming,' *Journal of the Royal Agricultural Society of England*, 2nd Series, I (1865), p. 108.

One cannot hope to understand this divergence between British and American experience during the third quarter of the nineteenth century without observing that the landscape of Britain's principal grain-producing districts – especially if viewed in juxtaposition to the United States' setting – was on balance inimical to the immediate successful introduction of the mechanical innovation that had been suddenly brought to the attention of her farmers through the medium of the Great Exhibition. While it has recently become less than fashionable to notice them, the nineteenth century offered significant contrasts between Britain and America in dimensions other than that of the relative prices of the factors of production. In the United States the broad, level and stone-free prairies of the Midwest, where grain production was becoming increasingly concentrated even before the Civil War, provided extensive regions whose topography was singularly well suited to the operation of horse-drawn machines in field work. And it was, of course, in precisely that physical setting that the cumbersome reapers designed by Hussey, McCormick and their followers, first gained popular acceptance. As I have elsewhere shown,[1] so technically hospitable an environment in itself was not sufficient to induce the farmers of the American Midwest to abandon hand methods of harvesting the small grains prior to the 1850s. At the half-way point in the century, American grain farming was actually not so much further down the road toward harvest mechanization than British agriculture: less than 1 per cent of the acreage under wheat, oats, rye, and barley in the United States appears to have been cut by reaping machines in 1849–50. On the other hand, the favorable physical conditions did mean that during the 1850s – when an alteration of the structure of Midwestern factor-prices combined with a rise in average farm acreage devoted to small grain crops, providing a stronger inducement to substitute machinery for labor on the region's farms, the mechanization of the harvest could proceed without the limitations that terrain problems would have imposed and actually did impose in sections of the Northeast and the seaboard South.

Two features of the farming landscape in Britain must be considered as having obstructed the progress of mechanical reaping: the character of the terrain – which is to say, the nature of the field surfaces across which the implements would have to be drawn; and secondly, the size, shape, and arrangement of the fields – or more generally speaking, the layout of farms' cereal acreage.

From the outset terrain problems were recognized. They occasioned the single note of caution Phillip Pusey felt obliged to sound in reporting

[1] Cf. David, 'The Mechanization of Reaping,' esp. pp. 26–8 for discussion of the relationship between the effects of the Midwestern economic environment and the effects of the physical environment on mechanization in the United States.

to the Royal Agricultural Society on the results of his trial of the McCormick reaper in 1851:[1] 'This trial was witnessed by many farmers, and no fault was found with the work. The land, I should say, however, being stock land is even; where ridges and water-furrows exist, some difficulties seem to arise.' The exact nature of the manifold 'difficulties' on ridge-and-furrow land soon became evident, and can be adumbrated from scattered reports of the performance of British, as well as of American machine designs, under trial and practical field conditions during the 1850s and 1860s.

The larger, heavier models – like Croskill's resurrection of the Bell reaper dating from 1826–8, and the Burgess and Key swath-delivery adaptation of McCormick's machine – were especially likely to be found 'now and then sticking fast into the uneven ground, especially when the off wheels descended into the furrow...'[2] Secondly, where the ground was not level, and particularly where high-ridge ploughing had left a pronounced corduroy field-profile, the reaping machines' draft might prove too great: a normal, easily managed two-horse team could not maintain a steady rate of forward motion sufficiently rapid to prevent the reciprocating knives of the cutter bar from clogging.[3] The larger the swath taken, the more serious was the problem of draft, and hence the poorer the performance of the machine was likely to be over uneven terrain. Smaller models, by cutting a swath narrower than the 5 ft 3 in. standard, 'were certainly enabled to do their work rather better: but, on the whole, not so thoroughly satisfactorily as we could have wished to witness.'[4] Thus, among the early machines that cut the standard-width swath, manual delivery reapers on Hussey's pattern won considerable favor in many parts of Britain for reasons other than their low cost:[5] 'They are lighter of draft, less liable to derangement... more easily managed, and thus more to be depended upon for the regular performance of a fair amount of daily work than their heavier rivals.' And when sturdy but lighter, more compact, and more efficiently geared reaping machines became available in the 1860s, they alone would be considered by farmers in hilly districts, or in areas like the North Riding of Yorkshire where the

[1] P. Pusey, 'On Mr. McCormick's Reaping-machine,' *Journal of the Royal Agricultural Society of England*, XII (1851), p. 160.

[2] *Farmer's Magazine* (London), September 1856, p. 189.

[3] Cf. Orwin and Whetham, *British Agriculture*, p. 112 on the problems caused by inefficient gearing, especially noticeable on Bell machines, where the cutter was placed far forward of the driving wheels. The use of additional horses would not, however, significantly increase speed in the case of 'pusher'-type machines on the Bell pattern, Cf. Wilson, 'Reaping Machines' (1864), p. 131.

[4] *Farmer's Magazine*, September 1856, p. 189.

[5] John Wilson, *British Farming* (Edinburgh, 1862), pp. 147–8.

old ridge-and-furrow system remained.[1] Finally, the pronounced undulations in such terrain also adversely affected the quality of the work. For, if the machine's cutter bar was set to take the grain properly at the crown of the ridge, it left the stubble too high in the furrows, causing losses of straw and troubles for the binders and stookers who would have to cope with the sheaves of varying lengths.

Elsewhere on the island, the progress of wheeled implements like reaping and mowing machines was obstructed by the presence of open irrigation trenches and deep drainage furrows, or 'water-cuts.'[2] This source of difficulty was especially noticed in Essex and Suffolk, where fields customarily were ploughed in long 'stetches' of varying width, each stetch being limited on its two sides by deep furrows that served as gutters to carry off the surface water.[3] On heavy lands, as in the Suffolk strong-loam belt, close spacing of water-cuts of this kind were required and dictated stetches as narrow as 2 yards across. Narrow stetches, in addition, were usually ridged up slightly further to improve the drainage. Such fields proved notoriously inhospitable to the early reaping machines, and at later dates to self-binders and combine-harvesters, for, 'the continual jolting over deep furrows soon put the most robust machine out of action.'[4]

Of course, in the absence of effective sub-surface drainage, grain fields adequately drained by high-ridge ploughing or water-cuts would at least have been freer from the hazards encountered in negotiating soggy ground with a horse-drawn reaping machine weighing upward of a half-ton. Additional horses could be used to overcome the greater friction on wet fields, although withal, as one farmer despairingly reported, 'at the turns the [reaping machine's] wheels sunk so deeply into the soft ground, that care and the occasional aid of a lever were required.'[5] In some districts these problems were especially serious. The fen arable in Huntingdon

[1] Cf. William Wright, 'On the Improvements in the Farming of Yorkshire since the date of the last Reports in the Journal,' *Journal of the R.A.S.E.*, XXII (1861), p. 127, on the contrast between the type of reaping machine which could be used in the North Riding and on farms in the Yorkshire Wolds. For parallel evidence primarily relating to Scotland, cf. Wilson, 'Reaping Machines' (1864), pp. 131, 135–9. On the progress of 'agricultural mechanics,' as applied to building mechanically more efficient reapers, cf., e.g., John Coleman and F. A. Paget, 'General Report on the Exhibition of Implements at the Plymouth Meeting,' *Journal of the R.A.S.E.*, 2nd Series, I (1865), pp. 373–405.

[2] Cf. W. J. Moscrop, 'A Report on the Farming of Leicestershire,' *Journal of the R.A.S.E.*, 2nd Series, II (1866), p. 321, on irrigation trenches preventing the use of grass mowing machines, although hay-makers and horse-rakes could be used.

[3] James Caird, *English Agriculture in 1850–51* (London, 1852); G. E. Evans, *The Horse in the Furrow* (London: Faber and Faber, 1960), pp. 30–1.

[4] Evans, *Horse in the Furrow*, p. 31.

[5] C. Lawrence, 'Letter on the Use of the Reaping Machine, and the Root-Crops in 1860,' *Journal of the R.A.S.E.*, XXI (1860), p. 551.

was said to be 'mostly too soft to carry reaping machines,'[1] while on Lincolnshire's heavy-clays – even after underdrainage was installed – in wet weather the horses' hooves sank deeply into the soil.[2] Such considerations had led to the double-furrows separating Suffolk's stetches being used 'in every operation, either in sowing the seed or after the seed was in the ground, as a trackway for the horses, which are thus never permitted to trample or injure the tender soil.'[3] The fact that the water-cuts formed impassable barriers to wheeled implements had contributed to this practice, by forcing cultivation to be done along the length of the stetch. Yet, where the movement of horses and machines was so constrained, the breadth of the swath cut by the generally available makes of reaping machines often turned out to be ill-matched to the width of the cultivated tracts. Indeed, as Caird observed in 1850–1, this was a more general problem and Suffolk farmers had overcome it only by providing themselves with drills and harrows designed to fit the dimensions of their particular stetches.[4] But the reaping machine, not being similarly built up from modular components, proved more difficult to adapt.

The lessons learned at the expense of those who purchased the new machines without heed to the state of their farm's terrain were succinctly communicated to the membership of the Highland and Agricultural Society by Jacob Wilson in 1844:[5]

> Too much, therefore *ought* not to be expected from the reaping machine: and where it *is* employed, care should be taken that the previous preparation and treatment of the land be such as will *allow* it to work to the greatest advantage. Such preparation should consist in the thorough drainage of the land and subsequent levelling of the surface, in the abolition of water-cuts and furrows, well as of large stones, and in properly finishing the process of cultivation by rolling, & c. By careful observance of these few points, the chief obstacles to reaping by machinery will thus be removed.

Farm layout was another matter. Alderman J. J. Mechi, that irrepressible lay apostle of High Farming, in a speech before the Coggleshall Institute in 1850, gave it as his opinion that a farm of 600 acres[6]

[1] J. H. Clapham, *An Economic History of Modern Britain* (Cambridge: Cambridge University Press, 1952), II, p. 269, citing the *Journal of the R.A.S.E.* Prize Essay for 1868.

[2] Cf. John Algernon Clarke, 'Farming of Lincolnshire,' *Journal of the R.A.S.E.*, XII (1851), p. 346.

[3] Caird, *English Agriculture*, p. 153.

[4] Cf. Evans, *Horse in the Furrow*, pp. 30–1; Caird, *English Agriculture*, p. 153. Where 8½ ft stetches called for uncommonly wide sowing and cultivating implements, the gateways into the fields also had to be correspondingly enlarged.

[5] 'Reaping Machines' (1864), p. 149.

[6] Reported in the *Farmer's Magazine*, XXII (July 1850), pp. 20–1, and quoted by Clark C. Spence, *God Speed the Plow, The Coming of Steam Cultivation to Great Britain* (Urbana, Illinois: University of Illinois Press, 1960), p. 121. Mechi's theme on this occasion was the

should represent a square mile with a farmery in its centre having half a mile diverging roads to its extremeties; whereas now, under the system of old custom and inalterability, a farm of that size generally involved the intricate threading of miles of almost impassable green and muddy lanes, with fields of every form except the right one.

Reaping machines did not require rectangular, and preferably square fields as did the steam-plowing rigs which were being tried in Britain – and notably in the Lothians and Northumberland – during these decades. Nevertheless, small and irregularly shaped fields did cause losses of time in turning, particularly when the lie of the grain called for the cutting to proceed in a direction orthogonal to the principal axis of an oblong enclosure. And where very small fields were common – such as those of the 160-acre Devonshire farm visited by Caird shortly after its average field size had been increased to 10 acres by the removal of 7 miles of hedgerows – the early reaping machines would have had to be set to work in two or more enclosures during the course of a single day. Devon was, of course, quite notorious for the smallness of its farms' enclosures, but was not so exceptional as is sometimes thought. In the parts of Durham which had been longest enclosed, average field sizes ran from 2 to 6 acres during the 1850s.[1] Considerable time might thus be spent leaving ancillary harvest laborers less than fully occupied whilst the machine was being resituated. For, it was complained that the reaping machines' form was 'unwieldy, and requires much time and labour in taking to pieces for removal from field or from farm to farm.'[2]

Additional, albeit only marginal, wastage of time was likely to be incurred in moving equipment between adjacent enclosures, due to the separation of fields by hedges through which only a single gate had been let. But the grain ripe for harvesting did not necessarily lie in adjacent enclosures on the farm. Hence, when small dispersed fields required a number of moves during the day, carefully taking the reaping machine along miserable

[1] On Devonshire enclosures, cf. Caird, *English Agriculture*, p. 52 who regarded the presence of hedgerows as terribly injurious even before the appearance of mechanical reapers and mowers: 'Every operation of husbandry is impeded, a constant shifting of implements from field to field occasions waste of time, and does positive damage to the implements and the fields through which they are passed....' Cf. also Thomas George Bell, 'A Report upon the Agriculture of the Country of Durham,' *Journal of the R.A.S.E.*, XVII (1856), p. 111.

[2] Wright, 'Improvements in the Farming of Yorkshire' (1861), p. 127. This criticism was lodged particularly against the versions of the McCormick reaper built and marketed in Britain by Burgess and Key.

---

physical requirements for the successful application of steam-power to cultivation, but with the exception of the special importance of field *shape* for efficient steam-ploughing, many of the same considerations are relevant to the case of mechanical reaping. Cf. *ibid.*, pp. 120, 143–4.

lanes that wound their way through the least well-drained portions of the farm, the overall expenditure of time might amount to a significant reduction of the acreage that could actually be harvested by the cumbersome early models.

Still, in comparison with the problems encountered on farms whose fields remained unimproved by sub-surface draining and leveling, the non-optimal aspects of typical farm layouts in mid-Victorian Britain posed rather subsidiary impediments to the efficient use of reaping machines. Certainly such was the implicit judgement of contemporaries, who fastened quickly upon drainage as the key to the matter. In 1852, James Slight – whose efforts to restore injured Scottish pride, in the aftermath of the American reapers' triumphant appearance, succeeded in retrieving Patrick Bell's swath-delivery machine from its long initial career of obscurity – thought he saw clearly the cause for the workable reaping machines invented in Britain during the first half of the century having 'fallen so much into disuse.' They had simply appeared 'before their time';[1]

> ... that is to say, before the subject on which they were to act had been prepared for their reception. In the first quarter of the present century, furrow-draining, levelling high ridges, and filling up the old deep intervening furrows, were only beginning to assume their due prominence in the practice of agriculture; and so long as these improvements remained in abeyance, the surface of the land was very ill suited for such operations as that of a reaping machine.

Actually, the truth of the matter[2] seems to have been that prior to the late 1840s the high purchase cost of the Bell reaper, relative to the wage of the harvest labor it could displace, would have prevented widespread adoption of the device on the large farms in the Scottish lowlands, as well as south of the Tweed, even if the land improvements described by Slight had not remained largely in abeyance. John Wilson's formulation, offered a decade later, comes closer to the position one is now inclined to maintain, in suggesting that changes in both the state of the terrain *and* rural labor markets explain the curious re-emergence of the neglected

[1] James Slight, 'Report on Reaping Machines,' *Trans. H.A.S.*, 3rd Series, v (January 1852), pp. 193–4.

[2] This is based upon a set of calculations of the 'threshold' acreage, at which it would begin to be profitable to adopt Bell's machine, given the factor-prices prevailing in Scotland during the period from *c.* 1828 to *c.* 1851, and assuming optimal terrain conditions such as those available to Patrick Bell's brother George, who continued to operate the machine on his farm in the Carse of Gowrie, Fofarshire, until sometime in the early 1840s. Slight's 'Report on Reaping Machines,' pp. 193–4, gives some of the details of George Bell's practice. The nature of the 'threshold calculation' for the case of simple mechanization is reviewed by Paragraphs 1–3 of the Technical Notes in Appendix A, below.

Bell reaper, which began to take its place alongside the American-style machines in Britain's grain fields during the 1850s:[1]

> Not only was manual labour then abundant and cheap...; but thorough draining had made little progress, and the land was everywhere laid into high ridges, presenting a surface peculiarly unfavourable for the successful working of a reaping machine. Now [1862] conditions are reversed.

Wilson rather exaggerated the historical contrast in making his point. Although contemporaries might properly be impressed with the progress of drainage and kindred improvements since the second quarter of the century, the physical transformation of Britain's farmlands was far from complete twenty years after the Great Exhibition opened. It is true that between 1846 and 1876 something on the order of £24 million was spent on drainage and related works, one third of it having been financed with public funds under the terms of the program of government loans for thorough drainage schemes that had been set up by Parliament in 1846–8. Yet Caird thought that more than four-fifths of the farmland in England and Wales which ought to have been drained still stood in need of that treatment in 1873. And on the greater part of the 4 or 5 million acres that had been thus improved by the mid-1870s, the work was executed outside the somewhat exacting conditions laid down by the Inclosure Commissioners as qualification for public loans.[2] Drainage, moreover, was a necessary but not sufficient preparation for the mechanization of field operations. The layout of the existing stetches often conflicted with the desirable arrangement of runs for a more carefully calculated system of sub-surface drains, which might well traverse both stetches and furrows, resulting in an awkward variation in the depth of soil above the installed pipes. Rather than taking the additional time and trouble then to plough out the ridges and level the field, the conflict could be resolved in favor of retaining the existing stetches: 'field drains were therefore sometimes run down the old furrows, leaving the ridges to plague another generation of farmers whose drills and reapers were not adapted to undulating surfaces.'[3] Thus, a quarter-century after the advent of the American reaping machines, there remained much to be done even in Britain's corn

[1] Wilson, *British Farming* (1862), p. 148. Cf. also *The Encyclopaedia Britannica*, Ninth Edition (Edinburgh, 1875), Vol. I (A–ANA), p. 322, for a reiteration of this argument in the article on 'Agricultural Implements and Machinery' contributed by John Wilson.

[2] Cf. Clapham, *An Economic History*, II, pp. 270–1 for contemporary estimates of the cost and volume of drainage improvement work. The reference to James Caird appears in Orwin and Whetham, *British Agriculture*, p. 101. On the operation of the government loan program, including estimates of expenditures during 1847–72 under the various Drainage Improvement of Land Acts, cf. *ibid.*, pp. 194–200.

[3] Orwin and Whetham, *British Agriculture*, p. 101.

counties if the landscape was to be thoroughly cleared of those vestiges of an older system of husbandry which obstructed the harvesting of grain and mowing of rotation grasses by horse power.

And yet this cannot be the whole of the story. While the extent of the mechanical reaper's diffusion in Britain in the period after 1851 may have been at any *given moment of time* effectively circumscribed by the serious difficulties encountered in seeking to operate it on 'unimproved' arable, the pace at which farm improvement – and specifically land improvements – proceeded cannot legitimately be taken as an exogenously determined feature of the British environment. It might be admissible to argue that the lack of technical advances in the design and manufacture of suitable drainage materials in the period before the Repeal of the Corn Laws, had held up extensive land improvement. But clearly, from the early 1850s onward, this technical barrier had been removed by the successful machine production of cylindrical agricultural pipe, which offered drains that were cheap, durable and available in standard sizes for ready installation.[1]

If, as appears to have been the case, mechanization of the corn-harvest would have been a profitable undertaking on a great part of Britain's cereal acreage even at the beginning of the 1850s, supposing only that the more serious among the terrain problems surveyed here could have been first removed, the logic of the argument demands a reason why this significant profit opportunity could have remained only partially exploited for want of the indicated modifications. Putting the matter another way, the proposed explanation of the comparatively limited diffusion of mechanical reaping in Britain during the 1850s and 1860s may be correct so far as it goes; but it is unsatisfyingly incomplete until we also know why the farming community did not more generally heed the advice of Jacob Wilson, and instead suffered the land to remain in a condition that did not permit the machine to be employed to advantage in place of harvesters wielding the sickle and the scythe.

TECHNICAL INTERRELATEDNESS AND FARM MECHANIZATION – RECASTING THE PROBLEM

In seeking an answer to the question just posed, it is illuminating to begin with the perception of most of the physical obstructions to mechanical reaping surveyed by the preceding section as having been manmade, rather than of 'natural' origins. They represented the legacy of past capital formation directed toward improvement of the British farm. Some

[1] Cf. Clapham, *An Economic History*, I, pp. 458–61 on the introduction of pipe-making machinery in the 1840s. See Appendix B, below, for the cost of drainage pipes, and their installation.

of these 'improvements,' such as the drainage afforded by ridge-and-furrow contours, were, to be sure, the durable by-products of cultivation activities extending over a very long span of time. Conceived of in this way, the topography of British agriculture at the mid-point of the nineteenth century bears an immediate resemblance to the structure of industrial plant that – at any given moment – remains as the heritage of investments in specific kinds of durable capital. The latter structure, in its turn, has sometimes been likened to the accumulation of strata deposited by historic geological action;[1] in the agricultural context, however, it proves instructive to reverse the usual direction of the analogy.

For, by drawing attention to the features of the British farm landscape which appears to have obstructed ubiquitous application of the novel harvesting technique, we are in effect acknowledging the importance of compatibility between the various capital goods available for employment in agriculture. The problem of the landscape is one of the 'interrelatedness' of the techniques of grain-harvesting on the one hand, and, on the other, the techniques used for drainage, wind-damage control, the management of livestock, and on-farm transportation of field equipment. These being mutually complementary activities of the Victorian arable farm, any capital goods embodying techniques for their execution within this general system of husbandry must have been, at least, mutually compatible.[2]

To recast the farmer's problem in this formal way may seem strained and too unfamiliar; it is nonetheless useful in calling forcibly to mind the existence of parallel situations in other, non-agricultural branches of production. For, it is widely recognized that attempts to introduce modern machinery into a plant composed of older equipment performing complementary functions may be utterly frustrated by the technical

[1] Cf., e.g., W. E. G. Salter, *Productivity and Technical Change*, Second Edition (Cambridge: Cambridge University Press, 1966), p. 52: '... the [industrial] plants in existence at any one time are, in effect, a fossilized history...'

[2] It is possible that when the advantages offered for application of new techniques in some sub-set of complementary activities are terribly strong, the attempt to introduce a process innovation will issue in the fragmentation and ultimate transformation of a previously interlocking set of operations, or system of husbandry. In Britain during the 1890s, the efforts of a few enterprising arable farmers to use tractors to mechanize their field work led to the demonstration that specialized grain-farming was feasible. Since the early tractors could not deal with row-crops, these were dropped from the rotation; and since the tractors were inefficient at haulage and often could not get into the cattle yards through the narrow entrances, as did the horses, stall-feeding of livestock and the carting of their dung were also dispensed with. The stockless grain farm in Britain was thus the logical consequence of the dogged effort to accommodate the early, non-versatile tractor, as a satisfactory replacement for the horse. Cf. E. H. Whetham, 'Mechanization of British Farming, 1910–1945,' *Journal of Agricultural Economics*, XXI, 2 (May 1970), pp. 6–7.

incompatibility of the new with the old.[1] The inadequacy of the floor layout, pillar spacing, and lighting arrangements in existing factory buildings appears to have hampered the installation of new, more automatic equipment in the British cotton textile industry – providing an oft-cited illustration of simple, yet frequently encountered problems of technical interrelatedness and incompatibility in the manufacturing sector.[2] Perhaps still better known is the instance which Veblen cited as illustrative of the penalties of Britain's having an early start: the impossibility of replacing the inefficiently small 10-ton railway coal wagons without also modifying the coal screens, weigh-bridges and sidings in the collieries, and altering the hoists and tips at the ports, to enable these facilities to accommodate the larger wagons.[3]

Now, in the case at hand the difficulty posed by the presence of interrelatedness is essentially one of cost: the newly available type of capital equipment, the reaping machine, would have higher unit costs of operation when used in the environment of an outmoded physical 'plant,' i.e., on a farm whose fields had not been rationally laid out, thoroughly underdrained and leveled. The implication this carried for the choice of harvest technique by the operator of such a farm is quite straightforward. Translated into the terms I have used in an earlier analysis of the adoption of mechanical reaping,[4] operation of the machine in small enclosures and under terrain conditions causing it to stick in furrows or bog down at the turns, obviously would reduce the acreage of grain it could cut in a normal working day. Since the crew responsible for the machine and the horses, and possibly also the laborers needed to bind up the sheaves and set them up in stooks, would remain on hand throughout the day, the number of man-days of labor saved through mechanization would be reduced – *per acre cut*. Time lost in frequent moves between enclosures would have the same consequence. Because sub-optimal farm conditions would thereby lower the realized per acre savings on labor costs which

[1] Cf. M. Frankel, 'Obsolescence and Technological Change,' *American Economic Review*, Vol. 45 (June 1955), p. 296. The term 'interrelatedness' is Frankel's, but the modifier ('technical') has been added here to distinguish the class of relationship discussed by Frankel from the same heading in C. P. Kindleberger, *Economic Growth in France and Britain 1851–1950* (Cambridge, Mass.: Harvard University Press, 1964), pp. 193–247.

[2] Cf. *Report of the Cotton Textile Mission to the U.S.A.* (London: H.M.S.O., 1944), and *The Jute Working Party Report* (London: H.M.S.O., 1948), both of which are discussed by Salter, *Productivity and Technical Change*, p. 85.

[3] Cf. Thorstein Veblen, *Imperial Germany and the Industrial Revolution* (New York: Macmillan, 1915), pp. 126–7. Veblen actually referred to the necessity of changing 'terminal, facilities, tracks and shunting facilities,' but cf. Kindleberger, *Economic Growth in France and Britain*, p. 142, n. 27.

[4] David, 'Mechanization of Reaping.' For convenience, the variables and parameters entering the determination of the choice between mechanical and hand methods of harvesting grain are defined in Appendix A to this chapter: see Paragraphs 1, 2 and 3.

the farmer could set against the fixed capital charge he incurred for the use of the machine, a larger acreage would have to be harvested in an average season before the break-even point was reached for the conversion from hand- to machine-harvesting. Rough terrain might also take a toll by increasing normal maintenance expenses, or by actually shortening the expected service life of the machine and, hence, raising the (imputed) annual rental cost of the equipment. Other things being equal, this too would serve directly to raise the scale of the annual harvest operation required just to cover the interest and amortization charges entailed by the more capital-intensive technique. Of course, once the higher break-even scale – which I have called the 'threshold acreage' – thus established for sub-optimal operating conditions was reached, for farms further and further up in the size distribution the adoption of mechanical reaping would be increasingly profitable.[1]

Comparing Britain's 'improved' farms with farms of equal extent existing in an already 'improved' state – improved either as a result of remedial outlays made in the past, or because, as on the prairies of Illinois in the 1850s, durable capital formation of a kind incompatible with subsequent mechanization had never been undertaken – the higher costs of harvesting with machinery on the former would effectively have raised the threshold acreage and thereby reduced the profitability of abandoning the old, hand method of reaping grain. The distinction here between improved and unimproved farms does not turn on whether or not the land was in virgin state; we may designate as 'unimproved' all farmlands which either in their original condition, or as a result of past investments that had become outmoded, possessed features incompatible with the efficient operation of reaping machines. Functionally, an unimproved farm in this context is one on which additional capital outlays would be required to remove existing incompatible elements, and 'install' wheeled implements like the early generations of mechanical reapers and mowers in technically optimal operating condition.[2]

[1] If a reaping machine employed on a farm of a size equal to the threshold (defining 'size' here in terms of grain acreage) must earn – by definition – the annual imputed rental rate which covers interest and amortization charges, on a farm with twice as much grain acreage the same machine would earn a gross rate of return equal to twice the rental rate. It seems equally obvious that for a farm of a *given* size, the gross rental rate earned on a single machine must vary *inversely* with the level of the threshold acreage. Nevertheless, since the point apparently eluded at least one reader of my earlier analysis (cf. George G. S. Murphy, 'Review of *Industrialization in Two Systems*,' in *Journal of Economic History*, XXVII (September 1967), p. 421 on the irrelevance of the concept of the threshold to any wealth-maximizing decision), a formal proof that the threshold acreage is the 'dual' of the gross rate of return is supplied by Paragraph 4 of Appendix A, below.

[2] Note the important theoretical implication of this definition: over time it would be possible for the proportion of farms in the 'unimproved' category to decline, without any change in

Of course it is possible that without some improvements being made to the land, the new technique simply could not be introduced; in the limiting case incompatibility with the terrain might push the threshold acreage for adoption of the mechanical reaper upward beyond the seasonal capacity of a single machine, thus restricting the commercial feasibility of harvest mechanization to the class of 'improved' farms. On undrained heavy clays, for instance, horse-drawn machinery might bog down so repeatedly that under no circumstances would it be worthwhile to rely upon them. More generally, however, the unsuitability of the terrain and farm layout did not constitute an absolute barrier to mechanizing the corn-harvest. For the purposes of explaining Britain's incomplete conversion to mechanical reaping during the third quarter of the century, it is quite enough that the efficiency of the machines was sufficiently impaired to render their use more costly than the hand-techniques on all but very large farms in many of the country's principal grain-producing districts.

One cannot deny the aesthetic appeal of the argument that in British agriculture the rapid adoption of new, mechanical methods (being taken up in America) was blocked by problems of technical interrelatedness akin to those perceived by Veblen in 1915. At very least, it offers a restoration of some measure of symmetry to discussions of developments in the industrial and non-industrial sectors of the British economy during the latter part of the nineteenth century. There is no *a priori* reason why the durable legacies of an 'early start' should impose obstacles in subsequent phases of the race for technological leadership in manufacturing, or in transportation, but not do so in agriculture. But by the same token, the argument raises the same questions irrespective of the sector to which it is applied: if old plants could not accommodate new machines, why were they not modified, or scrapped and replaced with new structures? Why did the use of antiquated labor-intensive methods on unimproved farms persist alongside progressive husbandry? Are we ultimately led to an explanation cast in terms of entrepreneurial poverty? Or can it be shown that rational profitability considerations – as well as the elements of ignorance, stupidity, inertia and economic irrationality, all of which un-

---

the total number of farms, or any physical improvements having been made. Progress in the design of machinery could increase its versatility in coping with a variety of terrain conditions. In such circumstances we would be justified in saying that outmoded farms had been directly 'improved' by subsequent technical progress in the farm machinery industry; and – so long as the manufacturers of machinery were unable to act as a discriminating monopolist would in setting different prices for their different farm customers – we might expect to find the 'improvement' reflected in a relative rise of the rental value of such formerly outmoded farms. The analogous point may be made in regard to the case of existing industrial buildings. Within the period with which the present analysis deals, however, there were no radical design advances of this sort in reaping equipment.

doubtedly were present in some degree – could have been responsible for the failure to modernize the physical overhead facilities and operations of existing high-cost farms?

According to the hypothesis advanced by W. E. G. Salter,[1] the continuing use of outmoded capital equipment is an indication not of inefficiency and entrepreneurial failure, but of relatively cheap labor, and/or high interest rates. The one – in Salter's model – keeps variable costs low in the old, ostensibly more labor-intensive production methods; the other makes for high fixed costs, and hence high full costs with the new and comparatively capital-intensive technique that might otherwise displace them. It is, then, just conceivable that outmoded farms and farming methods persisted in Britain because in relation to the prevailing cost of capital rural wage rates remained too low to warrant these facilities being 'scrapped,' or modified to accommodate labor-saving machinery. And were that the essence of the matter, would one not be compelled to acknowledge that the state of the British farming landscape – rather than furnishing a separate, quite independent reason for the restricted extent of the reaping machine's diffusion – is nothing but a less familiar incarnation of the 'relative labor abundance' explanation? Evidently there is much we may hope to learn by asking whether Jacob Wilson's admonition to prepare the land before undertaking to mechanize the corn-harvest – in addition to making obvious technical sense – represented sound economic advice for farmers in Victorian Britain.

INTERRELATEDNESS, TENANCY AND INTERDEPENDENT DECISION-MAKING

Physical improvements such as the draining and leveling of fields entailed capital expenditures. These outlays, it has already been suggested, could simply be regarded as installation costs required to assure efficient operation of a reaping machine. Thus, for the project to be at all worthwhile to the farmer, the sum of his annual fixed expenses arising from the purchase and installation of mechanical reaping equipment would have to be covered by the difference between the cost of harvesting his (unimproved) grain acreage by hand methods, on the one hand, and on the other, the labor costs of harvesting his grain with the aid of the reaping machine once the unsuitable features of the terrain had been removed.

Quite obviously the addition of a capital charge for 'installation' would reduce the rate of return on the farmer's investment in mechanization; the threshold adoption acreage for the reaping machine would be raised above the level applicable in the case of farms on which the existing

[1] *Productivity and Technical Change*, Second Edition (Cambridge: Cambridge University Press, 1966).

state of the terrain required no further improvement. Indeed, it is entirely conceivable that 'mechanization-*cum*-improvement' would be an attractive proposition only for very extensive grain farms, holdings even larger than those on which it would pay to leave the fields unimproved and use mechanical reapers in a less than fully effective manner.[1] Hence, the mere fact that terrain obstructions could be removed to permit efficient operation of mechanical reapers gives us no reason to presuppose it must have been sensible for farmers in general to have done so.

This, however, is an over-simplified formulation of the mechanization-*cum*-improvement problem. If it is not to prove misleading, it must be immediately amended in two important details. First, one would be wrong to depict investment in land improvements as literally nothing more than installation outlays facilitating the introduction of reaping machinery. Leveling and enlarging the size of fields allowed lower costs in cultivation and other field operations apart from the grain harvest, while sub-soil drainage – in addition to permitting the elimination of ridges and water-cuts – directly increased grain yields per acre and facilitated heavier stocking of the land. Regarded from this perspective, which is certainly the way contemporaries thought about it, the expense of removing physical concomitants of the old system of husbandry was only incidentally chargeable to the introduction of machinery in grain harvesting operations, and in its own right would yield economic benefits to the farmer. The cost savings afforded by mechanical reapers on an improved, rationally laid out farm must be seen to have constituted an *additional* incentive for the agricultural community to 'thoroughly reform the hedge-and-ditch-row-ism and topography of this kingdom.'[2] The question at issue is whether this *further* incentive was actually sufficient to induce a significant acceleration in the pace of such reform.

If we are then to consider the inducements to convert to a system of farming with which mechanical reaping was compatible, it is surely pertinent to specify who was to be induced. The need for a second emendation becomes evident at this point. For, no sooner is the question phrased in terms of the way the availability of reaping machines affected farmers' decisions about land improvements, we must acknowledge that the land

[1] Elements of indivisibility in the expenditures required to make the indicated improvements, leading to high fixed 'installation' costs, could give rise to this. Such indivisibilities need not have resulted from strict engineering considerations, and it should be noted that it was not *technically* impossible to drain and level, say, a single field, or even some portion of a field. Without indivisibilities, it is still possible that the fixed cost of preparing the land to permit effective mechanical reaping of a single acre of grain would largely offset the saving in labor costs per acre, thereby making it possible to cover the fixed annual rental charges for the machine only on big farms.

[2] *Farmer's Magazine*, XXII (July 1850), p. 20 (reporting a speech by Alderman Mechi to the Coggleshall Institute), quoted in Spence, *God Speed the Plow*, p. 120.

was typically not the farmer's to improve. Unlike the situation in American agriculture where owner-occupancy was the norm, nineteenth-century British grain-farming was typically conducted by tenants.[1] The form of land tenure greatly diminished not only the incentive, but also the financial ability of Britain's farmers to undertake substantial improvements of any sort. Admittedly, where custom and mutual trust had been firmly established, English annual tenancies were not necessarily less secure than the long leaseholds characteristic of Scottish farming; moreover, tenants might well prefer not to be bound to rent schedules and crop and grazing patterns by the terms of restrictive leases.[2]

But the issue here is not long leases versus tenancy at will; it is tenancy versus owner-occupancy. Contemporaries, like Caird, may have been prone to overstate the case for granting long leases, and too quick to ascribe the rapid pace of agricultural improvement taking place in Scotland during the 1840s and the 1850s to the fact that a large proportion of Scottish farmers held 19- or 21-year leases.[3] It is nonetheless true that before compensation for unexhausted improvements was made compulsory, in 1883, tenants were understandably loth to sink money into extremely durable improvements, the full benefits of which they would not capture should their occupancy of the farm be terminated, either voluntarily or at the landlord's insistence.[4] Furthermore, whereas owner-occupants were in a position to mortgage their land to secure funds needed in carrying out costly improvements, Britain's tenant farmers (having no comparable asset to hypothecate) would be obliged to seek the money in the same quarters, and pay the same overdraft rates, which provided their working capital: the country bank parlor.

[1] Cf. Caird, *The Landed Interest and the Supply of Food* (London, 1878), pp. 46–7, 58, 60; G. C. Brodrick, *English Land and English Landlords* (London, 1881), Pt II, Ch. IV.

[2] Cf. J. D. Chambers and G. E. Mingay, *The Agricultural Revolution 1750–1880* (London: B. T. Batsford, 1966), 160–1, 164–6; F. M. L. Thompson, *English Landed Society in the Nineteenth Century* (London: Routledge and Kegan Paul, 1963), pp. 203–4, 227–31, on the question of leases in the period under consideration. For an example of restrictions imposed in short, 3–7-year leases, cf. Bell, 'Agriculture of the Country of Durham' (1856), pp. 99–100.

[3] Cf. Caird, *English Agriculture*, pp. 503–10, and *The Landed Interest*; Alderman Mechi's address to the London Farmers' Club, *Farmer's Magazine*, XX (July 1849), p. 20; Léonce de Lavergne, *The Rural Economy of England, Scotland and Ireland* (Edinburgh, 1855), p. 294; Skirving, 'East Lothian Farming' (1865), p. 111.

[4] Cf. Clapham, *An Economic History*, II, pp. 256–7, on the permissive legislation of 1875 and the Agricultural Holdings Act of 1883. Even when resort to the Act of 1883 became possible, it is questionable how much security tenants actually derived on durable improvement investments. Claims under the Act required documentation, and exposed tenants to counterclaims for violations of restrictions and to substantial penalties. Cf. Kindleberger, *Economic Growth in France and Britain*, p. 246, n. 147.

Considerations of security and finances would have mattered less were the contemplated capital expenditures trifling in relation to the farmer's current outlays. But in the case of an improvement such as the laying of pipe drains, which was less disruptive to the routine of the farm and productive of more beneficial result when done all at once instead of piecemeal, very substantial sums would be required at the outset. Thus, in the event, it was left to the landowners to decide upon major estate improvements and arrange for their financing. Of course, this scarcely meant that tenants could expect landlords to bear the entire cost without recompense; improved farms would be expected to fetch higher rental incomes and, ultimately, higher market prices. Moreover, in the important instance of drainage projects it was usual for tenants to be required to pay a specific annual charge in addition to the agreed rent – normally a fixed percentage of the improvement outlay – to defray the interest and amortization costs of his landlord's investment.[1] For the landlord to profit under such an arrangement without otherwise increasing the rent of the farm, the annual improvement levies would have to be more than sufficient (over the amortization period) to recoup the principal and the actual or foregone interest. Caird, however, was to be found condemning such practice as extortionate – in the instance of these Yorkshire landowners who, after 1846, took government drainage loans repayable in twenty-two equal installments of 6·5 per cent of the principal, but charged their tenants 7·5 per cent of the initial improvement outlay.[2] Apparently, it would have been deemed more sporting for the landlord to assume the entire risk by trying to take his improvement-profit in the form of non-specific increases in the farm's rents. Implicitly, however, many landowners might still have done so whilst maintaining the fixed improvement charges: offsetting rent abatements frequently were granted when agricultural prices declined markedly.[3]

[1] Cf. Caird, *English Agriculture*, p. 328; Thompson, *Landed Society*, p. 250; Bell, 'Agriculture of Durham' (1856), pp. 99, 144. The typical charge in Durham seems to have been 5 per cent in the mid-1850s, but this was low by the standards that generally came to prevail. See further discussion below.

[2] Caird (*op. cit.*, p. 328) rather confusedly castigates the landlord who would do this, for 'putting into his pocket 1 per cent, besides [sic] securing a permanently higher value for his land by an outlay to which he does not contribute a single farthing.' Note that Caird also got the landlord's profit rate wrong: twenty-two annual installments of 6·5 per cent on the principal yielded the 3·5 per cent per annum interest rate charged under the government subsidized drainage loan program initiated by Peel in 1846, and renewed in 1849. Analogously, annual installments of 7·5 per cent would have yielded a 5 per cent per annum rate of return – a 1·5 per cent pure profit rate. On the government drainage loan schemes, cf. Orwin and Whetham, *British Agriculture*, pp. 195–6; Clapham, *An Economic History*, II, p. 271. Variants of these arrangements were set up by land improvement companies in the 1850s and 1860s (cf. *ibid.*, p. 272).

[3] Cf., e.g., Thompson, *Landed Society*, pp. 240–1.

The implications of these institutional arrangements are worth pursuing. Apparently mechanization-*cum*-improvement was not, in actuality, a subject for unilateral decision. It required, instead, a collaborative decision by two parties who normally would not be united by mutually shared hopes and fears. The problem was a recurring one, and would arise at later points, in connexion with the introduction of tractors, for example. Field gates, and entrances to cattle yards and cow sheds, were designed for horse work, and, as E. H. Whetham has recently observed, British landlords were not quick in volunteering new buildings merely to suit the tractor. While owner-occupiers might feel free to rip out hedges and knock larger holes in their walls, tenants were constrained to work with the existing buildings and gateways: 'Farmers will continue to ask their landlords to repair their old buildings and give them new ones, and landlords will continue to promise to do something next year.'[1]

Looking at the present problem from the viewpoint of the tenant, were his landlord to agree to make improvements which would increase a mechanical reaper's effectiveness he could anticipate that he would in one way or another be asked to bear the associated interest and amortization costs. Yet, could he be sure that the improvements would not occasion a readjustment in the farm's rent, which would effectively absorb whatever pure profit he might otherwise derive from his own investment in a reaping machine? Unless he secured a lease, which might be fraught with other dangers, what was to stop the landlord from leaving him no better off than he had been before he had gone to the trouble and fixed expense of changing his production methods? From the landowner's vantage point, sober consideration of the proposition would disclose parallel risks. Should he initiate improvements permitting a reaping machine to be employed to greater advantage, was there any assurance that his tenants would in fact seize that opportunity, prosper by it, and thus permit some profit to be extracted on the capital sunk into improvements? Were the landowner to contemplate setting a farm rent and improvement levy that would compel the occupant to offset the incremental burden by reducing harvesting costs through mechanization, such a coercive strategy might only end in his having to look for a new tenant. And, unless his fellow landowners could be depended upon to adopt the same stance, a suitable tenant might prove very hard to find.

On farms which would have to be physically improved before a reaping machine could become a paying proposition, the normal risks attending the adoption of a technical innovation were thus compounded with the uncertainties created by the characteristic system of land tenure in Britain. In the absence of owner-occupancy, a feature which, it must be stressed,

[1] C. I. C. Bosanquet, *Farm Structures, Farm Mechanization* (*1946/47*), p. 51, quoted in Whetham, 'The Mechanization of British Farming 1910–1945,' p. 3.

differentiated the American from the British agrarian scene, the problem of technical interrelatedness (between agricultural machinery and farm topography) necessitated joint decision-making by *at least* two parties who would have to reach some tacit or explicit understanding about the division of the prospective gains.[1] Reluctance on the part of landlords to grant long leases, and the disadvantages of tenants of restrictive leases, further complicated matters by casting doubt upon the reliability of such understandings.

The serious problems of investment coordination to which the system of tenancy gave rise have tended to go insufficiently noticed and analyzed. This oversight is not wholly surprising, in view of British agrarian historians' absorption with the (cricumstantial) association between the growth of tenancy and the economic benefits derived from the advance of larger-scale, commercial farming. When the alternative is taken to be the inefficiently small, imperfectly commercialized, owner-occupied holdings, 'large tenancies' become a 'good thing.' The preceding corrective observations have sought to shift attention from well-recognized issues of security of tenure and compensation for permanent improvements, to the less familiar difficulties which arose because the tenancy prevented automatic 'internalization' of the returns from complementary investments.

Though the import of these difficulties for the receptivity of British agriculture to technical innovations forms the present focus of concern, the basic point might be expanded into a more comprehensive argument concerning the change-inhibiting effects of the system of land tenure which had become so firmly established by the second quarter of the nineteenth century. 'In Britain,' as C. P. Kindleberger[2] has remarked, 'it is appropriate to regard owner, wheat-producing tenant, and other potential tenant or tenants as separate firms, and to observe that the gains from the change may involve economies that are external to a particular firm from which investment is required.' Having in mind the inadequately thorough conversion from grain-farming to dairying during the final

[1] Where realignment of field boundaries of neighboring tenant holdings was contemplated, such as was often the case in the conversion of farms to steam-ploughing, for example, the number of interested parties would exceed the landlord and a single tenant. Cf. Spence, *God Speed the Plow*, p. 121. And where technical 'externalities' in maintaining an efficient system of field drains made it necessary to consider other people's ditches, streams and outfalls, several neighboring landlords might have to combine. They were encouraged to do so by the Land Drainage Act of 1861 – which provided machinery for setting up drainage boards to deal with the improvement of a large area comprehensively. Cf. *Parliamentary Papers*, 1873, IX, House of Lords S.C. on Improvement of Land, pp. 344–52. This early legislative recognition of a classic 'non-pecuniary externalities problem' is noticed by Orwin and Whetham, *British Agriculture*, pp. 102, 198.

[2] Kindleberger, *Economic Growth in France and Britain*, p. 247.

quarter of the century, Kindleberger suggests that this failure fully to respond to profit opportunities was due in major part to the uncertainties created by the 'disarticulated' structure of British agriculture. Lack of an unambiguous relationship between respective efforts and gains diluted the incentives for each of the various parties who would have to have collaborated in the transformation. The parallel is apparent between this suggestion and the line of argument Kindleberger advances in explaining the failure of British mining companies to replace their coal wagons with larger vehicles which would have resulted in lower operating costs to the railways.

One must be careful, however, not to misapprehend the nature of this sort of argument and to ask too much of it. It is not suggested here that the problems of coordinating complementary investments were inherently insuperable. Far from it. If reciprocal externalities existed it remained entirely possible for the parties involved to 'internalize' the benefits flowing from their joint efforts, by entering a formal or informal 'profit-sharing' arrangements.[1] If within such arrangements, as we have pointed out, uncertainties remained concerning the precise compensation that each participant was going to receive, the problem boils down simply to the fact that the venture held some (perhaps appreciable) risks for the investors.

But neither in principle nor in practice can it be asserted that such uncertainties *alone* would have blocked the extension of mechanical reaping into areas where preparatory terrain modifications were indicated. Tenants and landowners were no strangers to risk-taking. Much would depend upon the magnitude of the expected gains that would be available to be shared among the investing parties. For, when great uncertainties cloud the future division of the pie, petty collaborative baking ventures are much more likely to be abandoned than are large ones.

Properly to assess the significance that should be attached to the uncertainty created by land tenure arrangements in Britain, one must, therefore, return to the original question of the profitability of investment in mechanization-*cum*-improvement. We should, in other words, first establish whether or not the whole of this particular pie

[1] This is an objection which must be leveled at Kindleberger's more clearly elaborated analysis (*Economic Growth in France and Britain*, pp. 141–4) of the failure of British railways to adopt larger coal cars. If the railways companies would have saved enough on operating costs to justify the change under a system in which coal cars and all related facilities would be owned by the railways, why could they not have arranged to compensate the mining companies (which owned the coal-cars) for the investment in making the switch? Kindleberger does not examine whether sufficient compensation could have been paid to make collaboration attractive to the colliery-owners, or even whether the return on the total investment would have justified the investment under conditions of complete integration of activities, i.e., internalization of all benefits.

was really big enough to make it worth baking under the most certain of conditions.

Providing a rigorous answer to this question would be no mean undertaking. That much should be clear from the first of the two emendations set out above. For, to insure that due allowance was made for interdependence among the myriad activities comprising an agricultural enterprise, and for the multiple elements of technical interrelatedness and joint production cost to which that interdependence gave rise, it first would be necessary to explore the full range of the choice of technique problems confronting farmers in the period under consideration. The aim would be to arrive at a cost comparison between the best mode of grain-farming in Britain and the most economically efficient among those combinations of techniques which would, nonetheless, restrict farmers to harvesting their grain by hand methods. The comparison would, moreover, have to be made for farms of varying sizes. More precisely: we should consider the difference between the unit operating costs (including interest on working capital) of (1) an existing unimproved farm restricted to using outmoded methods, including hand-harvesting of grain at very least; and (2) a thoroughly improved farm using the most superior techniques available, including mechanical reaping. Assuming the *natural* fertility of the soil to be the same in both cases, the difference in unit operating costs should be compared with the unit rental charges – i.e., annual interest and amortization costs per unit of output – for all the additional fixed reproducible capital, the livestock inventory being more conveniently treated as part of the farmer's working capital. Since less land would undoubtedly be needed to produce a given volume of output on an improved farm, it would be appropriate to subtract the land rents on the unnecessary acreage (at the pre-improvement rental rate) from the additional fixed capital costs.[1]

It is, indeed, a daunting prospect, one that would vastly extend the scope of the present inquiry and turn it into a full-scale study of mid-Victorian agricultural technology. Humbled before so imposing a task, it is permissible to ask if the relevant empirical point may not be established by pursuing a less ambitious course and focusing attention upon the core of the problem, the connexion between land improvements and harvest mechanization. The choice between machine methods and hand methods

[1] Alternatively, the land rent saving could be added to the difference in operating costs in determining whether or not the conversion would be profitable for a farm of the size initially stipulated. Either procedure would amount to treating the capitalized value of the land rents saved on a par with the resale value of any other assets which could be disposed of in 'modernizing' the farm without reducing its output level. Cf. Salter, *Productivity and Technical Change*, pp. 55–8, for discussion of scrapping and replacement criteria in the case of industrial plant and equipment.

of cutting grain was surely bound up with decisions regarding drainage and, consequently, the terrain profile. But to suppose it was equally entangled with decisions about the application of fertilizers, or the conduct of non-field operations such as grain threshing and the preparation of livestock fodder, does seem to be stretching the point. Similarly, while better drainage did permit a shift to mixed farming, and large additional outlays for 'elaborate new dairy parlours, cattle houses, pig-sties and barns' often did accompany land improvement projects,[1] these were hardly necessary concomitants in a technical sense. In fact, there are grounds for the suspicion that the incremental advantages derived by grain farmers from stall-feeding of livestock and greater recovery of animal manure failed, in the event, to justify the associated capital expenditures which were made for buildings and yards.[2] It is certainly clear that during the period 1847–79 as a whole, English landowners received miserably low returns from their investments in new farm buildings – as distinct from land improvements – and that the former thus served only to depress the overall rate of return on expenditures for comprehensive estate improvement schemes.[3]

THE RATE OF RETURN: A QUANTITATIVE EXERCISE

There is, then, some reasonable ground for narrowing the focus to consider only complementary improvements made upon the land itself. Doing so makes it feasible to try to gauge the profitability of mechanization-*cum*-improvement by examining the following illustrative investment project: (a) installing pipe drainage, (b) spreading the sub-soil removed from the drainage trenches, (c) leveling the tillage, and (d) purchasing a manual sheaf-delivery reaping machine. We shall posit that the land improvements specified offered a direct gain in the form of increased yields on the farm's total cereal acreage without any additional application of fertilizers. Further, it will be supposed they assured optimal terrain conditions, thereby guaranteeing the full saving on harvest labor costs obtainable by employing the reaping machine in place of scythesmen to harvest the crops of wheat and oats – and what little rye might be grown.

From a purely technical standpoint there was no reason why the cutting apparatus of the manual sheaf-delivery (or, in simpler contemporary American parlance, the 'hand-rake') reaper could not cope perfectly well with barely, as with the other small grains. But in the case of

[1] Cf. Chambers and Mingay, *Agricultural Revolution*, p. 176.

[2] On the changing balance of grain and livestock profitability within mixed farming enterprise, cf. E. L. Jones, 'The Changing Basis of Agricultural Prosperity, 1853–73,' *Agricultural History Review*, x (1962), pp. 102–19.

[3] Cf. Thompson, *Landed Society*, pp. 250–3.

the barley crop – which in Britain vied with oats in terms of the extent of acreage sown – a complication arose due to the common practice in handling the grain once it was cut. In the era with which we are concerned, English farmers most frequently regarded it as not worthwhile to bind barley into sheaves. Instead, they left it to dry lying in swath in the mown fields – before carting and stacking it in much the same way as hay or straw.[1] This practice, however, would not have been feasible where a rear-delivery machine of the American (Hussey–McCormick) type was employed: the gavels, swept from the platform by the raker who rode the machine, had to be quickly bound up and removed from the horses' path before the contraption came round on its next pass. It is therefore supposed that a farmer purchasing one of the comparatively small and moderately priced machines in this design class would not contemplate using it to harvest his barley.[2]

In reckoning the per acre saving of harvest labor costs (wL_s^*) for the representative acre cut by machine, one must take into account the differential advantage the mechanical reaper offered in wheat compared with oats; this may be done simply enough, by weighting the unit labor savings coefficient ($L_s^*[j]$) for each j-th crop in proportion to the relative acreage devoted to the two grains in the country at large.[3] To the unit saving in labor cost, it is then necessary to add – for each acre of wheat and oats harvested – the per acre value of the incremental cereal yield. The latter ($p\dot{y}G'$) depends upon the absolute increase in the number of bushels per acre sown with cereal crops ($\dot{y}$), the market price per bushel (p), and (G') the number of acres annually devoted to cereals of all kinds per acre sown with wheat and oats. Now it would be just profitable to undertake

[1] Cf. H. S. Thompson, 'Account of a subsequent Trial of the American Reapers' (Appendix to Implement Report), *Journal of the R.A.S.E.*, XII (1851), p. 648.

[2] The use of swath-delivery reaping machines – of native British design – in this connexion is referred to below. With manual delivery reapers fitted up with tipping platforms rigged for side-discharge, it was possible either to rake off the cut grain continuously, effectively laying it in swath, or to form sheaves. Such contrivances were developed during the 1860s as modifications of the original Hussey and McCormick devices, but unfortunately there is little information about their purchase costs and the per acre labor requirements associated with their use.

Self-acting sheaf-delivery reapers, which also began to be marketed in Britain in the early 1860s, would only make work in the barley harvest – though they might afford side-delivery; barley delivered automatically in sheaf would have then to be spread out by hand to dry. The limited appeal of the 'self-rake' reaping machines in Scotland appears to have been due in large measure to essentially the same problem: they gave farmers no flexibility in handling grain that was too wet when cut to be safely bound into sheaves. Cf. the letter of 20 September 1865 from E. Alexander, of Stirling, to C. H. McCormick, cited by William T. Hutchinson, *Cyrus Hall McCormick*, Vol. II (New York: The Century Co., 1930), p. 414, n. 19.

[3] See Appendix B, Table 16.

the complete investment project if, at the combined wheat and oats acreage being contemplated (S_T^{**}), the total monetary benefits ($[wL_s^* + p\dot{y}G']\, S_T^{**}$) began to exceed the sum of the fixed rental cost of the land improvements and the yearly interest and amortization cost of the reaping machine. Denoting the latter as the product of the machine rental rate and purchase cost, $R_m C_m$, the imputed annual charge for the land improvements may be analogously represented as $R_I[C_I A' S_T^{**}]$; for each of the S_T^{**} acres of grain harvested with the machine during the year, it would be necessary to allow for continuing crop rotation by draining and leveling A' acres of the arable, at a capital cost of C_I per acre of land thus improved. The interest rate at which finance could be obtained (r_I), and the conventional amortization period for such improvements (D_I), between them, determine the appropriate annual rental rate expressed as (R_I) a fraction of the initial capital expenditures.[1] Setting out the problem in this way facilitates arranging the pertinent quantitative information in an augmented threshold

TABLE 14 Determinants of the threshold acreage for harvest mechanization under illustrative conditions of interrelatedness

Implicit threshold function for mechanization-*cum*-improvement:

$$R_m C_m - (p\dot{y}G' + wL_s^* - R_I C_I A')S_T^{**} = 0$$

Fixed technical coefficients and parameters:

$L_s^* = 1{\cdot}096$,	for hand-rake machine vs. scythes, per acre of wheat and oats cut, bound and stooked;
$\dot{y} = 4{\cdot}0$,	the additional wheat yield in bushels per acre;
$\varepsilon = 0{\cdot}70$,	the wage-share in improvement costs;
$G' = 1{\cdot}60$,	the ratio of cereal acreage to wheat and oats acreage;
$A' = 2{\cdot}85$,	the ratio of arable to wheat and oats acreage;
$R_m = 0{\cdot}127$,	the machine rental rate, for $r_m = 0{\cdot}05$ and $D_m = 10$ years;
$D_I = 22$,	the land improvement amortization period, in years.

Variables subject to change over time (all prices expressed in shillings):

Price	*c.* 1851 = t_0	*c.* 1861	*c.* 1871
$w(t)$	3	$1{\cdot}30(w_0)$	$1{\cdot}50(w_0)$
$p(t)$	5	$1{\cdot}25(p_0)$	$1{\cdot}25(p_0)$
$C_m(t)$	660	$1{\cdot}00(C_{m0})$	$0{\cdot}76(C_{m0})$
$C_I(t)$	180	$[1+0{\cdot}30\varepsilon](C_{I0})$	$[1+0{\cdot}50\varepsilon](C_{I0})$

Source: See Appendix B for estimates of technical coefficients and price-variables; see Paragraph 3 of Appendix A for the relationship between rental rates (R) and parameters (r) and (D).

[1] For the use of the standard annuity formula in making the calculation, see Paragraph 3 of the Technical Notes in Appendix A.

function – from which it is a simple matter to calculate the wheat and oats acreage (S_T^{**}) beyond which this particular project, considered as a package, would be economically rational.

While the purpose of calculating the level of the mechanization-*cum*-improvement threshold thus determined must be regarded as primarily illustrative, it will be seen from Appendix B that a serious effort has been made to arrive at appropriate magnitudes for the parameters and variables which have thus been explicitly introduced into the discussion. The general expression defining the threshold, and the parameter estimates, are arrayed in summary form in the upper panel of Table 14. The lower panel of the same table carries the values assumed by the four price variables of the analysis – w, C_m, C_I, p – at the beginning, middle and end of the period 1851–71.

The proximate results of this exercise in quantification are displayed by the main portion of Table 15 and are subject to interpretation in either of two ways. The first is quite straightforward: each column in the body of the table shows how the mechanization-*cum*-improvement threshold would be altered by changes in (R_I) the yearly rental rate for the land improvements, given the structure of other input costs and output benefits prevailing at the date in question. As expected, one may see that if lower annual rental rates were charged for the land improvements, *ceteris paribus*, it would become worthwhile for tenants with smaller acreage in

TABLE 15 The mechanization-*cum*-improvement threshold acreage and the landlord's rate of return

Land improvement investment's		Threshold wheat and oats acreage for investment in mechanization-*cum*-improvement, S_T^{**}:		
Net per annum rate of return r_I	Gross annual rental rate R_I	*c.* 1851	*c.* 1861	*c.* 1871
0·020	0·0559	12·7	8·7	10·1
0·030	0·0620	24·1	14·4	31·2
0·035	0·0650	43·0	21·1	0
0·040	0·0684	$>S_{max}$	45·0	0
0·045	0·0716	0	0	0
Threshold wheat and oats acreage for simple mechanization:				
$S_T^* = (R_m C_m)/(L_s^* w)$		25·5	19·6	12·9

Source: See Table 14 for underlying parameters and variables.

wheat and oats to apply to their landlord to have these improvements made, and then to invest in the purchase of a reaping machine.[1]

An alternative interpretation of the same data, this time from the viewpoint of the landowner, is perhaps more illuminating. Given the prevailing price of grain, the level of agricultural wages, the fixed annual (imputed) rental cost of a reaping machine, the per acre cost of improvements, and a farm on which the combined wheat and oats acreage is that shown in the body of the table, the corresponding (r_I) entry in the left-most column reveals the 'maximum compatible rate of return' on the associated improvement outlay. This is the highest average annual rate of return, net of amortization charges over a 22-year period, which a landlord could extract on the land improvement expenditures without actually making it unprofitable for his tenant to purchase a hand-rake reaping machine on the date in question.[2] An improving landlord who was thus prepared to skim off any pure profits from mechanization – that is, any savings in harvest labor costs over and above the imputed rental costs of the reaping machine used by his tenant – as well as the entire incremental cereal yield on the improved arable, could obviously secure higher rates of return on (larger) outlays made to improve the land of larger individual farms. Here we have a nice concrete illustration, should one be needed, of the potential economic advantages that larger-scale tenant farming offered to the nineteenth-century landowner.[3]

The blessings of scale were not, however, boundless. In this case a proximate upper limit to the maximum compatible rate of return would be reached on farms whose area under wheat and oats equaled the

[1] Once the landowner had committed himself to a general policy of improvements, and announced the yearly improvement rental charge, his tenants would commonly have to apply to have specific changes made on the farms they occupied. Cf. Thompson, *Landed Society*, p. 255: 'It was very usual, as on the Northumberland estate, for the owner to fix the total annual appropriation for expenditure on his estate, but for its detailed farm by farm application to be determined by the requests made by individual farmers for the draining of a field, the erection of a barn or the removal of a hedge.' The argument being advanced here, of course, holds that the rational landowner would not fix an estate improvement budget without reference to the charges he could anticipate extracting from those of his tenants who would apply to have specific projects carried out. The alternative interpretation of Table 15 is thus to treat it as depicting the marginal efficiency of (landlord's) investment as a function of the size of his tenants' grain acreages.

[2] It is the maximum net rate of return on improvement capital which would be compatible with minimally profitable mechanization by that tenant.

[3] It certainly belongs in a category with the scale economies in the management of sheep flocks, and the utilization of horses and carts, described for eighteenth-century English farms by John Arbuthnot, of Mitcham, *An Inquiry into the Connection between the Present Price of Provisions and the Size of Farms. By a Farmer* (London, 1773), p. 8, to which reference is made in T. S. Ashton, *An Economic History of England: The Eighteenth Century* (London: Methuen, 1933), pp. 43–4.

100-or-so acres which represented the normal yearly limit on the cutting capacity of a single reaping machine (S_{max}). At still higher rental rates (R_I) designed to secure an even greater net rate of return on the landowner's improvement outlay, the tenant would not be left with enough to meet the yearly interest and depreciation costs of his reaper – no matter what the extent of his grain acreage.[1]

The low level of the maximum rates of return on improvement expenditures compatible with mechanization, in both the technical and the economic senses, constitutes a striking general feature of the findings presented by Table 15. It deserves further comment and amplification.

Under the terms of the government's Drainage Loan program initiated by Peel immediately following the Repeal Act, landowners were able to finance this class of improvements over a 22-year repayment period at an annual interest rate of 3·5 per cent – the annual interest and amortization charges worked out at 6·5 per cent of the principal. Perhaps for this reason, it seems that 3·5 per cent became accepted as a notional interest rate applicable to estate improvements of all descriptions, and not exclusively to drainage projects.[2] In the case of drainage loans, however, this rate did reflect an element of public subsidy. According to evidence given in the early 1870s, the interest rate at which the private land improvement companies had been accustomed to lending was 4·5 per cent; the sinking fund typically was calculated to repay the loan in twenty-five years, but when these companies were finished adding in interest and amortization of the 'preliminary expenses' of the loans, the average yearly payment on the 'effective outlay' was brought up 'to a little more than 7 per cent.'[3] Over a 25-year period the latter annual rental rate would in reality be generating a compound interest yield of 5 per cent per annum on the company's principal.

It is to be noted, then, that in the years immediately surrounding 1851 a yearly rental rate of 6·5 per cent, when applied to the expenditure required to drain, spread the sub-soil and level the arable of an average British farm, would apparently have imposed improvement rental costs upon the occupant exceeding the estimated value of the increase in cereal crop yields.[4] In these circumstances, for which the abnormally low price

[1] The situation just described is denoted by the zero entries for the mechanization-*cum*-improvement threshold in Table 15.

[2] Cf., e.g., the use made of this rate by R. J. Thompson, 'An Enquiry into the Rent of Agricultural Land in England and Wales during the Nineteenth Century,' *Journal of the Royal Statistical Society*, LXX (1907), p. 619.

[3] *Parliamentary Papers*, 1873, IX, House of Lords S.C. on Improvement of Land, 'Report,' Paragraph 9, Subsection 3, cited in Brodrick, *English Land* (1881), pp. 69–70.

[4] See Appendix B for further discussion. One may arrive at this inference directly, from the information presented by Table 15: at $r_I = 0{\cdot}035$ the mechanization-*cum*-improvement

of wheat during 1849–52 was in large part responsible, investments in this particular set of land improvements could provide a yield matching the 3·5 per cent interest rate only if other, indirect gains were captured – namely, the margin of supernormal profit derived by harvesting 43 acres of wheat and oats with a reaping machine instead of scythes.[1] Mechanizing the harvest of as much as 80 acres of wheat and oats – in the standard 57 : 43 proportions assumed throughout these calculations – would merely offer the landlord the prospect of extracting a 3·75 per annum net rate of return; for this, according to the present estimates, he would be obliged to expend more than £2,000 on the improvement of some 228 acres of tillage.[2] And the very most a landowner could hope for, with conditions as they were in *c.* 1851, appears to have been an annual pure profit rate less than one-half a percentage point above the (3·5 per cent) rate of interest on government improvement loans; unless, of course, his tenant was to be made worse off than he had been without the improvements and the reaping machine.

A 'maximum compatible rate of return' as high as 4·5 per cent per annum was not only unattainable when the Crystal Palace opened: Table 15 indicates that rates of that magnitude eluded the improving landlord's grasp over the course of the ensuing two decades. Moreover, the size distribution of existing agricultural holdings[3] would not have permitted even these terribly modest maximum feasible rates of return to be obtained on outlays for land improvements compatible with harvest mechanization. In England and Wales as a whole the average farm in 1851 was only 111 acres in extent, of which perhaps 23 acres were devoted to wheat, oats, and rye – the latter counting for little in the total. Over the next twenty years the size distribution of holdings remained much the same. Fully half of the agricultural acreage in England and Wales lay in holdings smaller

[1] As the last row of Table 15 reveals, in *c.* 1851 a manual sheaf-delivery reaper (in lieu of scythe-mowing of wheat and oats) would cover its own annual rental costs on the first 25·5 acres harvested, yielding pure profits – or quasi-rents – on the remaining 17·5 (= 43·0 – 25·5) acres up to the mechanization-*cum*-improvement threshold.

[2] For $r_I = 0{\cdot}0375$, $R_I = 0{\cdot}0667$, from which the data in Table 14 enables us to calculate the $S_T^{**} = 80{\cdot}0$ acres. Referring to the required improvement outlay per acre harvested in *c.* 1851, we then find that $S_T^{**}\,(C_I A') = £2{,}052$.

[3] The following statements draw upon an examination of the characteristics of the size distribution of British farms, farm acreage and grain acreage, based on the complete census of agricultural holding in 1851 and data for a sample of seventeen counties from the Census of 1871. The full details and documentation are too lengthy to give here, but will be provided in the larger monograph now in preparation.

---

threshold (S_T^{**}) exceeds the threshold applicable in the case of simple mechanization – S_T^{*}, shown by the last line of the table. This indicates that some of the pure gains from mechanization must be applied to cover the improvement rental charges.

than 200–250 acres, and it is quite likely that rather more than half of the country's acreage under wheat, oats (and rye) was being farmed in units of under 40–50 acres. Looking then at the rate of return which Table 15 indicates a landlord could hope to extract on his investment in improving a farm with 45 acres of these grain crops, we may therein observe the very highest rate attainable on the half of the country's grain acreage being farmed in holdings below that size class. It rose from a shade under 3·5 per cent in *c.* 1851 to 4·0 per cent a decade later, and then retreated to the neighborhood of 3·1 per cent in *c.* 1871, more or less paralleling the movement of the maximum feasible rate for acreage in the upper half of the farm size distribution – which remained under 4·0, 4·5, and 3·5 per cent per annum, respectively, on the same three dates. Giving equal weight to these two sets of maximum rates, in accord with the relative shares of the nation's grain acreage to which each refers, it appears appropriate to conclude that for England and Wales as a whole the maximum expected rate of return to a landlord's investment in the type of mechanization-*cum*-improvement project described here could not have been as high as 4 per cent on any of the three occasions in the period 1851–71.

Now these findings should not really provoke surprise. For, in this the line of the present argument converges with other, more familiar conclusions about developments in mid-Victorian agriculture. The £24 million spent on drainage and land improvements during the years 1846–76 notwithstanding, it was the rare and extremely lucky (or judicious) landowner who managed to secure an average annual rate of return as high as 4·5 per cent from land improvement outlays. Those wicked Yorkshire landlords, whose greed (in charging their tenants more than the rate implied by the government drainage loans) had drawn down upon them James Caird's wrath, were only aiming to extract a 5 per cent per annum rate of return. To most, the capital poured into their estates after the Repeal of the Corn Laws brought disappointingly meager financial rewards – well below those being earned from equally risky industrial investments.[1] The estate improvement expenditure of the great landowners recently studied by F. M. L. Thompson yielded – at maximum – a gross annual return of

[1] Cf. Thompson, *Landed Society*, Ch. IX, esp. pp. 248–55. 'Disappointingly' is used here with conscious intent, despite the view advanced by Thompson (*ibid*., p. 254) that landowners regarded their estates as assets whose very ownership constituted a luxury, and which *could not be expected* to yield financial returns of more than 2 or 2·5 per cent per annum on the purchase cost. There is a relevant distinction to be made between finding consolation for low monetary returns in the social advantages conferred by landownership, and proceeding to sink more money into the improving of estates whose ownership already conferred prestige. Before arguing that mid-Victorian landowners looked forward with equanimity to low rates of return on improvement expenditures, it is necessary to posit that in this era – and not only in eighteenth-century England – the improvement of estates, not simply their ownership, was considered socially prestigious.

3·6 per cent, and in many instances this high a rate appears to have been enjoyed only for a few years toward the close of the period 1847–79.[1] Less fortunate still were those who had begun improvements too late to clear off their debts before the rent reductions of the 1880s fell upon them.[2] Had the money thus deployed simply been left in Consols, it would have brought in a steady 3·2 per cent per annum – without the worry.[3]

In the light of the illustrative calculations presented here, one can well understand why in 1873 the House of Lords' Select Committee on Land Improvement believed that those who borrowed to finance drainage projects – let alone construction of new farm buildings and laborers' cottages – might readily find themselves out of pocket, despite the indirect pay-offs to facilitating the mechanization of field operations. Even were no allowance to be made for the uncertainty surrounding both the size of the total return and, owing to the prevailing system of tenure, its division between improving landlord and progressive tenant, the Committee's *Report* appears fully justified in its opinion 'that the improvement of land... as an investment is not sufficiently lucrative to offer much attraction to capital.'[4]

RICARDO EFFECTS, WICKSELL EFFECTS, AND THE PENALTIES OF AN EARLY START

A second broad facet of the findings reported by Table 15 concerns the way in which the profitability of mechanization-*cum*-improvement was affected by trends in grain prices and agricultural wages during the interval separating the opening of the Crystal Palace from the House of Lords' inquest on the nation's investment in land improvements. These movements, like the general absolute level of the rate of return, merit closer attention.

Previously, in discussing the determination of the profitability threshold for 'simple mechanization' – the introduction of mechanical reaping on a farm whose terrain already offered suitable operating conditions – it was proper to be exclusively concerned with the influence exerted by the level of agricultural wage rates relative to the rental cost, or (in the absence

[1] Cf. Thompson, *Landed Society*, pp. 250–1. Repairs of buildings are excluded from the underlying expenditure figures in this calculation.

[2] Cf. Chambers and Mingay, *Agricultural Revolution*, p. 177.

[3] The average annual yield on 3 per cent Consols during 1850–9 was 3·16 per cent; during 1860–9 it was 3·27 per cent; and during 1870–9, 3·19 per cent. Cf. Sidney Homer, *A History of Interest Rates* (New Brunswick, N.J.: Rutgers University Press, 1963), Table 19, pp. 195–7.

[4] *Parliamentary Papers*, 1873, IX, House of Lords S.C. on Improvement of Land, 'Report,' pp. iii–iv, quoted in Thompson, *Landed Society*, p. 251.

of interest rate and service life alterations) the purchase costs of the reaping machine itself. Now, however, it is important to recognize that through their influence upon the profitability of complementary land improvements, changes in the price of grain (p) also must have played a significant role governing the diffusion of the mechanical reaper in Britain. Moreover, it must be appreciated that although the profitability of 'simple mechanization' was enhanced by increases in the relative price of farm labor, paradoxically, the same upward drift of rural wage rates relative to the price of reaping machines *and grain* could simultaneously diminish the profitability of investment in harvest mechanization-*cum*-land improvement.[1]

One might be led instinctively to doubt the validity of the last contention, at least on the basis of some general acquaintance with W. E. G. Salter's notable theoretical work on the economics of scrapping and replacement decisions.[2] It is certainly more usual to find economists maintaining that, on the contrary, a rising real wage rate tends not only to hasten the scrapping of old machinery, but also works to insure that the introduction of new, more advanced labor-saving techniques will not be long obstructed by the existence of outmoded but extremely durable (industrial) plants. In the limit, 'interrelatedness' simply means that scrapping and replacement decisions must be taken with reference to complete plants, rather than individual pieces of equipment. But the 'Salter model' suggests that when wage rates rise relative to the price of an industry's product, the real costs of new capital goods available to the industry will have fallen in relation to the real savings in operating (labor) costs which their use would permit. Replacing out-moded plant and equipment with capital embodying superior techniques would, therefore, tend to be made more profitable; whatever 'penalty' had been imposed by the industry's 'early start' in accumulating durable capital would be sloughed off as soon as it grew burdensome. Clearly, to adopt such a position in the present connexion would be tantamount to reviving the germ of an all-embracing 'British labor abundance' explanation for the restricted diffusion of mechanical reaping: had the real wage of farm labor been higher in Britain, capital expenditure for reaping machines and complementary land improvements would have been more attractive.

I must confess to having brought this proposition forward only to reject it now the more firmly. It is not difficult to see why precisely the

[1] The reader may find it easier to grasp the essence of the following argument and counter-argument by turning at this point to a short mathematical treatment of the effects of price and wage changes upon S_T^*. Such a treatment is supplied by Paragraphs 5 and 6 of the Technical Notes in Appendix A, below. The verbal statement of the argument given in the text is, nonetheless, complete and self-contained.

[2] Cf. Salter, *Productivity and Technical Change*, pp. 50–4 and Ch. v *passim*.

opposite result could emerge from a rise in harvest wage rates relative to grain prices, when – as in the case at hand – the assumptions underlying the Salter model were not fulfilled.

As one might expect, if the fixed cost of reaping machines did not increase as rapidly as the level of wage rates facing the farmer, harvest mechanization *per se* would become a more attractive proposition on any given holding – whether otherwise improved or not. But concurrently, higher rural wage rates would also tend to force up the per acre costs of carrying out land improvements. The more labor-intensive this kind of capital formation was, and the more sluggish was the pace of productivity advance in such activities as trench-digging, the stronger the upward wage pressure upon the unit price of the improvements. Both conditions are especially pertinent here and, it might be added, also seem germane to the construction trades in general during the nineteenth century.[1] Nearly 60 per cent of the per acre cost of pipe drainage installations was accounted for by direct wage costs, materials making up the balance. Spreading the sub-soil and leveling the tillage, the two operations following the drainage work, were usually conducted in a manner which made wages the only significant cost item; thus, it is estimated that in *c.* 1851 labor's share comprised 70 per cent of the total unit price of this particular set of land improvements.[2] Moreover, there is little reason to think that the period under review saw any notable increase of productivity in on-farm capital formation activities of this nature. Note, however, that the degree of labor-intensity of improvements only affects the sensitivity of their price to changes in the level of agricultural wages; 'labor-intensity' itself – the description of the situation in which the wage share (strictly, the elasticity of output with respect to labor inputs) exceeds one-half – is not a necessary condition for a real wage increase to issue in a decline in the profitability of mechanization-*cum*-improvement.[3]

As improvement costs per acre rose in relation to the price of grain, and consequently in relation to the value of the direct benefits in the form of greater grain-yields per acre improved, investment in land improvement *per se* would become less and less profitable. Ignoring for the moment the possibility of a change in the grain price or the reaping machine, when the resulting addition to the cost of improvements reckoned in terms of grain turned out to exceed the incremental grain-value of the wage savings

[1] Evidence of the relatively slow rate of productivity growth in the British construction industry is, admittedly, rather shaky – a rather firmer case can be made in regard to the U.S. construction trades during the latter half of the nineteenth century. But cf., e.g., G. T. Jones, *Increasing Returns* (Cambridge: Cambridge University Press, 1933), pp. 58–100 on cost and price movements in the London building trades, 1845–1913.

[2] See Appendix B for discussion of the estimated wage share in improvement costs.

[3] This theoretical proposition is demonstrated in Paragraph 6 of the Technical Notes in Appendix A.

due to mechanization, the higher real wage rate would have the perverse effect of depressing the overall rate of return on the joint investment project. Restating the situation more concisely: although increasing real wage rates in agriculture produce the familiar 'Ricardo effect' as far as simple mechanization is concerned, in regard to the complementary land improvements the same wage movements result in an opposing 'Wicksell effect.' The latter could easily be strong enough to dominate the *net* influence exerted upon the rate of return to investment in the whole project.[1]

There is more to this than mere theoretical speculation. Looking at the influence of agricultural wage and price changes during the 1850s and 1860s, the situation just envisioned may be seen to have actually materialized well before the close of the 'Golden Age' of British farming. The upward course of grain prices had supplied vital impetus to the high farming movement of the 1850s which, through its emphasis on land improvement – if in no other way, was conducive to the more widespread diffusion of reaping machines.[2] Although with the cessation of hostilities in the Crimea grain prices dropped quickly from the heights touched in 1854–6, they remained well above the levels that prevailed when the decade opened. The rise of approximately 25 per cent registered by the *Gazette* price of wheat between 1849–52 and 1859–62 almost matched the rate of increase in average weekly wages of farm laborers, and therefore outpaced the estimated 21 per cent advance in the unit cost of land improvements, as may be seen from Table 14. Further, like drainage and

[1] Imposing the condition $\tilde{C}_m = \left[\frac{dC_m}{C_m} - \frac{dp}{p}\right] = 0$, equation (8) in Paragraph 5 of Appendix A may be rewritten as follows:

$$dS_T^{**} = \left(\frac{pS_T^{**}}{g_m + g_I}\right)[\tilde{C}_I(I/p) - \tilde{w}(g_m/p)]. \qquad (8a)$$

This expression mirrors the statement in the text above, showing that the threshold acreage will be raised if the term in the square brackets on the R.H.S. of the equation is positive – which it may well be even though real wages are rising in accord with the conditions

$$\tilde{w} = [(dw/w) - (dp/p)] > 0, \quad \text{and} \quad [\tilde{w} - \tilde{C}_m] > 0.$$

By imitating the proof given in Paragraph 4 of Appendix A, the reader may readily satisfy himself that the pure profit rate and the gross rate of return on a total investment for harvest mechanization and complementary improvement of a farm with G acres under small grains – an investment of $(C_m + IG)$ – must vary inversely with the mechanization-cum-improvement threshold S_T^{**}.

[2] The connexion between high farming's profitability and the adoption of new machinery is usually asserted in much looser fashion. It is suggested that 'prosperity' resulting from the level of grain prices in the 1853–73 era – in contrast with the plight of grain farmers in the last quarter of the century – encouraged (or, in some accounts, permitted) investment in farm equipment. Cf., e.g., Chambers and Mingay, *Agricultural Revolution*, pp. 177ff.

related land improvements, harvest mechanization *per se* was becoming more economically advantageous during the 1850s: the purchase price of a manual sheaf-delivery reaping machine remained essentially unaltered while agricultural money wages rose. Thus, we find that in *c.* 1861 mechanization-*cum*-improvement on a farm with 21 acres under wheat and oats held out the prospect of a 3·5 per cent annual net rate of return on the land improvement expenditures – the same rate which, ten years earlier, had only been attainable on farms where over twice the acreage of those crops was involved. Putting the position at the beginning of the 1860s another way, if new improvement outlays could be repayed over a 22-year period in annual installments of 6·5 per cent of the principal, the threshold wheat and oats acreage for mechanization-*cum*-improvement had become roughly coincident with the 19·6 acre threshold relevant in the case of farms whose terrain already was well suited to the use of the reaping machine.

But when the price of grains stopped moving upward, which is what happened for all intents and purposes during the 1860s, the precariousness of a high farming approach relying on rural labor-intensive capital formation to save land became more readily discernible. For, money wage rates in agriculture continued rising during the 1860s, and real farm wages reckoned in terms of grain increased about twice as rapidly as they had in the preceding decade. This was enough to drive up the threshold farm size required for barely profitable new investment in mechanization-*cum*-land improvement – even though the concurrent 35 per cent reduction of the reaping machine's price in relation to harvest wage rates served further to encourage substitution of mechanical for hand methods of cutting grain. By the opening of the 1870s, according to the evidence of Table 15, purchase of a manual sheaf-delivery machine and preparation of the land for its effective use had ceased to make economic sense if the improvements were to be financed at the 3·5 per cent rate of interest implicit in the terms of the government subsidized drainage loans.

For us to recognize the existence of a more complex set of connexions linking mechanization and the adoption of high farming practices, beyond the primitive interrelationship between land improvement and the effective use of the early reaping machines, can scarcely weaken the line of argument advanced here. In general it may be said that although the illustrative case examined in Tables 14 and 15 is rather generous in its evaluation of the direct grain-yield-raising effects of land improvement outlays, it does omit consideration of many potential secondary benefits – especially those connected with the mechanization of other field operations, such as sowing, and the cultivation of row-crops. Quite obviously, removal of obstacles to the movement of mechanical reapers would permit more effective use of grass mowers, hay-tedders, as well as of wheeled cultivators and grain drills, thereby generating labor cost savings that might be applied against the original capital charge for the improvements. It need not automatically follow, however, that the result was to reduce the size of the farm on which fuller mechanization became just profitable – bringing it below the threshold farm size defined for partial mechanization of the sort treated here.

There is even some reason to think that the opposite might have been closer to the truth of the matter, and that the replacement of sickles and scythes in the harvesting of wheat and oats comprised one of the more notably profitable among the various possibilities for mechanizing field operations which were available to British farmers during the 1850s and 1860s. Steam ploughing, however impressive a technical achievement, does not appear to have been a close contender.[1] Coming nearer to the specific activity examined here, we might briefly consider the possibility of a tenant of an improved farm pressing on beyond mechanization of the wheat and oats harvest, to replace hand-mowing of his barley as well. A swath-delivery machine – such as Crosskill's version of the Bell reaper, or the adaptations of McCormick's machine developed by the firm of Burgess and Key – would serve this purpose in a way in which, for reasons already mentioned, the rear-delivery hand-rake models could not.[2] Evidence available for the early 1860s, however, makes it apparent that swath-delivery reaping machines were not only somewhat more costly to purchase than the hand-rake models, but that they saved considerably less labor expense per acre harvested when set to work in wheat, or in wheat and oats. Indeed, it turns out that the unit labor saving per representative acre of wheat, oats, and barley – in the respective acreage proportions 42:31:27 observed nationally in the late 1860s and early 1870s – was greater for the manual sheaf-delivery machine than for the swath-delivery reaper, despite the latter's saving of labor in cutting barley where the former might be supposed to save nothing.[3]

[1] Cf. Spence, *God Speed the Plow*, pp. 103ff. An exact analysis of the economics of steam-ploughing remains to be carried out, but Spence's interesting study makes it clear that 'the simple matter of expense' was a problem that continued to plague steam plow enthusiasts. During the 1860s and 1870s the question whether steam cultivation was or was not economically advantageous remained a subject of controversy; the Royal Agricultural Society concluded that because of the initial expense entailed, for the equipment alone – leaving aside complementary land improvements, a steam plow would be an asset 'only on farms of more than 250 acres of heavy arable land.' (*Ibid.*, p. 118.)

[2] Cf. Orwin and Whetham, *British Agriculture*, p. 112, for a convenient, brief account of these machines. Further details are to be found in James Slight and R. Scott Burn, *The Book of Farm Implements and Machines*, ed. Henry Stephens (Edinburgh, 1858), esp. pp. 357–9.

[3] A comparison of average savings in mower-equivalent man-days for manual sheaf and automatic swath-delivery machines working in the several grain crops, and for repre-

These being the basic facts of the case, it follows that on an improved farm *of a given size* and representative mix of the major cereal crops, there was no way to proceed to mechanize the harvesting of all three grain crops using a 'swather' – either alone or in conjunction with a manual sheaf-delivery reaping machine – which would not reduce the farmer's rate of return on his total investment in reaping machinery, driving the latter below the rate he could obtain by remaining content to have partially mechanized the corn-harvest. This is tantamount to concluding that only the larger grain-farmers could employ swath-delivery reapers to cut their barley and still expect to earn a rate of return on their machinery outlay identical to the rate their lesser neighbors might obtain by investing in a device suited to mechanization of the wheat and oats harvest alone.

Such findings can be no more than suggestive. Were one to insist that harvest mechanization be evaluated in the context of the full array of new methods employed on improved farms, it should be remembered that the domain of profitable high farming remained quite rigidly circumscribed throughout the period under consideration. As Chambers and Mingay recently have pointed out, high farming was found to be economically advantageous only on large holdings, namely those of about 300 acres.[1] Yet medium and small units continued to predominate in the structure of British agriculture: the Census of 1851 had reported 16,840 farms in the size class above 299 acres in England and Wales, comprising about 7·5 per cent of the total number of holdings, and occupying approximately one-third of the agricultural land covered by the returns.[2] Over the next three decades, farm consolidation failed to augment this large-scale segment of the industry. Such concentration as did take place at the national level was slight, and appears to have resulted mainly from consolidations of unviable small-holdings – of less than 50 acres – into

[1] Cf. Chambers and Mingay, *Agricultural Revolution*, pp. 172–6. Farms of 300 acres, if they followed the pattern of land allocation characteristic of the agricultural sector in England and Wales as a whole, would have had approximately 60 acres under wheat and oats (and rye) – a figure quite consistent with the findings reported by Table 15.

[2] Cf. *Parliamentary Accounts and Papers*, 1852–3: XXXII, 1851 Census, 'Report,' p. lxxx. The totals cited by Clapham, *An Economic History*, II, p. 264, differ slightly from these, for reasons not altogether clear.

---

sentative mixes of wheat, oats, and barley suggested by the acreage figures for Great Britain in the late 1860s and early 1870s, is to be found in Table 16, Appendix B. Corresponding purchase prices for various designs and makes of reaping machinery in the early 1860s have been drawn from Wilson, 'Reaping Machines' (1864), pp. 131–9. The details of the threshold calculation are omitted here in the interest of brevity.

farms of under 100 acres.[1] Either in terms of absolute numbers, or absolute extent of acreage controlled, or relative share of the nation's agricultural acreage, the picture revealed by the available statistics for 1851 and 1885 – riddled as these are with elements of incomparability – is one of stability, and in some respects contraction of the potential segment within British agriculture which might profitably have taken up mechanical reaping as part of a thorough-going shift to high farming after 1850.

But the same conditions that were ultimately to spell the doom of high farming in Britain lay at the root of the problem exposed by the more restricted analysis pursued here. The price of grain prevailing in international markets simply was not high enough to warrant continuing British land improvement expenditures on the basis of the direct benefits conveyed in the form of higher cereal yields. Hence such investments could be recouped only when the indirect effects of the improvements, such as those obtained by reducing unit costs of field operations like the corn-harvest, were exploited to the fullest. And still, the expected rates of return to landlord and tenant were scarcely high enough to assuage the doubts which would beset even the mildly risk averse rational investor. That so much in the way of drainage and land improvement was accomplished in the face of such odds is remarkable. One is obliged to acknowledge that if the diffusion of mechanical reaping was at all held back by farmers' irrational resistance to innovation, surely it had been no less advanced by the monumental investment miscalculation on the part of British landowners.

To drive home the point that the heart of the problem lay in the long-run unprofitability of a post-Corn Law Repeal system of husbandry built around cereal production, it is worth stressing that the reason farm terrain and farm layouts in mid-Victorian Britain remained an obstacle to the further diffusion of the reaping machine is not properly traceable to the alleged 'abundance' of rural labor. At least not if that were meant to describe a situation in which farm laborers' wage rates remained so low in relation to prevailing grain prices that little effective inducement was afforded the substitution of other inputs in their place. As has been seen, rising real wage rates in agricultural districts, instead of promoting the replacement of Britain's old farms with modern ones, *à la* Salter, acted as an inhibitor of the preparatory investments which Jacob Wilson had urged upon all those who sought to mechanize their corn-harvest. By

[1] Cf. P. G. Craigie, 'Agricultural Holdings in England and Abroad,' *J.R.S.S.*, L (1887), pp. 86ff.: Clapham (*An Economic History*, II, pp. 264–5) thought it most likely that the throwing together of little hill farms in Wales figured significantly in these developments reflected by the farm size distribution statistics for 1885.

the close of the 1860s a substantial part of the arable farming sector was being forced to struggle on in awkward accommodation to the outmoded tangible legacies of its past, because within the then existing state of technology the transition to higher rural wage levels rendered it less and less economically worthwhile to erase the durable impress of history from the agricultural landscape and begin afresh.

EPILOGUE

In this tale the penalties of having had an early start in the accumulation of farm capital began to weigh heavily only when the advantages conveyed by taking the lead were already largely lost, when capital-intensive cereal production in Britain had ceased to be competitive with extensive grain-farming in the regions of recent settlement across the Atlantic. But is it not also plausible to think that the tug-of-war between 'Ricardo effects' and 'Wicksell effects,' whose outcome restricted the introduction of new labor-saving methods in agriculture, may have been still more directly responsible for the industrial dilemma of the late nineteenth-century British economy to which Veblen directed notice?

Certainly it now seems relevant to look again more closely at the difficulties created by the presence of strong elements of *technical* inter-relatedness between equipment and long-lived physical plant. For, by analogy with the case examined, we can see that limping productivity advance in the building trades would have made the costs of modifying structures, or actually tearing down and replacing plants inherited from earlier phases of the First Industrial Revolution, particularly sensitive to the upward movement of the urban-industrial wage structure in Britain during the second half of the nineteenth century. In the absence of strong product market incentives for rapid expansion of industrial capacity – which would have occasioned the building of new, 'best practice' plants suited to the most advanced machinery designs – an important segment of the domestic demand for capital equipment would thus be comprised of calls for replacements of old-style machinery compatible with the existing factories, mills, and mines. Equipment producers looking at the home market would thereby have been given a strong inducement to concentrate upon the perfection of older-vintage machine designs, rather than upon applications of the latest in technical knowledge. As real wage rates continued to rise, some of the more up-to-date equipment could perhaps be profitably installed even in sub-optimal operating conditions; but the fact that it was not being efficiently utilized – in the technical sense, at least – might well have had the perverse result of misinforming potential purchasers as to the true advantages of the equipment when installed in new facilities.

As a consequence of the direction in which machinery producers were encouraged to channel their energies, we might anticipate a tendency for the purchase costs of older styles of equipment to fall in relation to the prices of technically more advanced machinery – which would only serve to reinforce the rational decision of industrial firms to carry on with their outmoded physical plant instead of investing in thoroughly modernized facilities. And so on, in a vicious circle broken only by the eventually rising costs of maintaining increasingly decrepit structures, or the appearance of dramatically superior innovations in equipment designs – quite likely from abroad. We have here the core of an explanation of how, even in an era of rising real wages, an industry or sector that initially operated with the most up-to-date capital-intensive production processes could be guided by strict adherence to cost considerations into a position of protracted technological backwardness.

In the agricultural sector of the economy, with the route to more complete diffusion of mechanical reaping via renovation of the landscape being gradually sealed off, the immediate avenue of advance remaining open lay through the introduction of machinery on less-than-optimal terrain. Movement in this direction was, of course, fostered by the mounting real costs of labor in the harvest season – quite possibly receiving significant added impetus from the threat posed by Joseph Arch's organizing activities among the agricultural laborers in the 1870s. But after 1873 other forces were set in motion which also worked to reduce the proportion of the British grain harvest that was left for the sickle and scythe. The collapse of the level of grain prices under the onslaught of American wheat exports finally blocked further significant investment in land improvements, but by the same token it compelled a contraction of arable farming. And in the course of that painful readjustment it was the unimproved holdings, on which ridge-and-furrow remained pronounced, that were among the first to be left to 'tumble down to grass'; the nature of the terrain surface that had persisted on these sub-marginal arable farms is evident to this day in the undulating pastures of the East Midlands. The proportion of the British grain harvest that was machine-reaped thus tended to go on rising, in part because the segment of the now declining industry's capacity least well suited to mechanization was the first to be withdrawn from cereal production.

Then, too, technological progress continued in the development of farm machinery. The reaper-binder began to come into use in America in the mid-1870s, opening a new phase in the mechanization of harvest tasks, and it was not long in appearing on the British scene. Increased technical possibilities for saving labor promoted the use of new machinery under less than optimal conditions; and, in the longer run, the mounting potential benefits of farm mechanization – heightened by the exigencies of the nation's needs in wartime – became potent enough once again to initiate significant renovations of the landscape to make way for the tractor and the combine harvester. But in 1875 all that lay in the future, and would form part of a far more complicated story than the one I have sought to illuminate here.

Appendix A: Technical notes

Paragraph 1. Definition of the threshold acreage for simple mechanization, S_T^*.

Let the imputed annual money rental cost of a single reaping machine be M, and the money gain through the reduction of harvest labor costs by mechanization be $(g_m)_k$ for the k-th acre of grain harvested. The threshold acreage for adoption of mechanical harvesting is then defined by the equality,

$$M = \sum_{k=0}^{S_T^*} (g_m)_k. \tag{1}$$

As it is legitimate to treat the difference between the variable cost functions for the hand and machine methods of harvesting as linear over the relevant operating range, S_T^* in equation (1) may be estimated from

$$S_T^* = M/(g_m). \tag{2}$$

Paragraph 2. The saving in harvest costs per acre, g_m.

For the choice between harvesting with a manual delivery reaping machine and harvesting with scythes ('harvesting' including all operations from cutting to stooking), the following notation is relevant:

Hand-method labor coefficients, in man-days per crop acre,

n_1 = mowers;
n_2 = gatherers;
n_3 = binders and stookers.

Machine-method labor coefficients, in man-days per crop acre,

m_1 = machine drivers and rakers:
m_2 = binders and stookers.

Note that $m_1 = N_1/a$, where N_1 is the number of drivers and rakers per machine, and a is the normal daily cutting rate of the machine, in acres.

Relative wage rates exclusive of food, and food costs,

$v_1 = w_1/w_1 = 1$, mower's wage rates being the numeraire;
$v_i = w_i/w_1$, for hand-method wage rates in general;

$v_j = w_j/w_1$, for machine-method wage rates in general;
f, daily food cost per worker;
$f' =$ $(f)/(w_1+f)$.

Total labor costs per acre harvested, inclusive of food are therefore:

$$w_1\sum_i v_i n_i + f\sum_i n_i, \quad \text{for the hand method,}$$

and

$$w_1\sum_j v_j m_j + f\sum_j m_j, \quad \text{for the machine method.}$$

The difference in operating costs per acre harvested may then be represented as

$$g_m = w(L_s^*), \tag{3}$$

where

$w = (w_1+f)$, is the mower's daily wage rate including food,

and

L_s^*, the labor savings in mower-equivalent man-days per acre, is given by

$$L_s^* = (1-f')\,(Q_h\sum n_i - Q_m\sum m_j) + f'(\sum n_i + \sum m_j),$$

defining

$$Q_h = (\sum v_1 n_i)/(\sum n_i) \quad \text{and} \quad Q_m = (\sum v_j m_j)/(\sum m_j).$$

Paragraph 3. Determination of the imputed annual rental cost of a reaping machine.

Let C be the money purchase cost of a reaping machine, and R be its annual imputed rental cost to the farmer, such that

$$M = R(C). \tag{4}$$

Given the rate of interest r and the service life of the machine, in years, D, the standard annuity formula may be employed to determine

$$R = r(1-e^{-rD})^{-1}. \tag{5}$$

Note that this involves setting

$$C = \int_0^D M(t)e^{-rt}dt,$$

taking $M(t) = M$ for all $t = (0, \ldots, D)$, whence, upon evaluation of the definite integral we obtain equation (5).

Paragraph 4. To prove: The gross rate of return on a reaping machine is, for any given farm, inversely proportional to the threshold acreage.

Consider a farm with small grain acreage, G. After covering interest and depreciation charges on a reaping machine in the given year, the pure (or super-normal) profit rate on the original investment in the machine is

$$\pi = [(G - S_T^*)g_m]/C,$$

since, by definition, S_T^* is the break-even point in the farmer's harvesting operations. However, from equations (1) and (4) we have

$$S_T^* = RC/g_m,$$

which, upon substitution in the expression immediately above, yields,

$$\pi = [(G/S_T^*) - 1].$$

The *gross* rate of return on the original investment in the machine is $(R + \pi) = \rho$, from which it follows directly that

$$\rho = R(G)/(S_T^*). \tag{6}$$

Q.E.D.

Paragraph 5. Effects of price and wage rate changes on the mechanization-*cum*-improvement threshold acreage.

In the case of technical interrelatedness considered, we arrive at an augmented expression for the threshold acreage at which it would just begin to be profitable to mechanize the harvest by first making the required farm 'improvements':

$$R_m C_m - (p\dot{y}G' + wL_s^* - C_I R_I A')S_T^{**} = 0. \tag{7}$$

Note the following:

(i) The direct gain from improvements, per acre of small grain harvested, is a function of p, the price of grain and two parameters.

(ii) Sub-scripts m and I distinguish between the capital outlays for the reaping machine and those for 'improvements,' respectively.

(iii) The capital outlay on improvements, per acre of small grain harvested, is

$$I = A'C_I,$$

where A' is a parameter of proportionality, between acreage required to be improved for every acre harvested annually by the machine method. Furthermore,

$$C_I = C_I(w, z),$$

indicating that capital outlays on improvement are a function of agricultural wage rates and prices of materials, z.

Taking the total derivative of (7), under the following conditions,

$\left.\begin{aligned}\partial G'/\partial p = dG' = 0\\ \partial A'/\partial p = dA' = 0\end{aligned}\right\}$ there is no adjustment in crop mix within the arable;

$d\dot{y} = 0$ physical yield-effects of improvements are constant;

$\partial L_s^*/\partial w = dL_s^* = 0$ there is no autonomous or induced innovation in reaping machinery, or in hand methods of harvesting small grain;

$0 = dR_m = dR_I = dz$ rental rates, and the price of materials used for improvements are fixed parameters;

we therefore obtain the following expression:

$$\frac{dS_T^{**}}{S_T^{**}} = \frac{I}{(g_m+g_I)}\tilde{C}_I+\tilde{C}_m-\frac{g_m}{(g_m+g_I)}\tilde{w}, \tag{8}$$

in which the notation and definition given below are employed.

$\tilde{w} \equiv \frac{dw}{w}-\frac{dp}{p}$, the percentage change in the real wage rate;

$\tilde{C}_m \equiv \frac{dC_m}{C_m}-\frac{dp}{p}$, the percentage change in the real (grain) price of a reaping machine;

$\tilde{C}_I \equiv \varepsilon\frac{dw}{w}-\frac{dp}{p}$, the percentage change in the real (grain) cost of the improvement of an acre, due to the change in the wage rates of agricultural labor, $\varepsilon = (\partial C_I/C_I)/(\partial w/w)$;

$(g_m+g_I) = R_m C_m/S_T^{**}$ the initial monetary gain, per acre of small grains harvested, arising from mechanization-*cum*-improvement.

Paragraph 6. To prove: In the absence of other changes the perverse effect of a rise in real wage rates upon the mechanization-*cum*-improvement threshold is not restricted to the class of technically complementary improvements whose production is a labor-intensive process.

From equation (8) we can write the general condition for a perverse change in the threshold acreage:

$$\frac{dS_T^{**}}{S_T^{**}} > 0 \quad \text{when} \quad \tilde{w} > 0, \quad \text{if} \quad [(g_m + g_I)\tilde{C}_m + I\tilde{C}_I] > g_m\tilde{w}. \tag{9}$$

For simplicity let us impose the further restrictions, and consider the situation in which

$$\tilde{C}_m = 0 \quad \text{and} \quad \frac{dp}{p} = 0,$$

noting that $dw/w > 0$ still implies a rise in the price of labor relative to the price of the reaping machine and relative to the price of grain. The condition for a perverse movement in S_T^{**} then simplifies to:

$$(\tilde{C}_I/\tilde{w}) > (g_m/I).$$

But since under the assumed restrictions we now have $\tilde{w} = dw/w$, and $\tilde{C}_I = \varepsilon\, dw/w$, ε being the elasticity of C_I with respect to w, the condition for the perverse wage-effect is fulfilled for $(\varepsilon < \frac{1}{2})$ relatively capital-intensive 'improvements' whenever $(g_m/I) < \varepsilon$.

Q.E.D.

Appendix B: Sources of the estimates of parameters and variables (employed in Table 14)

1. COSTS OF IMPROVEMENTS PER ACRE OF WHEAT AND OATS HARVESTED, $C_I A'$, *c.* 1851.

(a) C_I, the cost of improvements per acre, represents the sum of unit costs (C_{I_t}) in each of three improvement activities: draining, spreading the sub-soil, and leveling.

The costs of drainage: At any given date the cost of installing sub-surface drains depended upon how widely apart the pipes were placed. The maximum spacing which seems to have been thought still worthwhile was 30–6 feet. Installation of pipes on this basis cost £5 per acre drained at the beginning of the 1850s, exclusive of the cost of carting the pipes – as carting was normally provided by the farmer rather than the landlord. Cf. Lord Wharncliffe, 'On Draining, Under Certain Conditions of Soil and Climate,' *Journal of the R.A.S.E.*, XII (1851), pp. 56–7, for estimates of cost of £5.7.9 per acre drained with 30-foot spacing, and £4.9.10½ with 36-foot spacing. On soggy land, however, this would prove inadequate and 24 feet between drains most likely would be required, which would run about £6 per acre – according to the information showing the variation of unit installation costs as a function of spacing, in the *Encyclopaedia Britannica* (1875), I, p. 334. Moscrop, 'The Farming of Leicestershire' (1866), pp. 306–7, refers to £6 as the usual, historical cost of draining an acre of clay land. Further contemporary testimony to this figure's appropriateness, drawn from the *Transactions of the Surveyors Institution, 1871–72*, pp. 66, 102, is cited by R. J. Thompson, 'The Rent of Agricultural Land' (1907). As one seeks contemporary evidence at later points in the century, e.g., in Albert Pell, 'The Making of the Land in England,' *Journal of the R.A.S.E.*, Ser. 2, XXIII, pp. 335ff., the representative historical drainage cost figures tend to drift upward – which is precisely what is to be anticipated.

Spreading sub-soil: Spreading the sub-soil raised by digging trenches for the drains was thought to be very beneficial in its effects upon grain yields, particularly when the sub-soil was mixed with chalk. Hamilton Fulton, 'Drainage of Hethel Wood Farm,' *Journal of the R.A.S.E.*, XII (1851), pp. 150–1, mentions 19 shillings per acre as the cost of spreading the sub-soil on a Norfolk farm on which the spacing between the trenches was closer to 30 feet than to the 24 feet assumed here. Adding 20 shillings to the drainage costs estimated above will bring the combined figure up to £7 per acre, without running a serious risk of overstatement.

Leveling: The cost of leveling the tillage once drained is more difficult to estimate. On the basis of statements as to the costs of ploughing up the topsoil and leveling the field by casting the soil into adjacent furrows, the unit price of the completed work may be put at £2 per acre. This makes use of Chandos Wren Hoskins, 'On "Ridge-and-Furrow" Pasture Land, and a Method of Levelling It,' *Journal of the R.A.S.E.*, XVII (1856), pp. 327–31, omitting the costs attributable to the operation of paring and rolling back the turf before ploughing on the crown of the ridge, and then rolling the turf back into place. On tillage land, obviously, the extra expense for the preservation of the turf during leveling would not arise. Note, however, that the cost estimate may well be on the low side, as the leveling of pasture land was undertaken to prevent burning of crown grass on drained land still lying in ridge and furrow, and would not have been carried out as thoroughly as improvement work undertaken to prepare a field for reaping machines.

Together, the cost of these operations *c.* 1851 comes to £9 per acre of land drained (C_I), a figure which is not out of line with the improvement expenditures per acre of *all* land – arable and pasture – reported for large landowners by Thompson, *Landed Society*, p. 250.

(b) A', the ratio of arable acreage to the acreage under wheat and oats is treated as a parameter in this analysis. From the *Agricultural Statistics* for the year 1867–72 it appears that approximately 35 per cent of arable was devoted to the two crops here assumed to be machine cut (cf. B. R. Mitchell and Phyllis Deane, *Abstract of British Historical Statistics* [Cambridge: Cambridge University Press, 1962], p. 78); the representative British farmer, whose arable was distributed like that of the country at large, would thus have to drain (1/0·35) = 2·85 acres of wheat and oats.

(c) ε, the share of C_I attributable to labor costs is a parameter which is relevant in estimating the influence exerted by changes in the level of agricultural wage rates upon unit improvement costs after 1851. It is found that labor costs comprised 57 per cent of the expense of installing pipe drainage, exclusive of haulage, and that this share was essentially invariant with respect to the spacing of the pipes. The *Encyclopaedia Britannica* (1875), I, p. 334, provides data from which we can compute that the labor share moved from 57 per cent to 58·4 per cent as the width of the spacing increased from 24 feet to 30 feet, and declined to 56·6 per cent when pipes were densely laid at 18-foot intervals. (Because it is argued that we can reasonably expect reductions in the relative price of drainage pipe *vis-à-vis* labor to have led to a higher pipe–labor input mix, and thus closer spacing of drains which permitted adequate draining of heavier soils, it is important that the labor share appears to have been an elasticity parameter which would remain unchanged in the face of the rise of wage rates relative to the price of drainage materials.) The other two operations – spreading the sub-soil and leveling – particularly the latter, would involve the use of horses: but as this work would be done at times during the year when the farmer's horses were not otherwise engaged, the two activities can justifiably be treated as having been purely labor-using.

Thus, for all three parts of the improvement work combined, the labor bill would amount to approximately 70 per cent of the estimated outlay, i.e., weighting the shares, we find: $= [0{\cdot}57(6/9)+1{\cdot}0(3/9)] = 0{\cdot}713$.

2. VALUE OF THE DIRECT BENEFIT FROM LAND IMPROVEMENTS PER ACRE OF WHEAT AND OATS HARVESTED ($p\dot{y}G'$), *c.* 1851.

(a) $\dot{y}$, the incremental wheat yield, in bushels per acre sown, is estimated on the assumption that the farmer would enjoy a 15 per cent improvement in the level of his cereal yields as a consequence of drainage and spreading the sub-soil – without any other changes in farm practice, such as manuring with solid or liquid fertilizers. The proportional improvement in yields is suggested by the few and scattered contemporary reports of what seem to have been noteworthy increases in yield following drainage: Caird (*English Agriculture*, p. 374) cites the case of a Northumberland strong-land farmer who was extremely pleased to find his farm 'had been increased 20 percent in its produce of wheat by pipe drainage.' There were other, still more optimistic figures mentioned, but they all are hedged with qualifications that lead one rather to discount them. For example, a Civil Engineer, writing in 1851, thought there was 'little doubt' that spreading the sub-soil 'in conjunction with perfect drainage' would raise yields 30 per cent. Cf. Fulton, 'Drainage of Hethel Wood Farm' (1851), p. 151.

Starting from an average wheat yield of 26·67 bushels per acre, as estimated for England as a whole in 1850–1 (cf. Caird, *English Agriculture*, p. 480), a 15 per cent gain in yield would provide $\dot{y} = 4{\cdot}0$, bringing the average above 30 bushels per acre. It can be argued that, historically, the reductions in the relative price of drainage pipe encouraged closer spacing of drains (requiring more pipe per acre) and thus permitted cheaper attainment of *adequate* draining on heavier land, rather than better and better drainage – with consequently greater yield increases – on moderate and light soils. These calculations therefore make no allowance for a rise in *average* cereal yields per acre in response to alterations in the price of drainage materials relative to rural labor wage rates.

(b) ($p\dot{y}$), the value of the additional grain yield per acre drained *c.* 1851, is determined by referring to the prevailing price of wheat per bushel, p. The average *London Gazette* price of wheat for 1849–52 stood at 41s. per imperial quarter, according to the data reproduced in Mitchell and Deane, *British Historical Statistics*, p. 488. Taking 40s. per quarter as a notional price, and reckoning 8 bushels of 63 lbs to the quarter, we find $p = 5$s., and $(p\dot{y}) = 20$s. at this date.

While ($p\dot{y}$) is estimated on the basis of the value of the yield increment when the drained land was sown with wheat, this does not imply that all the cereal acreage of the farm need have been devoted to that crop. Bushel yields of oats and barley per acre were higher than the average wheat yield, so that an appreciation of 15 per cent would imply larger absolute incremental bushels yields of those grains as a result of drainage. But the bushel price of oats and barley was correspondingly lower than that for wheat. Indeed, if the farmer had achieved an efficient allocation of his cereal acreage, the marginal value product of his land (and of land improvements) would be identical for all the cereals – relative quantity yields varying inversely with relative prices – and it would make no difference which of grains was considered for the purpose of estimating ($p\dot{y}$) at the margin. Note that this holds true among the cereals as these were, for all intents and purposes, pure substitutes in the rotation; it would not be correct to assert that in equilibrium the market value of an additional acre of root crops would equal the market value of giving that acre over to cereals. The shift away from the root crops, or rotation grasses under a regime of convertible husbandry, could be expected to lower the steady-state yield on the intra-marginal cereal acreage.

(c) A generous estimate of the value of the additional cereal yield per acre of wheat and oats harvested ($p\dot{y}G'$), *c.* 1851, is obtained by putting $G' = 1{\cdot}6$. The *Agricultural Statistics* for the late 1860s and early 1870s suggest that for Britain as a whole approximately 66 per cent of cereal acreage was devoted to wheat and oats, which would imply $G' = (1/0{\cdot}66) = 1{\cdot}5$. The stability of the relative position of wheat and oats, taken together, among all cereal crops in England and Wales throughout the last quarter of the nineteenth century is strikingly evident in the analysis of the acreage statistics presented by R. H. Best and J. T. Coppock, *The Changing Use of Land in Britain* (London: Faber and Faber, 1962), pp. 88–9.

(d) A check on the plausibility of the present set of estimates is provided by considering their implications regarding the anticipated rate of return to investment in land improvements *c.* 1851, in the absence of mechanization.

The government's program aimed at encouraging drainage investment was established during 1846–9, and, as noted above (pp. 252, 262), offered financing at 3·5 per cent per annum over a 22-year amortization period – with annual repayments running at 6·5 per cent of the principal. If Robert Peel and the other sponsors of the Drainage Act had in mind the draining and spreading of the sub-soil of the entire arable (omitting leveling) at a cost of £7 per acre, and the value of the incremental yield with wheat at 40s. per quarter, they were contemplating a subsidy to landowners of 2·0 or 2·5 per cent per annum. This would follow from consideration of the value of the annual incremental yield as a fraction of the total capital expense: (20s.)(1·6)/(140s.)(2·85) = (32s.)/(399s.) = 0·0802, which would be sufficient to amortize the principal in twenty-two years and yield a compound rate of return between 5·5 and 6·0 per cent per annum.

It is not inconceivable that in the late 1840s Peel and his friends had in mind an even more generous compensation to the landowning class for the injury so recently done them by the Repeal of the Corn Laws. The average price of wheat during the 1840s had stood above 50s. per quarter, so that were the latter taken as the notional price in valuing the incremental yield of grain – instead of the 40s. price which actually prevailed *c.* 1851 – the implied anticipated rate of return on the expenditures just considered would have been upward of 8·0 per cent per annum. Were the leveling of the drained fields to be undertaken, however, the initial outlay would have been larger, so that even with wheat at 50s. per quarter the annual proportional return [(4 bu.)(6·25s. per bu.)(1·6)/(180s.)(2·85)] = 0·078, would generate an average annual compound rate of return of only 5·5 per cent over the 22-year amortization period. As it was, *c.* 1851, with wheat fetching 5s. per bushel the annual value of the yield enhancement as a proportion of the corresponding total improvement outlay was (0·0624) insufficient to generate the 3·5 per cent per annum rate of interest made available to landlords by the Drainage Loan Acts.

3. THE PER ACRE SAVINGS IN HARVEST LABOR DUE TO MECHANIZATION (L_s^{**}).

The saving in labor realized through the use of mechanical reapers has been estimated on the basis of a comparison with the labor required when standing grain was mown by scythesmen, rather than cut by small hand-tools such as sickles or reaping hooks. [E. J. T. Collins, 'Harvest Technology and Labour Supply in Britain, 1790–1870,' *Economic History Review*, 2nd Ser., XXII, 3 (December 1969), pp. 453–73, provides a convenient description of the different hand techniques, and some of the available evidence indicating that per acre labor requirements were lower for mowing than for the other methods. On harvest cost differences between standing grain and laid grain, cf. P. A. David, 'Labour Productivity in English Agriculture, 1850–1914: Some Quantitative Evidence on Regional Differences,' *Economic History Review*, 2nd Ser., XXIII, 3 (December 1970), pp. 504–14.] The comparison between hand-method and machine-method labor requirements per acre of wheat and oats is made for the entire harvest operation – cutting, binding, and stooking, as described in Paragraph 2 of the Technical Notes in Appendix A.

Full details of the estimates made, with appropriate comments and corroborative evidence, cannot be presented in the space available here. The L^*-coefficient for the manual sheaf-delivery reaper which has been employed in Tables 14 and 15 is drawn from Table 16, below: it is the weighted average of separate coefficients of labor-saving estimated for wheat and for oats, on the basis of A. Hammond's Practice. Note from Table 16 that L_s^*-coefficient used is on the generous side of the estimate derived on the basis of P. Love's Practice with manual sheaf-delivery machines. Love, however, managed his harvest labor on a piece-work system known as 'thraving,' and appears to have been uncommonly economical in the use of ancillary workers in gathering, binding, and stooking behind the mowers; the overall labor-savings he reported having realized through instituting machine-cutting were rather smaller than might otherwise be anticipated. Cf. Peter Love, 'On Harvesting Corn,' *Journal of the R.A.S.E.*, XXIII (1862), pp. 217–26, for the data underlying estimated labor-savings in harvesting wheat with the hand-rake, and the swath-delivery machines, which are shown in Table 16.

The sources of the estimates based on A. Hammond's Practice may be briefly indicated, following the approach outlined in Paragraph 2 of the Technical Notes. The latter, of course, refers to the computation of L_s^* for a single homogenous grain crop.; thus,

Harvesting Wheat: the n_i– and m_j– coefficients are derived from Anthony Hammond, 'Use of Reaping-Machines,' *Journal of the R.A.S.E.*, XVII (1856), pp. 339–40; the relative wage-structures v_i and v_j are based on information in the same source, and the representative wage quotations for England, *c.* 1851 in Henry Stephens, *The Book of the Farm*, Second Edition (Edinburgh, 1855), II, pp. 345–6; daily food costs per workers, as a fraction of mowers' daily wages inclusive of food (f') are based on Stephens, *The Book of the Farm* (1855), II, p. 346 and *ibid.*, Third Edition (Edinburgh, 1871), II, p. 313.

Harvesting Oats: The n_i-coefficients derived for what from Hammond 'Use of Reaping-Machines' (1856) were adjusted downward using a factor of 0·667 derived from John Taylor, 'On the Comparative Merits of Different Modes of Reaping Grain,' *Trans. H.A.S.* (July 1844), pp. 262–3. According to the latter source, a force of 7 workers – 2 mowers, 2 gatherers, 2 binders and 1 boy, raking – in 10 hours cut 2·75 acres of wheat (2*A*.3*R*.0*P*.); working in oats and barley the same force cut 4·125 acres (4*A*.0*R*.20*P*.) in the same time. Consequently, the same downward adjustment of the n_i-coefficients provided by Love ['On Harvesting Corn' (1862)] for wheat were made in order to estimate the L_s^*-coefficients for barley harvested with the swath-delivery machine. Otherwise, all the data employed in estimating the L_s^*-coefficient for oats harvesting with the hand-rake reaping machine following Hammond's practice were drawn from the sources described above.

TABLE 16 Savings of mower-equivalent man-days of labor per acre for different grain-crops

	L_s^*-coefficient with a single machine		
	Manual sheaf-delivery reaper,		Swath-delivery reaper,
Per acre harvested	Hammond's Practice	Love's Practice	Love's Practice
WHEAT, cut (gathered), bound, and stooked	1·416	1·100	0·712
OATS, cut (gathered), bound, and stooked	0·667	0·413	0·440
WHEAT and OATS[a]	1·096	0·804	0·594
BARLET, cut	0	0	0·361
WHEAT, OATS, and BARLEY[b]	0·800	0·587	0·532

[a] Wheat and oats coefficients weighted in proportions of 57:43.
[b] Wheat, oats, and barley coefficients weighted in proportions of 42:31:27, respectively.

4. WAGES AND PRICES – w, p, C_m, AND C_I – FROM *c.* 1851 TO *c.* 1871

(a) Mowers' daily wage inclusive of the cost of food, *w*, is estimated as having been 3s. for England in general *c.* 1851, on the basis of Stephens, *Book of The Farm* (1855), pp. 345–6. This initial observation was extrapolated to later dates on the basis of an index of harvest labor wage rates in England's 'corn-counties' in 1850, 1860, and 1870. The index was formed by first computing an average of the daily farm money wage indexes for a representative group of twelve counties in England drawn from A. L. Bowley, 'Statistics of Wages in the United Kingdom during the Past Hundred Years,' *Journal of the R.S.S.*, LXI (December 1898), pp. 704–6.

The resulting average index numbers resemble the daily wage series for Northumberland quite closely, permitting the use of the relationship between the wage rates of 'ordinary' and 'harvest' workers observed in

Northumberland on these three dates to convert the index based on Bowley's data into a harvest wage rate index. Cf. A. Wilson Fox, 'Agricultural Wages in England and Wales during the Last Fifty Years,' *Journal of the R.S.S.*, LXVI (June 1903), p. 320, for the harvest differential in Northumberland. The resulting index, based on 1850 = 100, is: 131·3 in 1860, and 154·5 in 1870. The decadal movements have been put at 30 per cent and 50 per cent, respectively, for the 1850s and 1860s in Table 14.

(b) The price of wheat per bushel, *p*, *c.* 1851 was roughly extrapolated on the basis of the *Gazette* quotations of the price of wheat per imperial quarter averaged for 1849–52, 1859–62 and 1869–72: 41s., 51s. 8d., 52s. 2d., as computed from Mitchell and Deane, *British Historical Statistics*, p. 488. The decade averages show a slightly different course; for 1841–50, 53s. 3d.; for 1851–60, 54s. 7d.; for 1861–70, 51s. 1d.

(c) Estimated delivered cost of McCormick–Hussey type, manual sheaf-delivery, reaping machines (C_m); based on quotations of £30 *c.* 1851 and *c.* 1861, and £21 *c.* 1871, with 10 per cent delivery charge added. Cf. Slight, 'Report on Reaping Machines' (1852), p. 197; Hammond, 'Use of reaping machines' (1856), p. 49; Wilson, 'Reaping Machines' (1864), p. 135; Coleman and Paget, 'Exhibition of Implements' (1865), p. 389; Orwin and Whetham, *British Agriculture*, p. 110 (for price of two-horse machine quoted at the Manchester Exhibition of 1869).

(d) The unit cost of improvements (C_I) after *c.* 1851 has, for simplicity, been estimated in Table 14 on the assumption that the elements of the farm wage structure moved together. This allows the use of the percentage rate of rise in *w* to serve as a proxy for the percentage increase in wages paid workers engaged on land improvement projects. Table 14 further assumes that the money price of agricultural pipe and other materials used in drainage remained essentially unchanged. Thus the decadal changes in C_I are estimated from the percentage change in *w* and the estimate of the elasticity parameter ε.

Ramifications

6 Transport innovations and economic growth: Professor Fogel on and off the rails

I

Upon its appearance at the end of 1964, Robert W. Fogel's *Railroads and American Economic Growth*[1] drew wide notice as a prototype of the *nouvelle vague* approach to economic history; indeed, the ripples excited on that once quiet pond soon spread as far as the pages of *The Times Literary Supplement.*[2] And rightly so. Fogel's study is without question an important, extremely impressive contribution. Economists and historians concerned with the consequences of transportation improvements in developing economies especially ought to welcome this re-examination of the United States' nineteenth-century railroad experience, while its subtle integration of sophisticated theoretical argument with painstaking, highly ingenious quantitative analysis should find a considerably larger appreciative audience.

This is, however, a book that requires – as well as one that will repay – exceptionally careful reading. In it, Fogel has permitted himself a distractingly provocative style, particularly when summarizing his principal conclusions: 'despite its dramatically rapid and massive growth over a period of a half century, despite its eventual ubiquity in inland transportation, despite its devouring appetite for capital ... the railroad did not make an overwhelming contribution to the production potential of the economy' (p. 235). Such flights of radical revisionism, so boldly challenging to conventional wisdom and historiographic traditions concerning railroad transportation's role in American economic development, will on closer scrutiny be found to have been managed in large part by the device of affixing very special meanings to commonplace phrases. At times the prose edges perilously close to inviting casual misreading of both the nature of the argument being advanced and the substance of the evidence adduced in its support.

To be fair, however, it is not Fogel but some among his readers who must bear the immediate responsibility for actually saying that Fogel's work reveals that 'the railroads made some contribution to growth, but ... America would not be drastically different if they had never existed.'[3]

[1] Robert W. Fogel, *Railroads and American Economic Growth: Essays in Econometric History* (Baltimore: The Johns Hopkins Press, 1964).

[2] Cf. Keith Thomas, 'New Ways in History,' *Times Literary Supplement*, 7 April 1966; Peter Temin, 'In Pursuit of the Exact,' *ibid.*, 28 July 1966.

[3] Lance E. Davis, 'Professor Fogel and the New Economic History,' *Economic History Review*, 2nd ser. XIX (1966), p. 659.

Likewise, should a reader find it 'difficult to avoid the conclusion that we are unlikely to generate much economic growth in an underdeveloped country . . . by bringing about improvements in its internal transportation facilities,'[1] either he has begun playing deceptive games with the word 'much' – similar to the bits of semantic legerdemain that can be managed with 'drastically' and 'overwhelming contribution' – or he is seriously suggesting a policy conclusion that Fogel's book neither advances nor supports. Lamentably, misconstructions of this kind – one hopes they are not consciously ambiguous glosses – threaten to turn the work into a subject of heated, if not always illuminating, controversy.[2] The importance of some of the theoretical and empirical issues makes it worthwhile trying to bring them into clearer focus.

II

Odd as it may seem, the central idea in Fogel's book is really not especially concerned with the role railroads played in U.S. economic growth. It touches that question only by implication and by way of illustration. For, most simply stated, Fogel's main thesis is that 'no *single* innovation was *vital* for economic growth during the nineteenth century' (p. 234, my italics). To substantiate this thoroughly reasonable general view of the nature of technical progress and its relationship to economic growth, the author focuses upon one particular innovation, the railroad, whose introduction contemporaries and later analysis alike hailed as having wrought exceptionally profound economic changes. Fogel seeks to destroy the symbolic myth of the 'Iron Horse' by demonstrating that the railroad, considered as an innovation distinct from pre-existing modes of inland transportation, was not an 'indispensable' ingredient in the development process.

Elisions between form and function do turn up frequently in historical narratives. Although in employing them the historian's intent may be to evoke the drama of the past by momentarily depriving his readers of a detached analytical vantage-point, such elisions contain the unwarranted

[1] George W. Hilton, 'Review of *Railroads and American Economic Growth*,' *Explorations in Entrepreneurial History*, 2nd ser. III (1966), p. 237.

[2] For reviews and commentaries in addition to those already cited, cf. Albert Fishlow, *American Railroads and the Transformation of the Ante-Bellum Economy* (Cambridge, Massachusetts: Harvard University Press, 1965), esp. Ch. 2; Louis M. Hacker, 'The New Revolution in Economic History,' *Explorations in Entrepreneurial History*, 2nd ser. III (1966), pp. 159–75; Stanley Lebergott, 'United States Transport Advance and Externalities,' *Journal of Economic History*, XXVI (1966), esp. pp. 437–44; John Meyer, 'Review of *Railroads and American Economic Growth*,' *Journal of Political Economy*, LXXIV (1966), pp. 87–8; Marc Nerlove, 'Railroads and American Economic Growth,' *Journal of Economic History*, XXVI (1966), pp. 107–15; George R. Taylor, 'Review of *Railroads and American Economic Growth*,' *American Economic Review*, IV (1965), pp. 890–2.

assertion that specific institutional or technical arrangements were necessary to a process simply because they fulfilled one or more of its crucial functional requirements. The new-style economic history championed by Fogel shows scant charity toward such lapses from analytical precision, however compellingly they may be used in the telling of true stories.

It is no doubt salutary for us to be thus reminded that economic questions arise, by definition, only when there is more than one way to skin the same cat. Improved transportation service, which became available to American farmers, manufacturers, and merchants in the form of the railroads, could have been provided by other means. But by the same token, laboriously to erect and then topple the null-hypothesis of indispensability – or, as Fogel calls it, 'The Axiom of Indispensability' – must surely be a limited, if not redundant, way to evaluate the importance of any innovation in the *economic* domain. So long as possibilities exist for substituting other goods and techniques in order to fulfil material wants, can any single alternative truly be held to be indispensable? The state of the arts typically offers an array of production and distribution options among which may be found some method that is not a *dismally* inferior alternative to the best-practice solution under consideration. Not only does one have trouble conjuring up specific goods or production methods which would not be replaceable at a finite cost. It is hard to conceive of the entailed substitution costs being so great that they jeopardize a society's long-run economic viability, even in the case of products or processes that happen to have been universally selected in preference to available alternatives.

Even the most skeptical intelligence should be readily persuaded by Fogel's study that the provision of transportation by railroads in nineteenth-century America constituted no exception to the general state of affairs just described. But the book ultimately ventures beyond mere rejection of the notion that railroads were literally indispensable. As a means of judging the *importance* of the economic contribution made by this particular innovation, Fogel ambitiously undertakes to quantify the finite social costs of dispensing with rail transportation in 1890, a date which found the U.S. steam railroad system quite well developed.

The outcome is an estimate of the combined direct and indirect resource costs which turns out to be far from 'overwhelming' in comparison with the total productive capacity of the U.S. economy in 1890. Indeed, many readers will find it quite startling that the entailed social costs, in Fogel's terminology 'the social savings' attributable to the railroad, are put at less than 5 per cent of G.N.P. Adherents to the theory of strategic bombing may be especially upset, if not by the idea that eliminating all rail communication would just amount (in the not-so-long run) to bombing a country's G.N.P., then surely by the implication that although G.N.P. is an easy enough target to hit, denting it seriously by knocking out a 'key' sector is altogether another proposition. Fogel, however, extracts a quite different moral: the demonstration that in 1890 Americans could, at an apparently trivial cost in terms of final output foregone, have stopped using their extensive railroad network, 'clashes with the notion that economic growth can be explained by leading sector concepts' (p. 236). One is left with the tacit suggestion that, save for a mere difference in sign, leading sector hypotheses about economic growth can all be directly equated with the rationale of strategic bombing: a thought that quite possibly accounts for the sense of respectful exasperation which the book seems capable of provoking in otherwise patient men.

Whatever the ultimate interpretation placed upon these quantitative results, pursuit of the social savings does lead in the course of Fogel's study to efforts at answering new and interesting questions. The assessment of the direct and indirect advantages afforded by the availability of rail service is an intricate, multi-phased affair in which long-distance, or interregional, freight carriage is treated separately (in Chapter 2) from the question of intraregional transportation benefits (dealt with in Chapter 3). That separation proves practical because the empirical investigations are conducted in detail only for shipments of agricultural commodities which move, intraregionally, from farms to well-identified primary markets, and thence across regional boundaries to secondary markets for distribution to consumers. A reckoning of the direct and indirect social costs of dispensing with railroads as carriers of *all* classes of freight is subsequently provided (in Chapter 6) by rough extrapolation from the findings pertaining to freight shipments that originated on farms.

The specific steps in Fogel's calculations defy faithful description in brief compass. But, inasmuch as a detailed summary account by the author himself has already been provided in article form,[1] it may suffice here to indicate only the general approach, particularly the fashion in which the concept of social savings is used to gauge the 'primary' and 'derived' effects, i.e., the direct and indirect costs of the loss of railroads *as a medium of transportation.* For a single year, 1890, a measure of the direct social cost is provided by determining the additional resources that would have been required if one relied exclusively upon inland ship, barge, and wagon conveyance to execute the program of inter- and intraregional freight shipments actually handled by the railroads at that date. Included in this measure is an estimate of the added cost of holding enlarged inventories to compensate for the faster, year-round service that would have been sacrificed by keeping the 'Iron Horse' forever stabled.

[1] R. W. Fogel, 'Railroads as an Analogy to the Space Effort: Some Economic Aspects,' *Economic Journal*, LXXVI (1966), pp. 16–43.

It is pointed out that in principle such a reckoning is likely to overstate the direct social savings, since no deductions are made for adjustments (in the freight-mix and locational pattern of production and consumption) which would have permitted a less costly solution to the transportation problem in the absence of railroad facilities. Moreover, the type of social savings estimate just described is held to capture not only the direct social costs, those in transportation operations alone, but also all the indirect social costs of the loss of rail service. Fogel maintains (pp. 224–5) that as the calculation supposes an exact replacement of the transportation functions performed by railroad in 1890, there is no need to consider any indirect repercussions of the loss of rail service; to make an additional allowance for indirect effects, those upon other productive operations would, he says, amount to double counting.

The foregoing mode of accounting thus serves to dispose of all the generic transport-providing consequences of the innovation, leaving only the question of the indirect effects of the railroad sector's peculiar input demands upon other supplying industries. In treating the latter, Fogel puts aside the apparatus of the social savings calculus and, taking aim at W. W. Rostow's well-known views on the role which railroad investment demands played in the so-called 'American take-off,' examines the railroad sector's backward linkages with manufacturing industries in the two decades before the Civil War. The principal point which emerges (from Chapters 4 and 5) is that railroads remained a marginal purchaser of domestic industrial output during the 1840s and 1850s. *In toto*, railroad purchases during 1859–60 accounted for little more than a tenth of the value added in the manufacture of transportation equipment, lumber, machinery and castings, and blast-furnace and rolling-mill products – a group of industries which was responsible at the time for about 25 per cent of aggregate value added in U.S. manufacturing. In the specific instance of pig-iron production, to which Fogel devotes closest attention, the requirements for rails laid in the period 1840–60 represented a substantial *potential* claim on domestic blast-furnace capacity, one that on average exceeded 50 per cent of annual pig-iron production. Yet with the bulk of those rail orders being filled by importation, and with worn-out rails (notably the inferior 'American rail' turned out specially for that market by British ironmasters) increasing the supply of scrap for U.S. re-rolling mills after 1850, it is found that on average only 5 per cent of annual domestic pig-iron output actually flowed into the production of new rails during the 1840–60 period.

The attack mounted against the hypothesis that U.S. industrialization before the Civil War rested largely on the creation of additional capacity to provide railroads with capital goods and current inputs proceeds along rather conventional lines, but is still so overwhelming that one may now wonder how such a notion could ever have been seriously entertained by historians acquainted with the broad features of the ante-bellum economy. By contrast, however, Fogel's assessment of the railroad's impact as a comparatively efficient means of transportation, which forms the arrestingly novel centerpiece of his study, remains open to many criticisms that have thus far escaped systematic review. These fall under three major headings. The problems one encounters in the first category are rather diverse, but all bear immediately on the accuracy of the estimated social cost of dispensing with railroad service in 1890. Next, it is necessary to reconsider the basis on which the proportional social loss entailed in giving up railroads is held to equal, if not to exceed, the proportional net social gain attributable to the innovation's introduction earlier in the century. In the last category one must lodge the most strenuous objections to the study's reliance upon the ratio of social savings to G.N.P. as a meaningful index of the importance of the contribution railroads made to U.S. economic growth in the nineteenth century.

III

In the above schedule our three classes of criticism have been listed in ascending order of their gravity and general interest. Noting that the subject of passenger conveyance receives scant mention in Fogel's inquiry, the following adumbration contrives to do honor to that ancient railroad custom which informed prospective ticket-purchasers of the degree of discomfort they might expect to endure in each of the available classes of accommodation. This seems little enough attention to accord to the now-vanishing carriage of people, as well as freight, via rail, an activity that brought U.S. steam railroads more than 25 per cent of all their operating revenues in 1890 and still higher proportions at earlier dates.

DISCOMFORTS OF THE FIRST CLASS

The task of reckoning the direct social savings attributable to railroads in 1890 would seem to afford a splendid occasion for a thorough examination of the technical and economic aspects of late nineteenth-century transport and distribution arrangements in the United States. An undertaking of that sort undoubtedly would shed much light upon elements of interrelatedness among methods of shipping, handling, and storing goods. It would, moreover, be particularly germane to the thesis that the appearance of isolated 'great innovations' is less consequential for economic progress than the rate at which clusters of interlocking, mutually supporting techniques can be brought into use. Regrettably, the opportunity to pursue such an inquiry has not been seized in Fogel's book.

Consequently, technical problems that should concern transportation specialists are grossly simplified, or completely overlooked, in estimating the direct component of the social costs of dispensing with rail service. In 1890, in excess of 76 billion ton-miles of freight were handled by U.S. railroads. That canals could have carried a major portion of this additional traffic without an increase in the marginal social cost of transportation is a vital assumption in Fogel's argument. Yet only one page is devoted to supporting this assertion by examining the cost characteristics of canal transportation in the United States, and it is not a very satisfactory page at that. The short-run terms in which the problem is considered would be justified if existing inland water-transport facilities in 1890 were shown to be adequate to handle the additional traffic diverted from the railroads. But the sole concrete reference to this problem (p. 28, n. 28) points out that in the absence of rail carriage from the Midwest to the eastern seaboard, the agricultural products shifted to interregional water routes would have sorely taxed the normal tonnage capacity of the Erie Canal, forcing greater reliance on the (more expensive) New York State Barge Canal route. Furthermore, the railroads did perform other services besides moving farm products from west to east, and we are never told how the existing canals would have coped with the total transfer of east-bound and west-bound interregional freight, plus way-traffic carried by the railroads.

Even with the less efficient canals being utilized to a much fuller extent than they were in 1890, avoidance of bottlenecks would probably have necessitated building additional canals in areas where terrain and available water supplies entailed higher unit costs for operation at desired capacity. As an alternative, the prisms and locks of canals already situated along the more advantageous routes might have been enlarged, thereby enabling them to accommodate greater tonnage. However, supplementary capital outlays for reservoirs and feeders required to maintain an adequate, reliable supply of water cannot be overlooked when one reckons the social costs of expanding the capacity of existing canal facilities, since that solution would have increased the volume of water lost in lockage of vessels, especially those of smaller displacement, as well as increasing the rate of loss through evaporation and filtration.[1]

Similar skepticism is warranted in regard to Fogel's treatment of other components in his hypothetical non-rail system of freight transportation. Consider only the supplementary requirements for wagon-haulage to and from water shipment points in the country's interior, and the costs of holding enlarged inventories to compensate for the absence of all-year rail service. These are not inconsequential items: as it is, the latter accounts

[1] Cf., e.g., Harry N. Scheiber, 'Discussion of "United States Transport Advance,"' *Journal of Economic History*, Vol. XXVI (1966), pp. 462–5.

for 65 per cent of the interregional transportation social savings estimate, while the former is similarly dominant in the estimated social savings afforded by railroads in intraregional freight movements. Yet Fogel's calculations (esp. pp. 85–7) implicitly assume that all sites along navigable waterways in the United States would have been feasible and more or less equally efficient cargo trans-shipping points, and in that respect would have resembled sites along most railroad lines. In the absence of any allowance for additional capital outlays to create passable ubiquitous access to lake shores and river banks for both wagons and vessels, this simplification must lead to an understatement of the supplementary cost of intraregional wagon-haulage.

Fogel also fails to consider that even if social marginal costs of actually storing a vastly increased volume of goods remained strictly constant, as he implicitly assumes, the additional warehouses, grain elevators, coal bunkers, and stockyards required to hold the higher peak-level inventories when transportation was interrupted by snow on the roads and ice in the lakes and rivers, would all be absorbing productive resources in the form of average excess (storage) capacity. The magnitude of the problem is perhaps suggested by the following specific case: storage techniques at the Union Stock Yards in Chicago were such that in order to hold a peak inventory of livestock equal to half the Yards' annual receipts – the latter proportion being Fogel's estimate of the average compensatory inventory requirement for agricultural products – it would have been necessary in 1890 to set aside not some 200 acres of land for livestock pens but an area closer to 10,000 acres. That is to say, livestock pens would have taken up an area roughly equal to half of all the privately utilized land in the city of Chicago in 1890.[1]

[1] The calculation is primarily of heuristic value. Although specific numerical results ought not to be taken too literally, they do appear to be fairly reliable as an indication of the orders of magnitude involved. The basis for the calculation is as follows:

(a) The daily yardage capacity of the Union Stock Yards (in numbers of head of livestock) and the land area occupied by the yards remained essentially unchanged from 1868 to 1889. Cf. *Report of the Select Committee on the Transportation and Sale of Meat Products*, 51st Congress, 1st session, U.S. Senate Report no. 829 (1890), p. 17 (hereafter *Vest Committee Report*); B. L. Pierce, *A History of Chicago*, II (New York: A. Knopf, 1940), pp. 93–4. From these sources it is found that 200 acres of the yards were devoted to livestock pens, apportioned between uncovered cattle pens and covered hog and sheep pens in the approximate ratio of 1:2.

(b) Using the *Vest Committee Report* figures on daily cattle yardage capacity (25,000 head) and daily hog and sheep yardage capacity (135,000 head), and livestock receipts data for 1890 (*Chicago Board of Trade Report for 1896*, p. 42), we may separately compute the ratios of one-half annual receipts to daily yardage capacity for cattle (69·1), and for hogs and sheep (37·1).

(c) The total land area required for pens to carry a six-month livestock inventory in 1890 is therefore computed as $200 \times [(1/3)69{\cdot}2 + (2/3)37{\cdot}1] = 9{,}600$ acres, assuming the

Should not the same general point also be made in regard to the incremental social costs entailed in having a vastly augmented fleet of ships and barges, not to mention the increased canal capacity proposed by Fogel for the Midwest, standing idle during the five months of the year when northern inland water navigation was suspended? And what of the untold extra wagons and teams immobilized by snow and mud for want of all-weather roads? Now, the sum of all such cost items neglected by Fogel's estimates might still not look very impressive alongside G.N.P. Will waving them away on that account advance our understanding of the connexions between the availability of railroad transportation and the changing level of capital utilization and productivity observed throughout the economy?

The root of these and other kindred difficulties lies in Fogel's attempt to apply the sort of techniques developed in cost–benefit analysis to evaluate the long-run impact of the 'large-scale' investment decision represented by the construction of the U.S. railroad system. The deficiencies of partial analyses, which accept the existing structure of prices and production, in handling such large projects are, however, generally recognized. As Prest and Turvey observe:[1]

> If investment decisions are so large relatively to a given economy (e.g., a major dam project in a small country), that they are likely to alter the constellation of relative outputs and prices over the whole economy, the standard technique [of cost-benefit analysis] is likely to fail us, for nothing less than some sort of general equilibrium approach would suffice in such cases.

Fogel's unsubstantiated assumption that all non-rail transport and storage activities could be almost indefinitely expanded at constant social unit cost is no innocent lapse from the standards of relentless empirical inquiry prescribed in the book's methodological passages. It is, rather, a crucial tactical step taken as part of the author's grand design to finesse the staggering problems which would be encountered in implementing the requisite general equilibrium analysis. But since Fogel's awareness of what he is

[1] A. R. Prest and R. Turvey, 'Cost–Benefit Analysis: A Survey,' Ch. 13 of American Economic Association and Royal Economic Society, *Surveys of Economic Theory*, III (London: Macmillan Co., 1967), p. 157.

---

composition of the inventory, in terms of cattle v. hogs and sheep, was that actually observed in 1890.

(d) The latter figure may be compared with the 20,056 acres of land being utilized in the city of Chicago in 1890 for all private purposes (residential, manufacturing, business, educational, religious, cemeteries) other than steam railroads. Cf. Homer Hoyt, *One Hundred Years of Land Values in Chicago* (Chicago: Chicago University Press, 1933), p. 290. Of course, the 'removal' of railroads from the scene would have released urban land for other uses by Chicagoans: to the extent of approximately 4,500 acres in 1890.

about has not been adequately shared with his readers, the ingenuity of this essentially theoretical exercise does not render it any the less misleading with respect to either the historical realities of transportation in nineteenth-century America, or the relevance of its conclusions to the planning of major transportation projects in underdeveloped countries.

The foregoing unfortunately does not exhaust the fund of criticism pertinent to Fogel's estimate of the direct social savings. Cost–benefit analysis becomes arduous when market prices fail to reflect marginal social costs and benefits, and it is therefore not terribly surprising to find the present study running into troubles on this point. These, however, are nowhere adequately acknowledged by the author. In taking observed market prices as indicative of marginal *social* costs for the alternative modes of freight transportation available in 1890, he simply slides over some awkward historical circumstances: inland water-transport rates were affected by a record of past state borrowing at *differentially* low interest rates in order to finance canal construction, by federal tax-financed harbor improvements, and by outright subsidies granted to canal shippers. The rate on all-water shipments from Chicago to New York in 1890, for example, is used by Fogel in calculating the direct social savings in interregional transportation, although it certainly reflected the removal of toll charges in the Erie Canal by New York State after 1882.[1]

Had there not been a discrepancy between private and social costs, it would make little sense even to consider Fogel's estimates: how could the use of market equilibrium prices for effectively identical transportation services yield anything other than a zero estimate of the direct social savings? Moreover, quite apart from the question of empirical validity, Fogel's computational assumption that wagon- and water-transport techniques were characterized by constant, if not by declining, marginal costs is rather hard to reconcile with his concurrent supposition that alternative transport costs, and the division of traffic between railroads and other carriers observed in 1890, actually represented a close approximation to a long-run equilibrium position in the transportation sector.

So much for the quantification of the direct effects. We must now turn to the second component of the overall social savings estimate. Some reference has already been made to the contention that basing the direct social savings estimate on the actual pattern of freight movement in 1890 obviates the need to make any further allowance for indirect effects of the loss of rail service. That argument is unquestionably right, *if* final demands for all products embodying transport inputs were completely inelastic to a rise in their delivered prices – a rise implied by the positive figure found for the direct social costs. With such demand conditions (which might be relevant for some agricultural goods in the short run,

[1] Cf. Lebergott, *loc. cit.*

were the country concerned not engaged in competition with other agricultural producers in world markets), the incremental social cost of exactly replicating the services performed by the railroads in 1890 would simply be equivalent to a tax that all transport-using productive activities could shift completely to the ultimate consumers. Otherwise, unless one also rules out decreasing long-run marginal costs in all activities with positive and/or negative outputs of transportables, the rise in the social cost of transport services could subject the economy to a further, indirect, reduction in productive capacity as consumers substituted non-transportables for goods requiring the now more expensive inputs of transportation. On this question Fogel is not explicit, but appears to have in mind the former condition (demand inelastic to a price rise) rather than a justification based on the global absence of increasing returns to scale: at one juncture he states that his estimate of the direct social savings also 'includes all the increase in national income attributable to *regional specialization* in agriculture induced by the decline in shipping costs' (p. 225, emphasis mine). To say the least, the foregoing are rather strong, albeit convenient, demand/supply assumptions to leave unstated and uninvestigated in a serious empirical assessment of the indirect effects of improved transportation technology. Relying upon either of them, as we shall shortly see, leads to the neglect of crucial issues.

DISCOMFORTS OF THE SECOND CLASS

Most of the points made in the preceding section compel suspicion of Fogel's claim to having ensured that the overall bias of the estimates is toward overstatement, and not under-representation of the true social savings. Let us suppose, however, that the direct and indirect social costs of abandoning railroads as a transportation medium in 1890 actually would, on full accounting, amount to less than 5 per cent of U.S. productive capacity. What can we then conclude regarding the size of the net social gains attributable to this innovation's introduction? The inference Fogel draws (pp. 22–3) is that relative to G.N.P. the social cost of dispensing with rail service at dates prior to 1890 would have been no greater than the figure for the latter year. Although the argument offered in support of this assertion involves the questionable assumption of strict proportionality between the direct and total social savings per unit of transportation, and avoids reference to any evidence about comparative unit costs of rail and non-rail shipments in previous years, the general proposition remains plausible enough. The U.S. economy had become increasingly dependent upon rail transportation during the course of the century; thus, one finds that during the 1840–90 period, the combined ton-mile and passenger-mile output of the railroad sector expanded roughly a hundredfold in relation to real G.N.P.[1] But such considerations merely dodge the original question. They bear not upon the issue of the benefits accruing from the railroad's introduction but upon the relative size of the losses entailed in giving up its use at various points in time between 1890 and, say, 4 July 1828 – when, in a superb bit of technologically irrelevant ceremony, the building of the Baltimore and Ohio Railroad was inaugurated by the laying of a corner-stone.

To answer the question posed, one must consider the index-number problems that might vitiate the presumed direct equivalence between the measured effects of losing existing railroad service and the effects of gaining the transportation advantages afforded by the innovation. To demonstrate that such problems do not arise would be simple enough if the reality of historical experience corresponded to the peculiar demand supply conditions which would justify Fogel's treatment of the indirect social cost of dispensing with railroads. In a land of green cheese, a mouse might readily be king. Yet once one admits that the long-run demand schedules facing American producers of transportable goods were not completely price-inelastic during the nineteenth century, the question becomes rather more difficult to wave away.

Indeed, if we further consider – not unrealistically – that historical long-run marginal-cost curves for activities producing transportables were declining but not strictly reversible, the indirect social gain attributable to the initial introduction of a transport innovation might very well exceed the full indirect social loss occasioned by that innovation's subsequent withdrawal. One may argue, in effect, that the proportional social burden of eventually losing access to a superior technique would be found to be quite limited *because* the indirect net benefits deriving from its previous utilization had been, proportionally, far larger. Such asymmetries destroy formal parity between strategic-bombing theory and leading-sector theories of economic growth.

To arrive at such a conclusion, it is not necessary to suppose that some upward shifts occurred in the long-run demand for delivered goods, shifts that might be related to the introduction of rail transportation facilities. Nor is it essential to imagine a change in the cost characteristics of alternative transport techniques being brought about indirectly by the development of the railroad system. The previously stated conditions

[1] For estimates of combined ton-mile and passenger-mile railroad output (1840 to 1890), cf. A. Fishlow, 'Productivity and Technological Change in the Railroad Sector, 1840–1910,' in *Output, Employment and Productivity in the United States After 1800* (New York: Columbia University Press for National Bureau of Economic Research, 1966), pp. 584–9. Corresponding estimates of constant dollar G.N.P. were kindly made available by Professor Robert E. Gallman, 'Revised Annual Estimates of U.S. Gross National Product 1834–1909' (mimeographed, 1965).

alone are sufficient to give rise to two index-number problems which would render measured 'introduction gains' larger than 'withdrawal losses.' The first is nothing but the classic Laspeyres index-number effect, arising in this instance from the conjunction of a price-elastic final-demand schedule for transportable goods and their historically decreasing marginal cost of production. Essentially the same problem presents itself in regard to measurement of the direct social savings attributable to the cheapening of the cost of transportation services, at least to the extent that the latter is to be treated as a final good. Although, in taking the demand for transportation to be completely price-inelastic Fogel seeks to ensure that his estimate does not understate the *direct* additions to consumers' surplus, the question of properly measuring indirect increments in producers' surplus where transportation is an intermediate good has been suppressed.

The second problem is less familiar. It emerges because movements along the declining historical marginal-cost curve are non-reversible – in the sense that marginal costs, at production scales below the long-run optimum eventually attained, turn out to be smaller than the historical marginal costs of producing transportables at those sub-optimum scales. Learning by doing would give rise to such irreversibilities.

This argument lends itself to graphical elaboration in Figure 10, which has been constructed on the heuristic (Hotelling-like) assumption that successive units of the transportable good must be marketed to successive customers who are situated farther and farther away from the point of production at the origin *O*. Allowing some intra-locational price elasticity of demand would complicate the exposition without materially altering the conclusions. We consider first the initial historical observation: output *OB* is being produced at marginal cost (f.o.b. price) V_B and the last unit is delivered to purchasers at distance *OB* for the c.i.f. price P_B. The historical marginal production-cost curve is *LMC–LMC′*, and the relevant long-run market supply schedule is S_0S_0', because in this situation the available 'non-railroad' transport technology leads to the family of delivered price gradients paralleling *tw*. One such gradient exists for each (constant) f.o.b. price. The transport-cost gradient labeled *tw*, for example, is appropriate with marginal production costs at V_F; its progressively steeper slope indicates that the 'non-rail' technique is characterized by rising costs of delivering marginal units of the transportable good the (requisite) marginal distance, or, more simply, by rising marginal transport costs per ton-mile.

When 'railroad' service, a superior transport method assumed here to offer constant marginal delivery costs per ton-mile, becomes available to the economy at *OB*, a new long-run market supply schedule, *SS′*, is defined by *LMC–LMC′* and the new family of transport-cost gradients parallel to *tr*. Transportable goods production therefore expands from

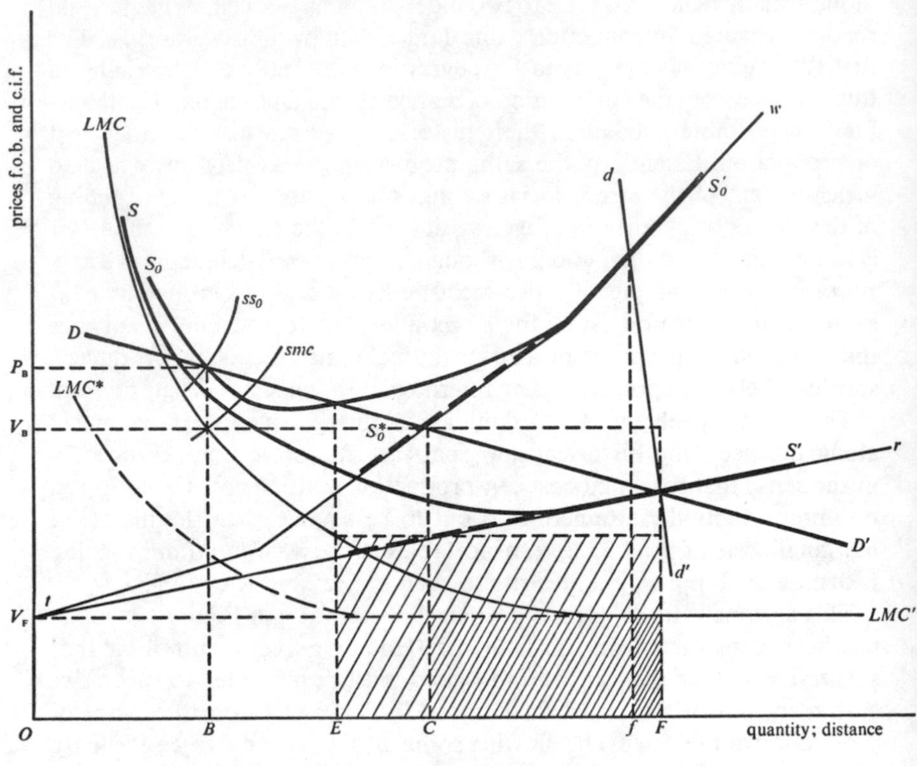

Figure 10

OB to *OF*, where the intersection of *SS'* and the long-run market demand schedule, *DD'*, establishes stable long-run equilibrium. It would be misleading, however, to regard the entire observed expansion of transportables output as a consequence of the availability of superior transport service in the form provided by 'the railroads.' The initial historical position, depicted as output level *OB* in the figure, was one of stable short-run but unstable long-run equilibrium. Thus, some event other than the actual lowering of transport charges, possibly a *temporary* upward shift of the demand for transportables connected with *building* 'railroads,' would have sufficed to disrupt the initial equilibrium and permit permanent realization of lower marginal costs of production at a scale exceeding *OB*. It is fully in keeping with Fogel's approach to delimit the problem by leaving out of account here any such indirect output gains attributable to the 'railroad's' advent as a (Schumpeterian?) unstable equilibrium-disrupting innovation. The figure does indicate, nonetheless, how the framework of the social-savings calculation could be extended to incorporate the latter class of indirect effects.

A merely transient disturbance of the initial unstable equilibrium could not, however, have given rise to the observed increase in production from *OB* to *OF*. So long as transportation continued to be provided by 'non-rail' techniques, no expansion of transportables output beyond *OE* would have been sustainable. For, as the figure shows, rising marginal ton-mile costs of transportation under that regime would serve to fix a position of stable long-run equilibrium at *OE*, in the range immediately beyond *OB*, despite the existence of decreasing long-run marginal costs of producing transportable goods. It is therefore appropriate to take the shaded rectangle above *EF*, rather than the entire area $BF \times V_B$, as the measure of the total indirect social benefit accruing from the introduction of the 'railroad' *qua* transportation technique.

Within the total shaded rectangle, however, lies a smaller area $CF \times V_F$ which represents the entire indirect social *loss* occasioned by falling back upon exclusive use of the inferior 'pre-rail' transport technique after having attained long-run equilibrium with output at *OF* and marginal cost of transportables production V_F. Measured from the vantage-point of *OF*, the indirect social 'withdrawal losses' thus understate the indirect social 'introduction gains' in both the relative output price and the relative quantity dimensions, prices and outputs of non-transportables being held constant in this partial equilibrium analysis. At *OF* the relative f.o.b. price of transportables (V_F) is below the marginal production cost that would have obtained at *OE*, while, equilibrium having been once attained at *OF*, the operative long-run marginal cost schedule for the transportables sector turns out to be LMC^*–LMC'. Thus, the relevant segment of the market supply schedule under 'non-rail' transport conditions is found to be $S_0^* S_0'$, and not $S_0 S_0'$. The hypothesized irreversibility of movement along LMC–LMC' therefore prevents the reimposition of 'pre-rail' transport-cost gradients (*tw*) from forcing the economy all the way back to *OE*. It leads, instead, to a shrinkage in the volume of transportables output indicated by the difference between *OF* and *OC*.

In the context of the foregoing illustration, Fogel's assumption that the available transportation alternatives to railroads were characterized by constant marginal social costs per ton-mile over the entire range of freight volumes contemplated may be seen to carry a dual significance. If the assumption is invalid, as we have reason to suspect, it would result in a progressively more serious understatement of the direct social savings afforded by the railroads in handling freight volumes greater than *OB*. In the second place, one would also be led to suppose that if a stable long-run equilibrium level of transportables production existed beyond *OB* under a regime of non-railroad transportation, the level of output at that equilibrium would lie much closer to *OF* than the actual solution at *OE* indicates. The social benefits derived from any policy, or random occurrence, which pushed transportables production above the initial unstable level (*OB*) might then be taken to be larger, and, correspondingly, the indirect social gains flowing from a superior mode of transportation – and the losses imposed by its subsequent withdrawal – would appear much smaller than those suggested by the preceding analysis. This last line of argument is not developed by Fogel, although his estimate of the indirect social loss would be smaller than the area given by $CF \times V_F$, and *a fortiori* less than the true indirect social benefits indicated by the shaded rectangle above *EF*. Indeed, Fogel's actual estimate is over-represented in the figure by the insignificant area $fF \times V_F$, since he takes the long-run demand for transportables to have been even less price-elastic than the schedule labeled *dd′* allows.

The conspicuous eschewal of any of the problems raised by increasing returns in treating the social benefits indirectly connected to the establishment of railroad service certainly increases the novelty of Fogel's study; but this break with tradition remains all the more puzzling in view of the recognition which is accorded essentially the same kind of questions when they arise in another context. In discussing the railroads' indirect effects as purchasers of industrial goods, specifically iron rails, Fogel does note that decreasing costs in activities supplying railroad inputs, coupled with price elasticity of aggregate demand for those industries' products, would considerably complicate the story (pp. 199–202). In such circumstances seemingly minor demand shifts would have far-reaching repercussions on marginal costs and the output of the industries concerned. The empirical investigation carried out in the study does not, however, deal with such questions; its purpose, we are told, is to provide evidence against a particular interpretation of the backward-linkage argument, namely the 'gross theory' that expansion of (constant cost) manufacturing activities depended in large measure upon rising railroad input demands in the ante-bellum era.

Thus, the importance of more subtle indirect influences transmitted via backward-linkages from the railroad sector – indirect effects involving scale economies, or learning effects dependent upon the volume of production of goods specifically needed by railroads – remain unexplored, and cannot be ruled out on the basis of Fogel's rejection of the 'gross theory.' 'Indeed,' he writes, 'one of the unfortunate effects of the reign of the gross theory is that it has choked off precisely such investigations. As long as one could [appear to] explain the growth of the iron industry by the sheer size of rail consumption relative to pig-iron production, there was little incentive to consider other alternatives' (p. 201). This may be all too accurate a historiographic diagnosis. It certainly does present a damning indictment, by suggesting that proponents of the 'backward-linkage hypothesis' have been thinking in terms of a world of strictly constant costs. After all, what theoretical basis can there be for the familiar balanced-growth arguments involving supply-side interdependence among industries, save that significant elements of indivisibility exist because decreasing cost techniques lead to comparatively large minimum efficient plant sizes? Ultimately, therefore, the complete case against the notion that backward linkages made the railroads pivotal in the early growth of U.S. manufacturing must be based upon the demonstrable inapplicability of the latter premise to the particular historical situation, and not by conveniently *postulating* that marginal costs remained strictly constant.

Be that as it may, who would seriously maintain that considerations of enhanced opportunities for realizing economies of scale and specialization have not been a crucial facet of traditional thinking in regard to the developmental consequences of the provision of cheaper transportation, however grossly such thoughts have been formulated? By ignoring those possibilities, by implicitly assuming constant returns everywhere, one may avoid becoming entangled in the difficult, perhaps intractable, index-number problems that impede unambiguous measurement of the railroad's indirect effects. Yet simplicity in quantification here is purchased at a high price: Fogel's study also avoids an open confrontation with economic issues that lie at the very heart of the ostensible subject of the inquiry.

DISCOMFORTS OF THE THIRD CLASS

Whatever reservations one may have about the way this study copes with the formidable problems of quantifying the direct and indirect social savings – and the preceding pages may suggest that such reservations are not lightly brushed aside – it should be apparent that a great deal of effort, and no little originality, have been poured into the task of securing concrete numbers. By comparison, the question of interpreting the social-savings figures obtained, of determining how they may best be used to shed light on 'the importance of the railroad's incremental contribution to economic growth,' has received far too little consideration. This is the most troubling aspect of the book, since it bears directly on the conclusions the author would have us draw from his empirical findings.

Beyond cavil, the estimated social savings (amounting to $560 million) are sufficiently small, in relation to the U.S. economy's productive potential in 1890, for us to accept Fogel's refutation of the railroad's 'indispensability' – if we adhere to the strict meaning of the term. But does a social saving of 5 per cent of G.N.P. warrant our also concluding that the railroads failed to make an important contribution to the economy's growth? Does it suggest that the significance of this innovation among nineteenth-century, or for that matter twentieth-century, innovations has been puffed up out of all true proportion by historians who foolishly relied upon the non-quantitative opinions of contemporary observers? Clearly it does not. Inasmuch as 'dispensability' and 'economic unimportance' cannot usefully be regarded as synonymous, there is no *a priori* basis for Fogel's implication that annual social savings below one-twentieth of G.N.P. represent a trivial magnitude on the scale of economic consequence. Just where does triviality end?

Comparing the social savings attributable to the railroads with the nation's productive capacity resolves nothing. It simply defers the question of evaluation. Without some relevant standard having first been established, the social savings fraction can be made the subject of all manner of conflicting appraisals and thus becomes, in the end, a meaningless number. Is 5 per cent to be judged an inconsequential fraction because it lies well within the range of the variations of *per capita* real output over the course of the nineteenth-century trade cycle? Perhaps we should, instead, find it impressively large: on the basis of such annual social savings, one would attribute to the introduction of the railroad about one-eighth of the total effect of all technological changes and other sources of conventional factor productivity growth in the United States during the half-century before 1890.[1]

Such attempts to impart meaning to the social savings–G.N.P. ratio presented by Fogel's study must remain bafflingly inconclusive. The exercise is jejune because it invokes comparisons which are not operationally relevant to the selection of a particular program of investment in transportation facilities, or, more generally, to the problem of choice among available alternative uses of resources.[2] One does not face a choice between having either railroads in place of canals or a half-century of technological change. Nor were Americans ever confronted with those options during the past century. The elimination of trade-cycle fluctua-

[1] Over the period 1839–89 the long-term rate of growth of real G.N.P. in the U.S. was roughly 4 per cent per annum, so that if the social savings attributable to the introduction of the railroads during the same time-period had the effect of making G.N.P. in 1889 5 per cent greater than it would otherwise have been, 'the railroad' as an innovation might be thought to have 'contributed' 0·12 percentage points per annum to the long-term output growth rate. Such a contribution would represent approximately one-eighth of the average annual rate of conventional factor productivity growth over the 1839–89 interval. I am indebted to Kenneth Arrow for suggesting that this sort of heuristic calculation might be made. The requisite data have been drawn from M. Abramovitz and P. A. David, 'Economic Growth in the United States,' Ch. 2 (mimeographed, Stanford Research Center in Economic Growth, 1965).

[2] This applies equally to Fogel's (*Railroads*, p. 47, n. 58) calculation that the loss of the railroad's services in the interregional transportation of major agricultural products would, in 1890, have represented only a three-month setback in the growth of U.S. real output.

tions is patently not, in any opportunity cost sense, a meaningful alternative to devoting resources to the provision of rail transportation.

The nature of the feasible, relevant alternatives to exploiting the technique of rail transportation suggests that we ought not to concern ourselves with the absolute size of the consequent social savings or their relationship to the aggregate volume of production. Rather, we should insist on knowing whether this particular innovation afforded opportunities for utilizing resources in a manner that yielded higher rates of return than those obtainable through investment in other directions. It should be clear in this that we are to consider the social rate of return, not simply the portion of the total yield that could be appropriated in the form of profits by railroad investors. Moreover, for these purposes the rate of return in question must be one that reflects the indirect social benefits as well as the direct: the logic of the Pigouvian argument for subsidizing a decreasing cost activity could also call for subsidization of a project supplying inputs to activities that were subject to increasing returns, even if consideration of the direct social rate of return alone did not justify making the outlay.

Were the railroads found to have offered a *differentially* high rate of return in the sense just defined, one would have to acknowledge that the innovation had made a positive contribution to the economy's growth. The importance of that contribution, however, can only be judged properly by reference to both the size of the differential social rate of return and the volume of real savings that could have been absorbed in railroad investment projects before the differential vanished. One really would want to be in a position to compare social rates of return to the allocation of successive equal lumps of resources for railroads and specific alternative projects.

It is curious that Fogel has not pursued this obvious formulation of the question, for, at the conclusion of his study, in the passage initially quoted here (p. 291 above), he appears to recognize its pertinence by suggesting that the railroad's 'contribution' was somehow incommensurate with 'its devouring appetite for capital.' But can such a view be supported with evidence regarding the social rate of return? To venture an answer, we must look elsewhere for information to supplement that which Fogel's work has made available.[1]

The essential bits of data have been set out in Table 17. Total social benefits from the operations of the U.S. railroad system in 1890 are taken

[1] Nerlove, *op. cit.*, pp. 112–15, has presented estimates of average and marginal social rates of return on U.S. railroad capital in 1890 which are quite at variance with those given here, and which therefore tend to accord with Fogel's position. Nerlove's calculations, however, are flawed by two serious errors which totally vitiate his findings. Those errors (details of which are noted below) aside, the present discussion owes much to Nerlove's general treatment of the problems. 'Tis not, then, a case of Nerlove's labor lost.

to be the sum of appropriable and unappropriable social benefits. Following a dodge common in practical cost–benefit accounting, we have associated the former with the private revenues generated by the sale of rail transport services, therein momentarily ignoring all possible sources of discrepancy between market prices and marginal social costs. For the present purpose, we have seized upon Fogel's final 'social savings' figure as a measure of the inappropriable social benefits afforded by the railroads, noting in so doing the previously stated reasons for thinking that this does not fully reflect the actual unappropriable benefits. Against the latter downward bias, however, must be set the likely overstatement of appropriable *social* benefits: in the case of this loosely regulated utility, strict adherence to marginal-cost pricing would have caused the railroads to run at a loss, which, collectively, they clearly did not do. (Of course, it is too much to expect the opposing biases to cancel out exactly in the estimate of the total social benefits.) Estimates of the total (gross and net) social returns to railroad capital have been analogously derived by adding unappropriable social benefits to the privately captured profits of the roads.

Referring now to the available alternative estimates of the 1890 value

TABLE 17 U.S. Class I–III railroads, 1890
(all figures in $ billions)

1. Unappropriable social benefits = 'social savings'	0·560	
	Gross	*Net*[a]
2. Appropriable social benefits = Operating revenues	1·052	0·785
3. Appropriable earnings = Operating earnings	0·627	0·360
4. Total social benefits = Social value of output	1·612	1·345
5. Total social return to capital	1·187	0·920
	Estimated value, net stock	*Reproduction costs, new*
6. Current value of capital, excluding land	5·830 (A)	7·500 (B)

Sources: *Line 1:* Fogel, *op. cit.*, p. 233. *Line 2:* Passenger and freight revenues, from U.S. Bureau of the Census, *Historical Statistics of the United States, Colonial Times to 1957* (Washington, D.C.: G.P.O., 1960), p. 434, series Q106. *Line 3:* Line 2 less operating expenses, *ibid.*, series Q107. *Line 4:* Line 1 plus line 2. *Line 5:* Line 1 plus line 3, or line 4 less operating expenses. *Line 6:* M. J. Ulmer, *Capital Formation in Transportation, Communications, and Public Utilities: Its Formation and Financing* (Princeton: Princeton University Press, for the National Bureau of Economic Research, 1960); Estimate A: p. 256; Estimate B: p. 288.

[a] Flows net of maintenance expenditures on freight and passenger railroads, as given by *U.S. Historical Statistics (1960)*, p. 434, series Q108, Q109.

of road and equipment which appear in the last line of Table 17,[1] it can be quickly calculated that the *average* social rate of return on a gross basis works out at 15·8 per cent to 20·4 per cent.[2] Thus, despite the massive program of railroad capital formation carried out in the United States after the Civil War, and in the face of the contemporaneous secular decline of money rates of interest throughout the economy, the average social rate of return on railroad capital appears to have held up remarkably well. In 1859, according to Fishlow's findings,[3] the average social rate of return on U.S. railroad capital was 15 per cent per annum, roughly twice the then prevailing money yield on low-risk bonds. In 1890, our figures suggest, the yield on low-risk financial assets was only one-third as high as the (12·3–15·8 per cent) average net social rate of return.[4]

The comparisons of rates of return just cited treat the American railroad system of 1890, and that of 1860, as indivisible investment projects, which

[1] Both estimates are the work of M. J. Ulmer (see the notes accompanying Table 17). The smaller estimate (A) is *conceptually* more appropriate in the present connexion, as it refers to the reproduction cost of stock net of physical depreciation and retirements, whereas estimate (B) refers to the reproduction cost of the stock net of retirements only. Ulmer's (type A) estimates of the railroad capital stock *in constant dollars* have been subjected to trenchant criticisms by Fishlow ('Productivity and Technological Change,' *loc. cit.*, pp. 589–94) who has prepared an alternative set of net stock figures, in constant 1909 dollars, for benchmark dates including 31 December 1889 = 1 January 1890. Cf. *ibid.*, p. 611. Fishlow, however, developed no cost of road and equipment (price) index which would permit translation of his constant dollar series into current dollar (reproduction cost) terms, and has criticized the cost index employed in that connexion (and others) by Ulmer. Most of Fishlow's comments throw particular doubt upon the reliability of Ulmer's cost index for the years prior to 1889, rather than the period from 1889 to 1909. It is therefore interesting to observe that if one uses this cost of road and equipment index (Ulmer, *op. cit.*, Table C.II, pp. 274–5) to express Fishlow's real net stock estimates (variant II) in current values, the figure obtained for end of year 1889 is $5·83 billion – which coincides exactly with Ulmer's estimates for beginning of year 1890! The same 'inflation' of Fishlow's variant I real net stock estimate yields a smaller current value figure for end of year 1889: $4·88 billion. The upshot is that, as far as we can determine, our use of the pair of capital stock estimates presented in Table 17 may lead to some systematic *understatement* of the true social rates of return. Such a bias would only strengthen the argument advanced in the following pages.

[2] The average social rate of return, gross of depreciation, reported by Nerlove (*op. cit.*, p. 114) for the same date is only half as high. But Nerlove neglected to add the gross operating earnings of the railroads to the estimated social savings in reckoning the gross social return. He simply divided the estimate of social savings by the value of the railroad capital stock employing the same alternative stock figures as have been used here (Table 17).

[3] Cf. Fishlow, *Railroads and the Transformation of the Ante-Bellum Economy*, Ch. 2.

[4] In contrast to the situation existing in 1859–60, by 1890 depreciation allowances on the railroad capital stock had assumed a sufficiently large size to make it more appropriate to compare financial asset yields with the social rate of return for railroads on a *net* basis. In 1890 the yield on railroad bonds averaged 4·55 per cent. Cf. *U.S. Historical Statistics* (1960), p. 656, series X332.

obviously they were not. In evaluating the railroad's significance it is perhaps instructive to ask, rhetorically: 'How many other discrete innovation-embodying projects can one find whose exploitation could similarly have absorbed more than an eighth of the nation's total reproducible tangible assets in 1890, yielding thereon a net social return in the neighborhood of 12 to 16 per cent per annum?' Still, average rates of return afford only the haziest impression of the possible social yield derived from the allocation of resources to rail transportation at various stages in the network's construction; explicit calculation of marginal social rates of return is clearly called for. (Of course, should it be assumed that the production function for *social* transport benefits was characterized by constant returns, average and marginal social rates of return would coincide. This assumption, however, seems very suspect indeed; in the following we eschew it, although we do entertain the assumption that the production function for *private* transport output was characterized by constant returns.)

The most immediately feasible path toward that goal unfortunately leads across some very treacherous ground: assuming there is an aggregate production function for the railroad industry, and further, that strict proportionality exists between the social and private output of that industry, the marginal social productivity of capital may be estimated as the product of the average social productivity of capital and the elasticity of (private) output with respect to capital. The data in Table 17 provide the means for computing the first variable, but we must now grapple with the problem of securing the appropriate elasticity coefficient. One solution, suggested by Nerlove,[1] is to make use of the results of Klein's regression analysis of cross-section data for U.S. steam railroads in 1936, in which an index of passenger and freight transportation serves as the output variable and inputs of capital are represented by train-hours. The partial elasticity estimate thus obtained is 0·28, as shown in Table 18. Aside from the dubious validity of using a cross-section production function coefficient in the present context, one is inclined to question the relevance of an estimate referring to a year nearly a half-century removed from 1890, and, moreover, to a year that found the U.S. railroad industry facing serious under-utilization of its operating capacity. The alternative elasticity estimates given in Table 18 seem at least as plausible as the 0·28 figure, even though their derivation from the railroad capital share (net and gross

[1] Nerlove, *op. cit.*, p. 114, employed Klein's estimate of the elasticity coefficient (see elasticity estimate (i), Table 18) in computing a marginal social rate of return, gross of depreciation, for 1890. Inexplicably, however, instead of multiplying the elasticity coefficient by the average (gross) social productivity of capital in the railroad industry, Nerlove multiplied it by his figure for the average gross rate of return – which is, as pointed out above, itself erroneous.

TABLE 18 Estimates of the marginal social rate of return on U.S. railroad capital in 1890

Elasticity of private output with respect to capital		Rate of return in percentage per annum: based on capital stock estimate B	A
Gross basis: (i)	0·28	6·0	7·7
(ii)	0·60	12·8	16·5
Net basis: (i)	0·28	5·0	6·5
(iii)	0·46	8·2	10·6

Sources: Elasticity estimates: (i) Lawrence R. Klein, *A Textbook of Econometrics* (Evanston, Illinois: Row, Peterson Co., 1953), p. 234. (ii) and (iii) Respective gross and net entries in Table 17, line 3, divided by corresponding entries in Table 17, line 2.

Marginal rate of return: Gross and net social output–capital ratios, from Table 17, lines 4 and 6, multiplied by corresponding elasticity coefficients shown in this table.

of depreciation) in 1890 implies, *inter alia*, a constant-returns production function for the industry.[1]

The single conclusion most strongly suggested by the array of alternative marginal rates of return in Table 18 is that, from a social viewpoint, by 1890 further investment in railroad transportation capacity did not constitute an especially advantageous way of using U.S. resources. On the one hand, the two pairs of marginal gross social rates presented in the table neatly bracket the estimated 9·3 per cent to 12·7 per cent gross marginal rate of return on the aggregate U.S. capital stock in 1890.[2] On the other hand, the pairs of marginal social rates net of depreciation span virtually the same range as the prevailing structure of yields on financial assets: the railroad bond rate was 4·5 per cent, New York call loans and four-to-six-month prime commercial paper fetched 5·8 per cent and 6·9 per cent

[1] Assuming the production function for private transport output of the railroad sector is first-degree homogeneous, and that the private return to capital is equal to the private marginal value product, the elasticity of (private) output with respect to capital in 1890 is given by the share of capital at that date. Capital shares (gross and net of depreciation) in 1890 are found from lines 2 and 3 of Table 17, by expressing earnings as a fraction of revenues. (See notes accompanying Table 18.)

As a check on the plausibility of the use of the capital share in this connexion, the share net of depreciation can be used to compute a marginal net *private* rate of return, which may be compared with the current yield on railroad bonds. The agreement is found to be reasonably close: bond yields averaged 4·5 per cent per annum in 1890, whereas the computed marginal net private rate of return to railroad capital is 4·8 per cent to 6·2 per cent.

[2] The estimate for the United States is due to Nerlove, *op. cit.*, p. 115, and is free from the errors which mar his marginal rate of return calculations for the railroad sector.

respectively, while farm mortgage rates ran from 5·9 per cent in the northeastern seaboard region to 7·7 per cent in the Midwest and 9·1 per cent in the western states.[1]

If it appears, not unexpectedly, that by 1890 the margin of railroad capital formation in the United States had already been pushed close to the point of exhausting any differential social returns afforded by exploitation of the innovation, what of the situation at earlier dates, when the railroad system was much less fully articulated? Here, alas, we can only guess, although as a by-product of Fogel's pioneering efforts we are provided with at least one clue to the possible orders of magnitude that may be involved. As a substitute for railroads, Fogel (pp. 92–100) contemplates the construction of some 5,000 miles of canals and feeders to serve the Midwest, at an estimated cost of roughtly $161·1 million. For such an incremental outlay, if the benefits to agriculture alone are considered among the indirect effects, his estimates indicate a gross social rate of return somewhere between 50 and 85 per cent per annum. On the railroad investment which obviated the need for this hypothetical canal system, even Fogel must concede that the social rate of return must have been still more spectacular.

Only a reckless man would refuse to admit the shakiness of the statistical basis for the preceding exercise, or the crudity of the edifice of social rate of return calculations erected upon those foundations. But surely it would be no less reckless to disregard the fact that when the available quantitative information is put into an analytically appropriate form, it simply will not justify the disparaging tenor of Fogel's implied assessment of the social profitability of nineteenth-century railroad investment in the United States.

[1] Cf. *U.S. Historical Statistics* (1960), pp. 645–6, series X306, X308, X332; *U.S. Eleventh Census* (1890), 'Report on Real Estate Mortgages,' XII, p. 248, cited in Fogel, *op. cit.*, pp. 82–3.

References

BOOKS

Arbuthnot, John, *An Inquiry into the Connection between the Present Price of Provisions and the Size of Farms. By a Farmer*, London, 1773.
Arrow, K. J. and Hahn, F. H., *General Competitive Analysis*, Edinburgh, Oliver and Boyd, 1971.
Asher, Harold, *Cost Quantity Relationships in the Airframe Industry*, Santa Monica, The Rand Corporation, R-291, July 1956.
Ashton, T. S., *An Economic History of England: The Eighteenth Century*, London, Methuen and Co. Ltd., 1955.
Baines, E., *A History of the Cotton Manufacture in Great Britain*, London, 1835.
Best, R. H. and Coppock, J. T., *The Changing Use of Land in Britain*, London, Faber and Faber, 1962.
Bidwell, Percy W. and Falconer, John I., *History of Agriculture in the Northern United States* (Publication No. 358), Washington, D.C., Carnegie Institute of Washington, 1925.
Brodrick, G. C., *English Land and English Landlords*, London, 1881.
Caird, James, *English Agriculture in 1850–51*, London, 1852.
The Landed Interest and the Supply of Food, London, 1878.
Calvert, Monte A., *The Mechanical Engineer in America, 1830–1910*, Baltimore, The Johns Hopkins University Press, 1967.
Cawley, E. Hoon (ed.), *The American Diaries of Richard Cobden*, Princeton, N.J., Princeton University Press, 1952.
Chambers, J. D. and Mingay, G. E., *The Agricultural Revolution 1750–1880*, London, B. T. Batsford, 1966.
Clapham, J. H., *An Economic History of Modern Britain*, Cambridge, Cambridge University Press, 1952.
Clark, V. S., *History of Manufactures in the United States*, I, New York, McGraw-Hill, 1929.
Davis, Lance E. *et al.*, *American Economic Growth*, New York, Harper and Row, 1972.
Deane, Phyllis and Cole, W. A., *British Economic Growth, 1688–1955*, Cambridge, Cambridge University Press, 1962.
Evans, G. E., *The Horse in the Furrow*, London, Faber and Faber, 1960.
Feller, William, *An Introduction to Probability Theory and Its Applications*, Third Edition, New York, John Wiley and Sons, 1968.
Fishlow, Albert, *American Railroads and the Transformation of the Ante-Bellum Economy*, Cambridge, Mass., Harvard University Press, 1965.
Fite, Emerson D., *Social and Industrial Conditions in the North during the Civil War*, New York, Macmillan Company, 1910.
Fite, G. C. and Reese, J. E., *An Economic History of the United States*, Second Edition, Boston, Houghton-Mifflin, 1965.
Fogel, Robert W., *Railroads and American Economic Growth: Essays in Econometric History*, Baltimore, The Johns Hopkins Press, 1964.
Gates, Paul W., *The Farmer's Age: Agriculture 1815–1860, The Economic History of the United States*, Vol. III, New York, Holt, Rinehart and Winston, 1960.
The Illinois Central Railroad and Its Colonization Work, Harvard Economic Studies, 42, Cambridge, Mass., Harvard University Press, 1934.

Georgescu-Roegen, *The Entropy Law and the Economic Process*, Cambridge, Mass., Harvard University Press, 1971.

Gerschenkron, A., *Economic Backwardness in Historical Perspective*, Cambridge, Mass., Harvard University Press, 1962.

Gibb, G. S., *The Saco–Lowell Shops, Textile Machinery Building in New England, 1813–1849*, Cambridge, Mass., Harvard University Press, 1950.

Habakkuk, H. J., *American and British Technology in the Nineteenth Century*, Cambridge, Cambridge University Press, 1962.

Hicks, J. R., *The Theory of Wages*, London, Macmillan, 1932.

Hirschman, A. O., *The Strategy of Economic Development*, New Haven, Conn., Yale University Press, 1958.

Homer, Sidney, *A History of Interest Rates*, New Brunswick, N.J., Rutgers University Press, 1963.

Hoyt, Homer, *One Hundred Years of Land Values in Chicago*, Chicago, Chicago University Press, 1933.

Hutchinson, William T., *Cyrus Hall McCormick*, Vols. I, II, New York, The Century Company, 1930.

Intrilligator, M. D., ed., *Frontiers in Quantitative Economics*, Amsterdam, North-Holland Publishing Co., 1971

Johnston, J., *Econometric Methods*, New York, McGraw-Hill, 1963.

Jones, G. T., *Increasing Returns*, Cambridge, Cambridge University Press, 1933.

Kindleberger, C. P., *Economic Growth in France and Britain 1851–1950*, Cambridge, Mass., Harvard University Press, 1964.

Klein, Lawrence R., *A Textbook of Econometrics*, Evanston, Ill., Row, Peterson Co., 1953.

Landes, D. S., *The Unbound Prometheus*, Cambridge, Cambridge University Press, 1969.

de Lavergne, Léonce, *The Rural Economy of England, Scotland and Ireland*, Edinburgh, 1855.

Layer, Robert G., *Earnings of Cotton Mill Operatives, 1825–1914*, Cambridge, Mass., Committee on Research in Economic History, 1955.

Lebergott, Stanley, *Manpower in Economic Growth: The United States Record since 1800*, New York, McGraw-Hill Book Co., 1964.

Lundberg, Erik, *Produktivitet öch Rantabilitet*, Stockholm, P.A. Norstedt and Söner, 1961.

Lyaschenko, P. I., *History of the National Economy of Russia to the 1917 Revolution*, New York, Macmillan, 1949.

Malinvaud, E., *Statistical Methods of Econometrics*, Chicago, Rand McNally, 1966.

McCloskey, D. N. (ed.), *Essays on a Mature Economy: Britain after 1840*, London, Methuen and Co. Ltd., 1971.

McGouldrick, Paul F., *New England Textiles in the Nineteenth Century, Profits and Investment*, Cambridge, Mass., Harvard University Press, 1968.

Meade, J. E., *Trade and Welfare*, Oxford, Oxford University Press, 1955.

Mitchell, B. R. and Deane, Phyllis, *Abstract of British Historical Statistics*, Cambridge, Cambridge University Press, 1962.

Navin, T. R., *The Whitin Machine Works Since 1831*, Cambridge, Mass., Harvard University Press, 1950.

North, D. C., *The Economic Growth of the United States, 1790–1860*, Englewood Cliffs, N.J., Prentice-Hall, 1961.

Growth and Welfare in the American Past, Englewood Cliffs, N.J., Prentice-Hall, 1966.

Orwin, C. S. and Whetham, E. H., *History of British Agriculture 1846–1914*, London, Archon Books, 1964.

Pierce, Bessie L., *A History of Chicago*, Vol. II: *From Town to City, 1848–1871*, New York, Alfred A. Knopf, 1940.

Richards, J., *A Treatise on the Construction and Operation of Woodworking Machines*, London, 1872.

Robertson, Ross M., *History of the American Economy*, Second Edition, New York, Harcourt, Brace, and World, Inc., 1964.

Rogin, Leo, *The Introduction of Farm Machinery in Its Relation to the Productivity of Labor in the Agriculture of the United States during the Nineteenth Century*, Berkeley, University of California Press, 1931.

Rosenberg, Nathan, *Technology and American Economic Growth*, New York, Harper and Row, 1972.

(ed.), *The American System of Manufactures*, Edinburgh, Edinburgh University Press, 1969.

Rostow, W. W., *The Stages of Economic Growth*, Cambridge, Cambridge University Press, 1960.

Salter, W. E. G., *Productivity and Technical Change*, Second Edition, Cambridge, Cambridge University Press, 1966.

Samuelson, Paul A., *Foundations of Economic Analysis*, Cambridge, Mass., Harvard University Press, 1947.

Saul, S. B. (ed.), *Technological Change: The United States and Britain in the Nineteenth Century*, London, Methuen and Co. Ltd., 1970.

Saville, John, *Rural Depopulation in England and Wales 1851–1951*, London, Routledge and Kegan Paul, 1957.

Schmookler, Jacob, *Invention and Economic Growth*, Cambridge, Mass., Harvard University Press, 1966.

Shannon, Fred A., *The Farmer's Last Frontier: Agriculture 1860–1897* (*The Economic History of the United States*, Vol. V), New York, Holt, Rinehart and Winston, 1945.

Slight, James and Burn, R. Scott, *The Book of Farm Implements and Machines*, ed. Henry Stephens, Edinburgh, 1858.

Smith, T. C., *Political Change and Industrial Development in Japan: Government Enterprise, 1868–1880*, Stanford, Calif., Stanford University Press, 1955.

Spence, Clark C., *God Speed the Plow, The Coming of Steam Cultivation to Great Britain*, Urbana, Illinois, University of Illinois Press, 1960.

Stephens, Henry, *The Book of the Farm*, Second Edition, Edinburgh, 1855.

The Book of the Farm, Third Edition, Edinburgh, 1871.

Strassman, W. Paul, *Risk and Technological Innovation*, Ithaca, N.Y., Cornell University Press, 1959.

Taussig, F. W., *Some Aspects of the Tariff Question*, Cambridge, Mass., Harvard University Press, 1915.

The Tariff History of the United States, Eighth Edition, New York, G. P. Putnam's Sons, 1931.

Temin, Peter, *Iron and Steel in Nineteenth Century America*, Cambridge, Mass., M.I.T. Press, 1964.

ed., *New Economic History*, Harmondsworth, Middlesex, Penguin Books, 1973.

Thompson, F. M. L., *English Landed Society in the Nineteenth Century*, London, Routledge and Kegan Paul, 1963.

Toulmin, Stephen and Goodfield, June, *The Discovery of Time*, London, Pelican Books, 1967.

Ulmer, M. J., *Capital Formation in Transportation, Communications, and Public Utilities: Its Formation and Financing*, Princeton, N.J., Princeton University Press for the National Bureau of Economic Research, 1960.

Veblen, Thorstein, *Imperial Germany and the Industrial Revolution*, New York, Macmillan, 1915.

Walras, Léon, *Elements of Pure Economics or the Theory of Social Wealth*, William Jaffé, transl., London, George Allen and Unwin, 1954.

Wilson, John, *British Farming*, Edinburgh, 1862.

Woodbury, Robert S., *History of the Grinding Machine*, Cambridge, Mass., M.I.T. Press, 1959.

History of the Milling Machine, Cambridge, Mass., M.I.T. Press, 1960.

Studies in the History of Machine Tools, Cambridge, Mass., M.I.T. Press, 1972.

ARTICLES AND CONTRIBUTIONS TO BOOKS

Abramovitz, Moses and David, Paul A., 'Economic Growth in America: Historical Parables and Realities,' *De Economist*, Vol. 121, 3 (May/June 1973).

'Reinterpreting Economic Growth: Parables and Realities of the American Experience,' *American Economic Review*, Vol. 58, 2 (May 1973).

Ahmad, S., 'On the Theory of Induced Innovation,' *Economic Journal*, Vol. LXXVI (June 1966).

Alchian, Armen, 'Reliability of Progress Curves in Airframe Production,' *Econometrica*, Vol. XXXI, 4 (October 1963).

Ames, Edward and Rosenberg, Nathan, 'The Enfield Arsenal in Theory and History,' *Economic Journal*, Vol. LXXVIII, 312 (December 1968).

Arrow, Kenneth J., 'Classificatory Notes on the Production and Transmission of Technological Knowledge,' *American Economic Review*, Vol. LIX, 2 (May 1969).

'Economic Welfare and the Allocation of Resources for Invention,' in *The Rate and Direction of Inventive Activity*, Universities-National Bureau Committee for Economic Research, Princeton, N.J., Princton University Press, 1962.

'The Economic Implications of Learning by Doing,' *Review of Economic Studies*, Vol. XXIX, 2 (June 1962).

Asher, Ephraim, 'Industrial Efficiency and Biased Technical Change in American and British Manufacturing: The Case of Textiles in the Nineteenth Century,' *Journal of Economic History*, Vol. 32, 2 (June 1972).

Atkinson, Anthony B. and Stiglitz, Joseph E., 'A New View of Technological Change,' *Economic Journal*, Vol. LXXIX, 315 (September 1969).

Baldwin, Robert E., 'The Case Against Infant-Industry Tariff Protection,' *Journal of Political Economy*, Vol. LXXVII, 3 (May/June 1969).

Bell, Thomas George, 'A Report upon the Agriculture of the Country of Durham,' *Journal of the Royal Agricultural Society of England*, Vol. XVII (1856).

Bhagwati, Jagdish, 'The Pure Theory of International Trade: A Survey,' in *Surveys in Economic Theory*, II, American Economic Association–Royal Economic Society, New York, St Martin's Press, 1967.

Bogue, Allan G., 'Farming in the Prairie Peninsula, 1830–1880,' *Journal of Economic History*, 23 (March 1963).

Bowley, A. L., 'Statistics of Wages in the United Kingdom during the Past Hundred Years,' *Journal of the Royal Statistical Society*, Vol. LXI (December 1898).

Bruni, L., 'Internal Economics of Scale with a Given Technique,' *Journal of Industrial Economics*, Vol. 12 (July 1964).

Carman, Harry J., 'English Views of Middle Western Agriculture, 1850–1870,' *Agricultural History*, 13 (January 1934).

Chenery, H. B., 'Process and Production Functions from Engineering Data,' in W. W. Leontief *et al.*, *Studies in the Structure of the American Economy*, New York, Oxford University Press, 1953.

Chicago Board of Trade, *Annual Report of the Trade, and Commerce of Chicago for 1896*, Chicago, 1896.

Clarke, John Algernon, 'Farming of Lincolnshire,' *Journal of the Royal Agricultural Society of England*, Vol. XXI (1851).

Coleman, John and Paget, F. A., 'General Report on the Exhibition of Implements at the Plymouth Meeting,' *Journal of the Royal Agricultural Society of England*, 2nd Series, I (1865).

Collins, E. J. T., 'Harvest Technology and Labour Supply in Britain, 1790–1870,' *Economic History Review*, 2nd Series, XXII, 3 (December 1969).

'Labour Supply and Demand in European Agriculture 1800–1880,' in *Agrarian Change and Economic Development, The Historical Problems*, eds. E. L. Jones and S. J. Woolf, London, Methuen and Co. Ltd., 1969.

Craigie, P. G., 'Agricultural Holdings in England and Abroad,' *Journal of the Royal Statistical Society*, Vol. L (1887).

Danhof, C., 'Agriculture,' in *The Growth of the American Economy*, Second Edition, ed. H. F. Williamson, New York, Prentice-Hall Inc., 1951.

'Farm-Making Costs and the "Safety-Valve": 1850–1860,' *Journal of Political Economy*, Vol. 49 (June 1941).

David, Paul A., 'A Contribution to the Theory of Diffusion,' Stanford Center for Research in Economic Growth, *Memorandum* No. 71 (June 1969).

'Industrialization and the Changing Labor Supply in a Region of Recent Settlement,' Stanford Center for Research in Economic Growth, *Memoranda*, Nos. 25, 26 (August 1963).

'Labour Productivity in English Agriculture, 1850–1914: Some Quantitative Evidence on Regional Differences,' *Economic History Review*, 2nd Series, Vol. XXIII, 3 (December 1970).

'Learning by Doing and Tariff Protection: A Reconsideration of the Case of the Ante-Bellum United States Cotton Textile Industry,' *Journal of Economic History*, Vol. XXX, 3 (September 1970).

'The Landscape and the Machine: Technical Interrelatedness, Land Tenure and the Mechanisation of the Victorian Corn Harvest,' in *Essays on a Mature Economy: Britain after 1840*, ed. D. N. McCloskey, London, Methuen and Co. Ltd., 1971.

'The "Horndal Effect" in Lowell, 1834–1856: A Short-Run Learning Curve for Integrated Cotton Textile Mills,' *Explorations in Economic History*, 2nd Series, Vol. 10, 2 (Winter 1973).

'The Mechanization of Reaping in the Ante-Bellum Midwest,' in *Industrialization in Two Systems*, ed. H. Rosovsky, New York, Wiley and Sons, 1966.

'The Use and Abuse of Prior Information in Econometric History: A Rejoinder to Professor Williamson on the Antebellum Cotton Textile Industry,' *Journal of Economic History*, Vol. XXII, 3 (September 1972).

and van de Klundert, Th., 'Biased Efficiency Growth and Capital–Labor Sustitution in the U.S., 1899–1960,' *American Economic Review*, Vol. 55, 3 (June 1965).

Davis, Lance E., 'Professor Fogel and the New Economic History,' *Economic History Review*, 2nd Series, XIX (1966).

'Mrs. Vatter on Industrial Borrowing: A Reply,' *Journal of Economic History*, Vol. XXI (June 1961).

'The New England Textile Mills and the Capital Markets: A Study of Industrial Borrowing 1840–1860,' *Journal of Economic History*, Vol. XX (March 1960).

'Stock Ownership in the Early New England Textile Industry,' *The Business History Review*, Vol. XXXII, 2 (Summer 1958).

and Stettler, H. Louis, 'The New England Textile Industry, 1825–60: Trends and Fluctuations,' in *Output, Employment and Productivity in the United States After 1800*, National Bureau of Economic Research Studies in Income and Wealth, Vol. XXX, New York, Columbia University Press, 1966.

Dovring, F., 'The Transformation of European Agriculture,' *Cambridge Economic History of Europe*, eds. M. M. Postan and H. J. Habakkuk, Vol. 6, Part II, Cambridge, Cambridge University Press, 1965.

Drandakis, E. M. and Phelps, E. S. 'A Model of Induced Invention, Growth, and Distribution,' *Economic Journal*, Vol. LXXVI (December 1966).

Drummond, Ian M., 'Labor Scarcity and the Problem of American Industrial Efficiency in the 1850's: A Comment,' *Journal of Economic History*, Vol. XXVII, 3 (September 1967).

Fellner, William, 'Does the Market Direct the Relative Factor-Saving Effects of Technological Progress?' in *The Rate and Direction of Inventive Activity*, Universities-National Bureau Committee for Economic Research, Princeton, N.J., Princeton University Press, 1962.

'Specific Interpretations of Learning by Doing,' *Journal of Economic Theory*, Vol. I, 2 (August 1969).

'Two Propositions in the Theory of Induced Innovations,' *Economic Journal*, Vol. LXXI (June 1961).

Fisher, Franklin M., 'Embodied Technical Change and the Existence of an Aggregate Capital Stock,' *Review of Economic Studies*, Vol. 32 (October 1965).

Fishlow, A., 'Productivity and Technological Change in the Railroad Sector, 1840–1910,' in *Output, Employment and Productivity in the United States After 1800*, New York, Columbia University Press for National Bureau of Economic Research, 1966.

Fogel, R. W., 'Railroads as an Analogy to the Space Effort: Some Economic Aspects,' *Economic Journal*, Vol. LXXVI (1966).

'The New Economic History: Its Findings and Methods,' *Economic History Review*, 2nd Series, Vol. LXX (December 1966).

'The Specification Problem in Economic History,' *Journal of Economic History*, Vol. XXVII (September 1967).

and Engerman, Stanley, 'A Model for the Explanation of Industrial Expansion during the Nineteenth Century: With an Application to the American Iron Industry,' *Journal of Political Economy*, Vol. LXXVII, 3 (May/June 1969).

Fox, A. Wilson, 'Agricultural Wages in England and Wales during the Last Fifty Years,' *Journal of the Royal Statistical Society*, Vol. LXVI (June 1903).

Frankel, M., 'Obsolescence and Technological Change,' *American Economic Review*, Vol. 45 (June 1955).

Fulton, Hamilton, 'Drainage of Hethel Wood Farm,' *Journal of the Royal Agricultural Society of England*, Vol. XII (1851).

Hacker, Louis M., 'The New Revolution in Economic History,' *Explorations in Entrepreneurial History*, 2nd Series, III (1966).

Hammond, Anthony, 'Use of Reaping-Machines,' *Journal of the Royal Agricultural Society of England*, Vol. XVII (1856).

Hilton, George W., 'Review of *Railroads and American Economic Growth*,' *Explorations in Entrepreneurial History*, 2nd Series, III (1966).

Hirsch, Werner A., 'Firm Progress Ratios,' *Econometrica*, Vol. XXIV (April 1956).

'Manufacturing Progress Functions,' *Review of Economics and Statistics*, Vol. XXXIV (May 1925).

Hirshmeier, J., 'Shibusawa Eiichi: Industrial Pioneer,' in *The State and Economic Enterprise in Japan*, ed. W. W. Lockwood, Princeton, N.J., Princeton University Press, 1965.

Horie, Y., 'Modern Entrepreneurship in Meiji Japan,' in *The State and Economic Enterprise in Japan*, ed. W. W. Lockwood, Princeton, N.J., Princeton University Press, 1965.

Hoskins, Chandos Wren, 'On "Ridge-and-Furrow" Pasture Land, and a Method of Levelling It,' *Journal of the Royal Agricultural Society of England*, Vol. XVII (1856).

Houthakker, Hendrik S., 'The Pareto Distribution and the Cobb-Douglas Production Function in Activity Analysis,' *Review of Economic Studies*, Vol. XXIII(1), 60 (1955–6).

Johnson, H. G., 'Optimal Trade Intervention in the Presence of Domestic Distortions,' in *Trade, Balance of Payments, and Growth: Papers in International Economics in Honor of Charles P. Kindleberger*, eds. Jagdish Bhagwati, Ronald W. Jones, Robert A. Mundell, and Jaroslav Vanek, Amsterdam, North-Holland Publishing Company, 1971.

Jones, E. L., 'The Agricultural Labour Market in England, 1793–1872,' *Economic History Review*, 2nd Series, Vol. 17 (1964).

'The Changing Basis of Agricultural Prosperity, 1853–73,' *Agricultural History Review*, Vol. X (1962).

Jones, Ronald W., 'A Three-Factor Model in Theory, Trade and History,' in *Trade, Balance of Payments, and Growth: Papers in International Economics in Honor of C. P. Kindleberger*, ed. Jagdish Bhagwati, Ronald W. Jones, Robert A. Mundell, and Jaroslav Vanek, Amsterdam, North-Holland Publishing Company, 1971.

Jorgenson, D. W. and Griliches, Z., 'The Explanation of Productivity Change,' *Review of Economic Studies*, Vol. XXXIV (July 1967).

Kaldor, Nicholas, 'The Irrelevance of Equilibrium Economics,' *Economic Journal*, Vol. 82 (December 1972).

Kennedy, Charles, 'Induced Bias in Innovation and the Theory of Distribution,' *Economic Journal*, Vol. LXXIV (September 1964).

Landes, David S., 'Factor Costs and Demand: Determinants of Economic Growth,' *Business History*, Vol. VII, 1 (January 1965).

Lawrence, C., 'Letter on the Use of the Reaping Machine, and the Root-Crops in 1860,' *Journal of the Royal Agricultural Society in England*, Vol. XXI (1860).

Lebergott, Stanley, 'United States Transport Advance and Externalities,' *Journal of Economic History*, Vol. XXVI (1966).

'Wage Trends, 1800–1900,' in *Trends in the American Economy in the Nineteenth Century*, National Bureau of Economic Research Income and Wealth Conference, XXIV, Princeton, N.J., Princeton University Press, 1960.

Lerner, A. P., 'The Symmetry Between Import and Export Taxes,' *Economica*, Vol. III (August 1936).

Long, W. Harwood, 'The Development of Mechanization in English Farming,' *Agricultural History Review*, Vol. II (1963).

Love, Peter, 'On Harvesting Corn,' *Journal of the Royal Agricultural Society of England*, Vol. XXIII (1862).

McGouldrick, Paul F., 'Comment,' in *Output, Employment and Productivity in the U.S. After 1800*, National Bureau of Economic Research, Vol. XXX, New York, Columbia University Press, 1966.

McKinnon, R. I., 'Intermediate Products and Differential Tariffs: A Generalization of Lerner's Symmetry Theorem,' *Quarterly Journal of Economics*, Vol. LXXX (November 1966).

'On Misunderstanding the Capital Constraint in LDC's: The Consequences for Trade Policy,' in *Trade, Balance of Payments, and Growth: Papers in International Economics in Honor of Charles P. Kindleberger*, eds. Jagdish N. Bhagwati, Ronald W. Jones, Robert A. Mundell, and Jaroslav Vanek, Amsterdam, North-Holland Publishing Company, 1971.

Meyer, John, 'Review of *Railroads and American Economic Growth*,' *Journal of Political Economy*, Vol. LXXIV (1966).

Montgomery, James, *A Practical Detail of the Cotton Manufacture of the United States . . . Compared with that of Great Britain*, Glasgow, 1840.

Moscrop, W. J., 'A Report on the Farming of Leicestershire,' *Journal of the Royal Agricultural Society of England*, 2nd Series, II (1866).

Murphy, George G. S., 'Review of *Industrialization in Two Systems*,' *Journal of Economic History*, Vol. XXVII (September 1967).

Nelson, Richard R. and Sidney G. Winter, 'Towards an Evolutionary Theory of Economic Capabilities,' *American Economic Review*, Vol. LXIII, 2 (May 1973).

Nerlove, Marc, 'Empirical Studies of the CES and Related Production Function,' in *The Theory and Empirical Analysis of Production*, ed. M. Brown, National Bureau of Economic Research Studies in Income and Wealth, XXXI, New York, 1967.

'Railroads and American Economic Growth,' *Journal of Economic History*, Vol. XXVI (1966).

Parker, William N., 'From Old to New to Old Economic History,' *Journal of Economic History*, Vol. XXXI (March 1971).

'Review of *American and British Technology*,' *The Business History Review*, Vol. XXXVII (Spring/Summer 1963).

and Klein, Judith L. V., 'Productivity Growth in Grain Production in the United States, 1840–60 and 1900–10,' in *Output, Employment, and Productivity in the United States After 1800*, New York, Columbia University Press for National Bureau of Economic Research, 1966.

Pasinetti, Luigi, 'Switches of Technique and the "Rate of Return" in Capital Theory,' *Economic Journal*, Vol. LXXIX (September 1969).

Pell, Albert, 'The Making of the Land in England,' *Journal of the Royal Agricultural Society of England*, 2nd Series, Vol. XXIII (1887).

Prest, A. R. and Turvey, R., 'Cost–Benefit Analysis: A Survey,' in American Economic Association and Royal Economic Society, *Surveys of Economic Theory*, Vol. III, London, Macmillan Co., 1967.

Primack, M., 'Land Clearing under Nineteenth Century Techniques: Some Preliminary Calculations,' *Journal of Economic History*, Vol. XXI (1962).

Pusey, Ph[illip], 'On Mr. McCormick's Reaping-machine,' *Journal of the Royal Agricultural Society of England*, Vol. XII (1851).

Rapping, Leonard, 'Learning and World War II Production Functions,' *Review of Economics and Statistics*, Vol. 47, 1 (February 1965).

Rosenberg, Nathan, 'The Direction of Technological Change: Inducement Mechanisms and Focusing Devices,' *Economic Development and Cultural Change*, Vol. 18, No. 1, part 1 (October 1969).

'Technological Change in the Machine Tool Industry, 1840–1910,' *Journal of Economic History*, Vol. XXIII, 4 (December 1963).

Rothbarth, E., 'Causes of the Superior Efficiency of U.S.A. Industry as Compared with British Industry,' *Economic Journal*, Vol. 56 (September 1946).

Samuelson, Paul A., 'A Theory of Induced Innovation Along Kennedy–Weisacker Lines,' *Review of Economics and Statistics*, Vol. XLVII, 4 (November 1965).

'Parable and Realism in Capital Theory: The Surrogate Production Function,' *Review of Economic Studies*, Vol. XXIX (June 1962).

Sandberg, Lars G., 'American Rings and English Mules: The Role of Economic Rationality,' *Quarterly Journal of Economics*, Vol. LXXXII, 4 (November 1968).

Sheshinski, Eytan, 'Tests of the Learning by Doing Hypothesis,' *Review of Economics and Statistics*, Vol. 49, 4 (November 1967).

Skirving, R. Scot, 'Ten Years of East Lothian Farming,' *Journal of the Royal Agricultural Society of England*, 2nd Series, Vol. I (1865).

Slight, James, 'Report on Reaping Machines,' *Transactions of the Highland and Agricultural Society of Scotland*, 3rd Series, Vol. V (January 1852).

Solow, Robert M., 'Investment and Technical Change,' in *Mathematical Methods in the Social Sciences*, eds. K. J. Arrow, S. Karlin and P. Suppes, Stanford, Calif., Stanford University Press, 1960.

Spaventa, Luigi, 'Realism Without Parables in Capital Theory,' in *Recherches Récentes sur la Fonction de Production*, Namur, Centre D'Etudes et des Recherches Universitaires de Namur, 1968.

Taylor, George R., 'Review of *Railroads and American Economic Growth*,' *American Economic Review*, Vol. IV (1965).

Taylor, John, 'On the Comparative Merits of Different Modes of Reaping Grain,' *Transactions of the Highland and Agricultural Society of Scotland* (July 1844).

Temin, Peter, 'In Pursuit of the Exact,' *Times Literary Supplement*, 28 July 1966.

'Labor Scarcity and the Problem of American Industrial Efficiency in the 1850's,' *Journal of Economic History*, Vol. XXVI (September 1966).

'Labor Scarcity in America,' *Journal of Interdisciplinary History*, Vol. I, 2 (Winter 1971).

'The Relative Decline of the British Iron and Steel Industry,' in *Industrialization in Two Systems: Essays in Honor of Alexander Gerschenkron*, ed. Henry Rosovsky, New York, John Wiley and Sons, 1966.

Thomas, Keith, 'New Ways in History,' *Times Literary Supplement*, 7 April 1966.

Thompson, H. S., 'Account of a subsequent Trial of the American Reapers' (Appendix to Implement Report), *Journal of the Royal Agricultural Society of England*, Vol. XII (1851).

Thompson, R. J., 'An Enquiry into the Rent of Agricultural Land in England and Wales during the Nineteenth Century,' *Journal of the Royal Statistical Society*, Vol. LXX (1907).

Vatter, Barbara, 'Industrial Borrowing by the New England Textile Mills, 1840–1860,' *Journal of Economic History*, Vol. XXI (June 1961).

Whetham, E. H., 'Mechanization of British Farming 1910–1945,' *Journal of Agricultural Economics*, Vol. XXI, 2 (May 1970).

Wilson, Jacob, 'Reaping Machines,' *Transactions of the Highland and Agricultural Society of Scotland*, 3rd Series, XI (January 1864).

Wilson, John, 'Agricultural Implements and Machinery,' *The Encyclopaedia Britannica*, Ninth Edition, Edinburgh, 1875.

Wright, William, 'On the Improvements in the Farming of Yorkshire since the date of the last Reports in the Journal,' *Journal of the Royal Agricultural Society of England*, Vol. XXII (1861).

Wright, T. P., 'Factors Affecting the Cost of Airplanes,' *Journal of the Aeronautical Sciences*, Vol. 3 (1936).

Young, Allyn, 'Increasing Returns and Economic Progress,' *Economic Journal*, Vol. 38 (December 1928).

NEWSPAPERS AND PERIODICALS

Chicago *Daily Democratic Press*. Chicago, Illinois.
Farmer's Magazine. London.
Prairie Farmer. Chicago, Illinois.

GOVERNMENT DOCUMENTS

Great Britain

The Jute Working Party Report, London, H.M.S.O., 1948.

Parliamentary Accounts and Papers, 1852–3, XXXII, *Report of the Census of 1851*.
1854, XXXVI, 'Special Reports of Mr. George Wallis and Mr. Joseph Whitworth on the New York Industrial Exhibition.'
1854–5, L, 'Report of the Committee on the Machinery of the United States.'
1873, LXXI, part 2, *Census of England and Wales, 1871*.
1873, IX, House of Lords Special Committee on Improvement of Land.

Report of the Cotton Textile Mission to the U.S.A., London, H.M.S.O., 1944.

United States: Federal Government

DeBow, J. D. B., *Compendium of the Seventh Census*, Washington, A. O. P. Nicholson, Public Printer, 1854.

Edwards, Everett E., 'American Agriculture – The First 300 Years,' in *Yearbook of Agriculture 1940*, Washington, Government Printing Office, 1940.

American State Papers, Finances, Vols. I and II, Washington, Gales and Seaton, 1832.

U.S. Bureau of the Census, *Eighth Census of the United States: 1860*, Vol. II, *Agriculture of the United States in 1860*, Washington, Government Printing Office, 1865.
Eighth Census of the United States: 1860, Vol. III, *Manufactures of the United States in 1860*, Washington, Government Printing Office, 1865.
Historical Statistics of the United States, Colonial Times to 1957, Washington, Government Printing Office, 1960.

U.S. Federal Reserve Bank of Boston, *A History of Investment Banking in New England*, Annual Report for 1960, Boston, 1961.

U.S. Senate, *Select Committee on the Transportation and Sale of Meat Products*, Report No. 829, 51st Congress, 1st Session, 1890.

United States: State Government

Ohio State Board of Agriculture Report, 1857.

Index